AF614799

METHODS IN MOLECULAR BIOLOGY™

Series Editor
John M. Walker
School of Life Sciences
University of Hertfordshire
Hatfield, Hertfordshire, AL10 9AB, UK

For further volumes:
http://www.springer.com/series/7651

Plant Salt Tolerance

Methods and Protocols

Edited by

Sergey Shabala

School of Agricultural Science, University of Tasmania, Hobart, TAS, Australia

Tracey Ann Cuin

Biochimie et Physiologie Moléculaire des Plantes, INRA, Montpellier, France

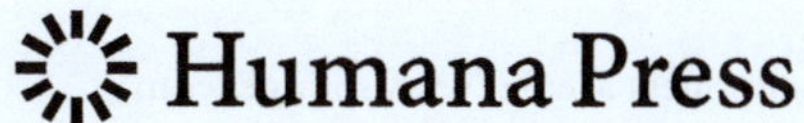

Editors
Sergey Shabala
School of Agricultural Science
University of Tasmania
Hobart, TAS, Australia

Tracey Ann Cuin
Biochimie et Physiologie Moléculaire des Plantes
INRA, Montpellier, France

ISSN 1064-3745 ISSN 1940-6029 (electronic)
ISBN 978-1-61779-985-3 ISBN 978-1-61779-986-0 (eBook)
DOI 10.1007/978-1-61779-986-0
Springer New York Heidelberg Dordrecht London

Library of Congress Control Number: 2012943159

Printed on acid-free paper

Humana Press is a brand of Springer
Springer is part of Springer Science+Business Media (www.springer.com)

Preface

Soil salinity caused the collapse of the world's first Sumer civilization and presently is destroying several hectares of arable land every minute. As estimated by the FAO, the global cost of irrigation-induced salinity is around US$11 billion per year. In addition, a much greater agricultural area is at high risk of "transient salinity," significantly reducing agricultural crop production. At the same time, global food production needs to increase by approximately 50% by 2050 to match the projected population growth. Because remedial land management cannot completely solve the problem, salt tolerant crops or plant species able to remove excessive salt from the soil could contribute significantly to managing the salinity problem. The key to engineering crops for salt tolerance lies in a thorough understanding of the physiological mechanisms underlying the adaptive responses of plants to salinity.

Salinity tolerance is a complex multigenic trait. Salt tolerance is also multifaceted physiologically with numerous tissue- and age-specific components involved. As such, a number of sub-traits will determine salt tolerance and each trait will be determined by a large number of genes. To add to this complexity, only a few of these genes are regulated specifically by salinity; most alterations in the regulation of gene expression and metabolic adjustment in response to salinity share common elements to other abiotic stresses. This makes sense in physiological terms, but it makes it extremely difficult to elucidate the causal relationships in the highly branched and complicated network of metabolic and signalling events. It is becoming increasingly evident that various "omics" on their own will not be able to deliver the "holy grail"—practical solutions of how to increase plant salinity tolerance without any associated yield penalties. Achieving this goal will only be possible when the intricacy and multitude of plant responses to salinity is matched by the complexity and a broad range of experimental techniques used to study this trait.

This volume describes recent advances and techniques employed by researchers to understand the molecular and ionic basis of salinity tolerance and to investigate the mechanisms of salt stress perception and signalling in plants. Written in a traditional "Methods in Molecular Biology" style, each chapter contains a comprehensive review on the use of a specific technique, before providing a detailed description of the laboratory protocols related to its application. Altogether, nearly 30 different methods are covered, including microelectrode and molecular methods, imaging techniques, and various biochemical assays. Written by leading international scientists, this volume is undoubtedly a "must have" book for every student or researcher dealing with various aspects of the salinity problem.

In conclusion, we would like to thank all the contributing authors for their great work and enthusiastic support of this project. We would also like to acknowledge financial support from the Australian Research Council and Grain Research and Development Corporation for research on plant stress physiology and, specifically, salinity tolerance at the University of Tasmania.

Enjoy reading!

Hobart, TAS, Australia — *Sergey Shabala*
Montpellier, France — *Tracey Ann Cuin*

Contents

Part V Other Methods

Contributors

N.A. Akram • *Department of Botany, University of Agriculture, Faisalabad, Pakistan; King Saud University, Riyadh, Saudi Arabia*

M. Ashraf • *Department of Botany, University of Agriculture, Faisalabad, Pakistan; King Saud University, Riyadh, Saudi Arabia*

Asmini Athman • *School of Agriculture, Food and Wine and Waite Research Institute, University of Adelaide, Glen Osmond, SA, Australia*

Olga Babourina • *School of Earth and Environment, University of Western Australia, Crawley, WA, Australia*

K.C. Bansal • *National Bureau of Plant Genetic Resources, New Delhi, India*

Bastiaan O.R. Bargmann • *Center for Genomics and Systems Biology, New York University, New York, NY, USA*

Ute Baumann • *Australian Centre for Plant Functional Genomics, University of Adelaide, Glen Osmond, SA, Australia*

Diane M. Beckles • *Department of Plant Sciences, University of California, Davis, CA, USA*

Begoña Benito • *Centro de Biotecnología y Genómica de Plantas, Universidad Politecnica de Madrid, Madrid, Spain*

Bettina Berger • *The Plant Accelerator, University of Adelaide, Urrbrae, SA, Australia*

Kenneth D. Birnbaum • *Center for Genomics and Systems Biology, New York University, New York, NY, USA*

Michael R. Blatt • *Laboratory of Plant Physiology and Biophysics, University of Glasgow, Glasgow, UK*

Dev T. Britto • *Department of Biological Sciences, University of Toronto, Toronto, ON, Canada*

Olivier Cagnac • *Dpto de Bioquimica, Biologia Celular y Molecular de Plantas CSIC, Estacion Experimental del Zaidin, Zaidin, Spain*

Martin Canny • *Research School of Biology, The Australian National University, Canberra, ACT, Australia*

Shao-Liang Chen • *College of Biological Sciences and Technology, Beijing Forestry University, Beijing, China*

Zhong-Hua Chen • *School of Natural Sciences, University of Western Sydney, Sydney, NSW, Australia*

Renato Colaço • *Department of Plant Sciences, University of Cambridge, Cambridge, UK*

Simon J. Conn • *European Molecular Biology Laboratory, Grenoble Outstation, Grenoble, France*

Tracey Ann Cuin • *Biochimie et Physiologie Moléculaire des Plantes, INRA, Montpellier, France*

Julia Davies • *Department of Plant Sciences, University of Cambridge, Cambridge, UK*

Bas de Regt • *The Plant Accelerator, University of Adelaide, Urrbrae, SA, Australia*
Petre I. Dobrev • *Institute of Experimental Botany AS CR, Prague, Czech Republic*
Richard Doyle • *School of Agricultural Science, Tasmanian Institute of Agricultural Research, University of Tasmania, Hobart, SA, Australia*
Jelena J. Dragišić Maksimović • *Institute for Multidisciplinary Research, University of Belgrade, Belgrade, Serbia*
Cornelia Eisenach • *Laboratory of Plant Physiology and Biophysics, University of Glasgow, Glasgow, UK*
René Alberto Enríquez-Figueroa • *Centro Universitario de Investigaciones Biomédicas, Universidad de Colima, Colima, Mexico*
Aurelie Evrard • *Australian Centre for Plant Functional Genomics, University of Adelaide, Glen Osmond, SA, Australia*
M.R. Foolad • *Department of Horticulture, The Pennsylvania State University, University Park, TX, USA*
Wieland Fricke • *School of Biology and Environmental Science, University College Dublin, Dublin, Ireland*
Anja Thoe Fuglsang • *Department of Plant Biology and Biotechnology, University of Copenhagen, Copenhagen, Denmark*
Matthew Gilliham • *School of Agriculture, Food and Wine and Waite Research Institute, University of Adelaide, Glen Osmond, SA, Australia*
Marcus Hardie • *School of Agricultural Science, Tasmanian Institute of Agricultural Research, University of Tasmania, Hobart, TAS, Australia*
Adrian Hills • *Laboratory of Plant Physiology and Biophysics, University of Glasgow, Glasgow, UK*
Qian Hu • *Department of Genetics and Biochemistry, Clemson University, Clemson, SC, USA*
Richard A. James • *CSIRO Plant Industry, Canberra, ACT, Australia*
Haiyan Jia • *Department of Genetics and Biochemistry, Clemson University, Clemson, SC, USA*
Alexander A.T. Johnson • *ACPFG and School of Botany, University of Melbourne, Melbourne, VIC, Australia*
Herbert J. Kronzucker • *Department of Biological Sciences, University of Toronto, Toronto, ON, Canada*
Anuphon Laohavisit • *Department of Plant Sciences, University of Cambridge, Cambridge, UK*
Dayong Li • *Department of Genetics and Biochemistry, Clemson University, Clemson, SC, USA*
Zhigang Li • *Department of Genetics and Biochemistry, Clemson University, Clemson, SC, USA*
Anette Lund • *Department of Plant Biology and Biotechnology, University of Copenhagen, Copenhagen, Denmark*
Hong Luo • *Department of Genetics and Biochemistry, Clemson University, Clemson, SC, USA*
Frans J.M. Maathuis • *Department of Biology, University of York, York, UK*
Gwenda M. Mayo • *Australian Centre for Plant Functional Genomics and School of Agriculture, Food and Wine and Waite Research Institute, Glen Osmond, SA, Australia*

MARGARET MCCULLY • *Division of Plant Industry, CSIRO, Canberra, ACT, Australia*
MEHBOOB-UR-RAHMAN • *National Institute for Biotechnology and Genetic Engineering, Faisalabad, Pakistan*
ANTHONY J. MILLER • *Department of Disease and Stress Biology, John Innes Centre, Norwich, UK*
IAN NEWMAN • *School of Mathematics and Physics, University of Tasmania, Hobart, TAS, Australia*
CSABA PAPDI • *Institute of Plant Biology, Biological Research Center, Szeged, Hungary*
D. MARSHALL PORTERFIELD • *Agricultural and Biological Engineering, Purdue University, West Lafayette, USA*
IGOR POTTOSIN • *Centro Universitario de Investigaciones Biomédicas, Universidad de Colima, Colima, Mexico*
SHANE REIGHARD • *Department of Genetics and Biochemistry, Clemson University, Clemson, SC, USA*
ZED RENGEL • *School of Earth and Environment, University of Western Australia, Crawley, WA, Australia*
GÁBOR RIGÓ • *Institute of Plant Biology, Biological Research Center, Szeged, Hungary*
ANDRÉS ALBERTO RODRÍGUEZ • *INTECH-CONICET, Chascomús, Argentina*
ALONSO RODRÍGUEZ • *Centro de Biotecnología y Genómica de Plantas, Universidad Politecnica de Madrid, Madrid, Spain*
UTE ROESSNER • *ACPFG Centre for Plant Functional Genomics and Metabolomics Australia, University of Melbourne, Melbourne, VIC, Australia*
STUART J. ROY • *Australian Centre for Plant Functional Genomics and School of Agriculture, Food and Wine and Waite Research Institute, Glen Osmond, SA, Australia*
BEKIR SAN • *Department of Genetics and Biochemistry, Clemson University, Clemson, SC, USA*
PRASAD SENADHEERA • *Department of Botany, The Open University of Sri Lanka, Nugegoda, Sri Lanka*
LANA SHABALA • *School of Agricultural Science, University of Tasmania, Hobart, TAS, Australia*
SERGEY SHABALA • *School of Agricultural Science, University of Tasmania, Hobart, TAS, Australia*
WEN-BIAO SHEN • *College of Life Sciences, Nanjing Agricultural University, Nanjing, China*
A.K. SINGH • *Department of Plant Pathology, University of Kentucky, Lexington, KY, USA*
XAVIER R.R. SIRAULT • *High Resolution Plant Phenomics Centre, Canberra, ACT, Australia*
SUSAN SMITH • *Department of Disease and Stress Biology, John Innes Centre, Norwich, UK*
JIAN SUN • *College of Biological Sciences and Technology, Beijing Forestry University, Beijing, China*
LÁSZLÓ SZABADOS • *Institute of Plant Biology, Biological Research Center, Szeged, Hungary*
EDITH L. TALEISNIK • *CIAP-INTA, Córdoba, Argentina*
MARK TESTER • *Australian Centre for Plant Functional Genomics, Urrbrae, SA, Australia*
RADOMIRA VANKOVA • *Institute of Experimental Botany AS CR, Prague, Czech Republic*

ANA MARÍA VELARDE-BUENDÍA • *Centro Universitario de Investigaciones Biomédicas, Universidad de Colima, Colima, Mexico*
S.H. WANI • *Central Institute of Temperate Horticulture, Srinagar, India*
LARS H. WEGNER • *Karlsruhe Institute of Technology, Institute of Botany I and Institute of Pulsed Power and Microwave Technology, Eggenstein-Leopoldshafen, Germany*
DEZHI WU • *Department of Agronomy, Zhejiang University, Hangzhou, China*
YAN-JIE XIE • *College of Life Sciences, Nanjing Agricultural University, Nanjing, China*
NING YUAN • *Department of Genetics and Biochemistry, Clemson University, Clemson, SC, USA*
SHUANGRONG YUAN • *Department of Genetics and Biochemistry, Clemson University, Clemson, SC, USA*
GUOPING ZHANG • *Department of Agronomy, Zhejiang University, Hangzhou, China*
MAN ZHOU • *Department of Genetics and Biochemistry, Clemson University, Clemson, SC, USA*
BRANKA D. ŽIVANOVIĆ • *Institute for Multidisciplinary Research, University of Belgrade, Belgrade, Serbia*

Part I

Microelectrode Techniques

Chapter 1

Patch-Clamp Protocols to Study Cell Ionic Homeostasis Under Saline Conditions

Ana María Velarde-Buendía, René Alberto Enríquez-Figueroa, and Igor Pottosin

Abstract

The patch-clamp technique was designed to measure any electrogenic transport across the whole cell and organelle (vacuolar) membranes and excised membrane patches. Here, we describe preparation of protoplasts and vacuoles, as well as patch-clamp assays, to detect the functional expression of K^+ and cation channels of plasma membrane and tonoplast, as well as plasma membrane anion channels and vacuolar and plasma membrane H^+ pumps. All of these contribute to the intracellular ionic homeostasis under saline conditions.

Key words: Patch-clamp, Ion channel, Electrogenic pump, Protoplast, Vacuoles, Saline stress, Ion transport, Ionic homeostasis

1. Introduction

Under saline conditions, ionic gradients across cell membranes are altered due to the influx of Na^+ and Cl^- into the cell and the loss of intracellular K^+. Potassium can be substituted for Na^+ in some functions such as osmotic adjustments, underlying the regulation of cell turgor and expansion, but activation of intracellular enzymes and protein synthesis rely on a high and stable, ~100 mM, K^+ concentration within the metabolic compartments (1). Stress-resistant plants are able to minimize the accumulation of salt in the cytosol, have efficient K^+ retention, and compartmentalize Na^+ in vacuoles or restrict its uploading into the xylem (2–8). Sodium enters the cell through non-selective cation channels (NSCCs) (9) and is exported from the cell and/or imported into the vacuole due to a

Sergey Shabala and Tracey Ann Cuin (eds.), *Plant Salt Tolerance: Methods and Protocols*, Methods in Molecular Biology, vol. 913, DOI 10.1007/978-1-61779-986-0_1, © Springer Science+Business Media, LLC 2012

concerted activity of H^+ pumps and electroneutral Na^+/H^+ exchangers (3, 10). Potassium is lost from the cell through plasma membrane (PM) outward-rectifying K^+ channels (KORC) or almost every available NSCC (11, 12), whereas K^+ leak from the vacuole through vacuolar K^+ (VK) channels may assist in improving the cytosolic K^+/Na^+ ratio (13). Lower expression and/or inhibition of NSCCs on the plasma membrane and tonoplast often correlate with a higher salt tolerance (12–16), although some NSCCs may be involved in the stress-induced Ca^{2+} signaling (17).

Patch-clamp technique was developed to record ion currents in whole cell mode and on cell-free membrane patches (18). This technique, as a pre-requisite, requires the achievement of a high-resistant (GOhm range) seal between a tip of a glass microelectrode and the membrane. Thus, it can be applied to isolated protoplasts or vacuoles only, not to intact plant cells. Another limitation is that patch-clamp in principle cannot detect electroneutral transport. Thus, the activity of Na^+/H^+ antiporters, although being of major importance for the plant performance under saline conditions, cannot be recorded by patch-clamp. Further, ion channels conduct currents in a pico-Amper (1 pA = 6×10^6 elementary charges per second) range. Thus, the activity of individual ion channels can be analyzed on small membrane patches. Yet, some important characteristics of ion currents, like activation and deactivation kinetics, and steady state current–voltage relations are more easily obtained from integrative responses in the whole cell mode. Besides, measurements in the whole cell mode involve the estimate of the membrane area based on the whole cell capacitance, which may not be accurately determined in small patches. Capacitance/area estimate allows a quantitative presentation of channels' functional expression, as a specific current (in pA/pF) or mean number of ion channels per unit membrane area. On the other hand, currents, generated by individual ion pumps, are well below the resolution limit of the patch-clamp, and only the activity of thousands of pumps in the whole cell mode may be detected. H^+ pumps current density is normally much smaller than channels-mediated currents, about or less than 1 pA/pF, with the exception of the guard cell PM, where it is 2–10 pA/pF (19–22). Thus, separation of H^+ pumps-mediated currents requires specially designed media, supplemented with large and impermeable cations and anions.

Summarizing, patch-clamp technique can be used to record ionic currents, generated by following key components of cell and vacuolar membranes, mediating Na^+, K^+, Cl^-, Ca^{2+}, and H^+ transport during plant responses to salinity (1) potassium-selective PM-bound KIRC (inward-rectifying K^+) and KORC, and tonoplast VK channels; (2) NSCCs in plasma membrane and tonoplast slow and fast vacuolar (SV and FV) channels; (3) PM anion channels; (4) PM H^+-pumping ATPase; and (5) tonoplast H^+ pumps, ATPase and pyrophosphatase (PPase).

2. Materials

Prepare all solutions using ultrapure (deionized, with a resistivity of 18 MOhm cm at 25°C) water. All chemicals should be analytical grade (we purchased most of chemicals from Sigma). Store all solutions at 4°C (unless indicated).

2.1. Growth of Barley Plants and Root Protoplasts Isolation

1. Plant material: barley (*Hordeum vulgare* L.) (salt-tolerant or salt-sensitive genotypes, see Note 1).
2. Growth solution: 0.5 L hydroponic solution (0.5 mM KCl and 0.1 mM $CaCl_2$).
3. 1 L plastic container.
4. Enzyme solution containing (w/w): 2% cellulase, 1.2% cellulysin, 0.1% pectolyase, 0.1% bovine serum albumin, 10 mM KCl, 10 mM $CaCl_2$, and 2 mM $MgCl_2$. Adjust pH 5.7 with 2 mM MES-KOH and osmolality to 750 mOs with D-sorbitol. Mix and filter the solution through a 0.22 μm Millipore filter. Store at −13°C.
5. Wash solution as enzyme solution, minus enzymes.
6. Release solution: 10 mM KCl, 2 mM $CaCl_2$, 1 mM $MgCl_2$, 2 mM MES-KOH, pH 5.7. Adjust osmolality to 380 mOs with D-sorbitol. Mix and filter the solution.
7. Bath solution: 5 mM KCl, 2 mM $CaCl_2$, 0.5 mM $MgCl_2$, 2 mM MES-KOH, pH 5.7. Adjust osmolality to 560 mOs with D-sorbitol. Mix and filter the solution.
8. EDTA-bath solution: the same as bath solution plus 5 mM EDTA. Mix and filter the solution.
9. Pipette solution: 100 mM KCl, 3 mM $MgCl_2$, and 0.8 mM $CaCl_2$, 2 mM K_2EGTA. Adjust pH 7.4 with 5 mM HEPES-KOH and osmolality to 560 mOs with D-sorbitol. Mix and filter the solution.
10. 10% bleach (NaClO), a common house bleach. Store at room temperature.

2.2. Media for Isolation of Sugar Beet Root Vacuoles and Patch-Clamp Electrophysiology

1. The release solution: 2 mM $CaCl_2$, 100 mM KCl, 15 mM HEPES-KOH pH 7.5. Adjust osmolarity to 680 mOs with D-sorbitol.
2. Solutions for slow vacuolar (SV) and vacuolar K^+ (VK) channels assays: Pipette and bath solutions containing symmetrical 100 mM KCl, 2 mM $CaCl_2$, 15 mM HEPES-KOH (pH 7.5). Adjust osmolality to 680 mOs with D-sorbitol. Mix and filter the solution.

3. Solutions for fast vacuolar (FV) channel assay: Pipette and bath solutions containing 2 mM K_2EGTA (~2 nM free Ca^{2+}) instead of 2 mM $CaCl_2$ to abolish Ca^{2+}-activating ion channels and prevent the FV current inhibition by Ca^{2+}. Mix and filter the solution.
4. Connect the reference AgCl electrode to the bath via a 3.5% agar bridge filled with 100 mM KCl (see Note 2).

2.3. Patch-Clamp Media for Measurement of Currents Mediated by H^+ Pumps in Barley Root Protoplasts and Vacuoles

1. Bath solution for the PM H^+ pump currents measurements: 5 mM Ca-gluconate, 8 mM $MgCl_2$, and 10 mM MES-Tris, pH 5.8.
2. Standard intracellular (pipette) solution; 10 mM Mg^{2+}-ATP (see Note 3), 0.04 mM Ca-gluconate (5 μM free Ca^{2+}), 10 mM $MgCl_2$, 10 mM Tris-MES, pH 7.2 (see Note 4). Adjust osmolality of bath and pipette solutions to that of the cell sap (500–580 mOs) with D-sorbitol.
3. For recordings of currents, generated by vacuolar H^+ pumps, ionic solutions in the bath and pipette: 250 mM *N*-methylglucamine-glutamate, 50 mM HEPES-imidazole (see Note 4), 1 mM $CaCl_2$, 2 mM $MgCl_2$ and pH 7.3, with or without Mg-ATP in the bath (see Note 3).

3. Methods

Conduct all experiments at room temperature (23–25°C).

3.1. Growth Conditions

1. In order to eliminate fungal and bacterial contaminations, sterilize barley seeds with 10% bleach for 10 min and thoroughly rinse with tap water, follow by a minimum five times with ultrapure water (18 MOhm cm).
2. Insert sterile seeds into the wet filter paper and transfer them into a growth cylinder (1 L plastic container) that contains 0.4 L of the growth solution.
3. Maintain seeds in a dark growth cabinet at 25°C.
4. Ensure that germinated seeds are not contaminated every 24 h.

3.2. Isolation of Protoplasts from the Root Mature Zone

This protocol was developed by modifying the previously described protocol by Chen et al. (11) and is used for a quick isolation of barley root epidermal protoplasts, but it can be adopted to obtain root protoplasts from other plants (see Note 5).

1. Select 3–5 days old hydroponically grown barley seedlings (see Note 6) with a root length 70 ± 10 mm for all electrophysiological experiments.

2. Cut the roots about 5 mm below the seed and their apical parts (7 mm from the tip) and discard those parts.
3. Cut the remaining mature parts into approximately 10 mm long segments and split them longitudinally under a dissecting microscope.
4. Split root segments are placed into 5 mL flask. Cover the flask openings with Parafilm and incubate tissues with 3 mL of the protoplast enzyme solution.
5. After 30 min of incubation in the enzyme solution (in the dark at 30°C; agitated on a 90 rpm rotary shaker), transfer root segments to the wash solution and agitate them on a shaker for another 4 min.
6. Transfer the segments into the measuring chamber filled with the release solution.
7. Release the protoplast by gently shaking into the measuring chamber used for patch-clamp experiments (see Note 7).
8. Wait 5–10 min until most of protoplasts settle to the bottom. Then perfuse the chamber with a fresh portion of bath solution, removing cellulose debris and protoplasts not firmly attached to the bottom glass.
9. Next, wash the preparation with the EDTA-bath solution to clean the cell membrane surface (5 min) (see Note 8) and wash again with normal bath solution.
10. The above protocol provides protoplasts preferentially from the cortical zone. Epidermal protoplasts can be further selected by size (see Note 9). Protoplasts form better seals with a patch-pipette within the first hour after their isolation.

3.3. Vacuoles Isolation for Patch-Clamp Measurements

1. Vacuoles from some tissues (e.g., taproot) can be isolated mechanically, otherwise the protoplasts need to be isolated first from the tissue of interest.
2. Obtain fresh beet (*Beta vulgaris* L., whole plants) from a local market and kept at +4°C. Prior to isolation of vacuoles, incubate slices of a taproot (~300 mg weight) 30 min in a Petri dish containing 3 mL of a release solution (see Note 10).
3. Cut the tissue by preparation needles (see Note 11). Collect a few released vacuoles with diameters 25–50 μm using a 10 μL micropipette with a cut tip and transfer them to the experimental chamber (~0.5 mL volume).
4. After adhesion of vacuoles to the bottom, pass 3 mL of fresh solution through the chamber to remove the contamination. The preparation containing a few vacuoles can be used for about 1 h. It should be renewed.
5. To obtain barley mesophyll (23) or root vacuoles, first prepare protoplasts from respective tissues (see Note 12).

3.4. Patch-Clamp Recording of Ion Channels in Protoplasts

1. Place the chlorinated silver wire in contact with the pipette solution, fix patch-pipette in the holder, and connect it to the patch-clamp amplifier. Connect the reference AgCl electrode to the bath via a 3.5% agar bridge filled with 100 mM KCl. The resistance of fire-polished patch-pipettes filled with a standard pipette solution and inserted in a standard bath should be 7–15 MOhm.
2. Cancel offset potential between patch and reference electrodes in a current-clamp (zero current condition), and run rectangular voltage test protocol (5 mV, 200 Hz).
3. Locate patch-pipette onto selected protoplast and press its tip lightly against the cell membrane. Then apply a gradual suction to assist the formation of a high-resistance seal between the glass and the cell membrane (a "gigaohm seal," typically 5–20 GOhm range). After achieving a tight seal in a cell-attached configuration (18), the membrane patch under micropipette needs to be ruptured. This provides low-resistance access to the protoplast interior (whole cell mode (18)), but a tight seal condition has to be maintained throughout the recording (see Note 13).
4. Switch amplifier to a voltage-clamp mode, measure whole cell capacitance and access resistance, compensate access resistance, using respective circuit of the patch-clamp amplifier. Suitable access resistance values are <50 MOhm and mean whole cell capacitance is 5–10 pF (see Note 14). Specific current is often expressed in pA/pF; 1 pF is approximately equivalent to 100 μm^2 of the membrane surface.
5. Switch off voltage test and apply a standard sequence of voltage steps (see Fig. 1) to evoke ionic currents. Individual currents may be distinguished by their voltage and time dependence and by reversal potential values (Fig. 1) (see Note 15). For time-dependent currents (KIRC, KORC, NORC-1, and NORC-2), a specific steady state current is the difference between the steady state current and the current recorded at the beginning of the voltage step. Instantaneous currents could be evaluated at any time, but unspecific leak contribution needs to be taken into the account (see Note 16).

3.5. Patch-Clamp Protocols and Recordings for Vacuolar Ion Channels

1. Select large (few tens of μm diameter) clean vacuoles for measurements.
2. Use a relatively wide (2–3 μm tip opening, $R = 3–5$ MOhm) fire-polished patch-pipette. Poise the pipette tip at the edge of vacuole, slightly pressing tonoplast from the top. Apply a gentle suction, achieve (normally, within a second or few) a GOhm seal between the pipette tip and tonoplast, and immediately break the tonoplast patch by application of voltage (zap) pulse, shortly followed by a strong and short suction. This leads to a

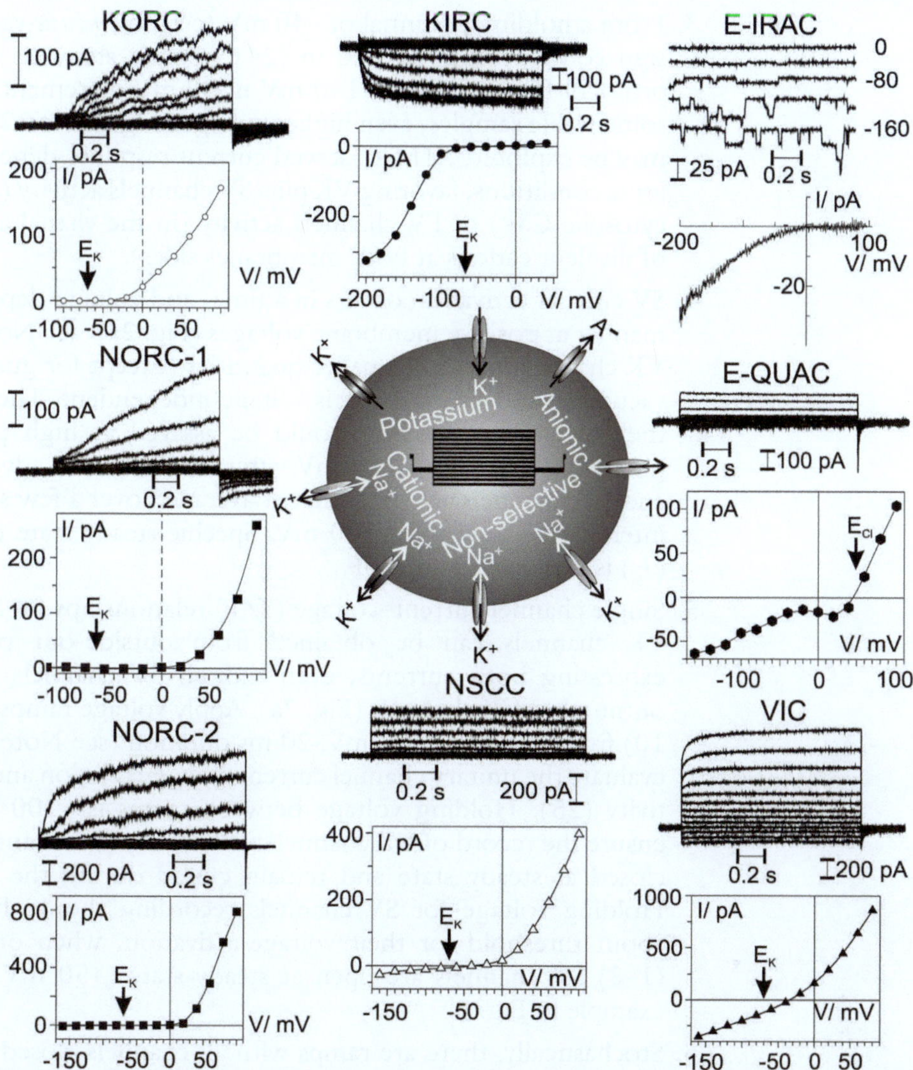

Fig. 1. Individual ionic currents expressed in the plasma membrane of protoplasts isolated from barley roots. Typical current traces recorded in the whole cell mode using standard bath and pipette-filling solutions, and steady state current–voltage relations derived from these experiments, are presented. From a holding potential of −100 mV, voltage was changed stepwise from −160 mV to 80 or 100 mV in 20 mV increments; at the end of each step the voltage was switched to −100 mV. At least three types of conductances are present: K^+-selective (time-dependent outward-rectifying KORC and inward-rectifying KIRC), non-selective cation (outward-rectifying time-dependent NORC-1 and NORC-2, and instantaneous VIC and NSCC), and anionic (E-QUAC and E-IRAC) ones. These conductances can be distinguished by reversal (zero current) potential values. K^+-selective currents show current reversal close to E_K approximately −72 mV, non-selective cationic currents reverse at slightly negative potentials, and anionic currents reverse at positive voltages close to E_{Cl} ~58 mV. Further currents classification is in accord with their voltage dependence and kinetic patterns.

whole vacuole configuration, essentially similar to the whole cell one (18) (see Note 17). Alternatively, a rapid pipette elevation after achieving a whole-vacuole configuration results in a formation of a right-oriented tiny (membrane capacitance C_m << 1 pF) outside-out patch (18) (see Note 18).

3. From a holding potential of −40 mV (cytosol-*minus*-vacuole), sign convention as advised in (24), apply a stepwise voltage protocol from −160 to +140 mV in 20 mV increments (with some stable samples, even higher voltage range up to ±200 mV may be explored). The observed current response depends on ionic conditions, favoring VK plus SV channels activity (at high cytosolic Ca^{2+}) or FV channels activity (in the virtual absence of divalent cations at both membranes sides).
4. SV current activation occurs in a time- and voltage-dependent manner at positive membrane voltages (Fig. 2a) (see Note 19). VK channels occur at smaller quantities (except for guard cell vacuoles), and their activity is voltage-independent. Therefore, the SV channel activity should be assayed at high positive potentials (+100 to +160 mV, after reaching a steady state), and VK channels activity (current averaged over a few seconds interval) is assayed at −100 mV. Specific steady state current (I_{ss}) is expressed in pA/pF.
5. Single channel current–voltage (I/V) relationships for SV and VK channels can be obtained from outside-out patches, expressing both currents, even though SV channels greatly outnumber VK channels (Fig. 2a). Apply voltage ramps (set of 10) from −150 to +150 mV, 20 ms duration (see Note 20) to evaluate the unitary channel current–voltage relation and selectivity (25). Holding voltage between ramps is −100 mV to ensure the record of VK channel current only (SV channels are closed at steady state and remain closed during the ramp). Holding voltage for SV channels recording should be kept about threshold for their voltage-activation, when only few (1–3) SV channels are open at steady state (+50 mV in the example in Fig. 2b).
6. Stochastically, there are ramps with all channels closed (background current only) and with one channel open. Subtract the averaged current of several ramps of the first type from that for the ramps of second type, yielding unitary channel current–voltage (I/V) curve, as ones shown in Fig. 2b.
7. This approach allows the accumulation of a large number of individual I/V relations for SV or VK in a short time. These can be further averaged to obtain a smooth and low-noise mean I/V relation. Once it is determined, divide whole vacuole steady state current (I_{ss}) through single channel current (I). This procedure yields a mean number of open channels at chosen test voltage (NP_o, where N is a total number of channels of certain type in the sample, and P_o is open probability). Express the SV or VK channels density as the (mean number of open channels)/pF.

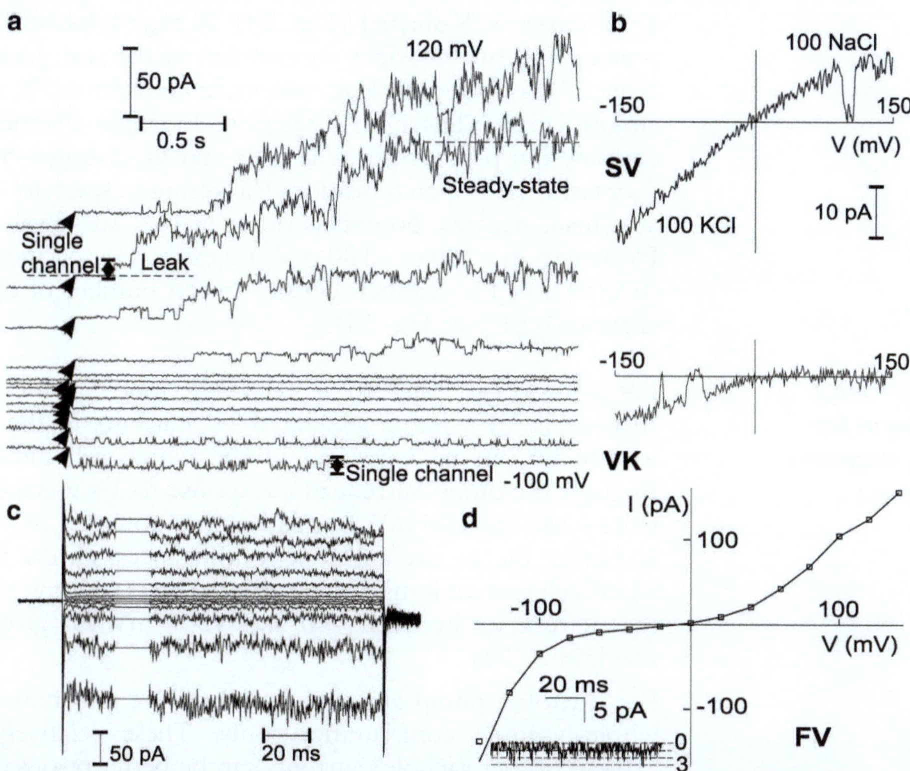

Fig. 2. Cationic currents of higher plant vacuoles. (**a**, **c**, **d**) Records obtained in symmetrical 100 mM KCl, pH 7.5. (**a**) Small ($C = 1$ pF) tonoplast vesicle contains at least 12 large conductance SV channels, which open with a delay after application of steps to $V > +40$ mV and a single voltage-independent VK channel (open 50% of time, as an average, 0.5 open VK channels/pF), whose activity could be seen at negative voltages (−80 and −100 mV), where SV channels are closed at all times. Voltage steps from a holding potential of 0 mV to more positive ones in 20 mV increments are indicated by *arrowheads*. (**b**) Single channel *I*/*V* relations, obtained in bi-ionic conditions (100 mM KCl vs. 100 mM NaCl) by application of rapid (15 mV/ms) voltage ramps for the sample, presented in A. While the SV has a slightly higher relative permeability for Na^+ as compared to K^+ (negative reversal potential), the VK does not conduct Na^+ at all (current at positive potentials asymptotically approaching *V*-axis). (**c**) In the absence of divalent cations on both membrane side, the tonoplast conductance is dominated by the FV current, which is activated instantaneously by the application of positive and negative voltage steps (same pulse protocol as in **a**, holding potential −40 mV). Steady state current I_{ss} can be determined at any time during the test pulse: here part of the current record is substituted by a mean current for a given interval. Around −40 mV the activity of FV channels is minimal, producing a high-resistance (fixed 20 mV voltage increments produce lesser current changes) region in the I_{ss}/V relation (see also **d**). Record from a small ($C = 2.3$ pF) right-oriented tonoplast vesicle. (**d**) Steady state FV current–voltage relationship obtained by plotting the mean current vs. voltage for the record in (**c**). The inset shows closures and openings of several FV channels in a small outside-out tonoplast patch clamped at −75 mV, current levels from 0 up to 3 open FV channels are indicated at the *right*. Mean single channel current is −2.0 pA at −75 mV. For the I_{ss}/V relation in **d**, mean current at −75 mV is equal to −16 pA (−7 pA/pF), which corresponds to 3.5 open FV/pF as an average at this voltage.

8. FV channels maximal activity is observed after the removal of all divalent cations from both tonoplast sides (see Note 21). At symmetrical 100 mM KCl, FV channels display a minimal activity around −40 mV. Steps from this holding voltage value to higher positive or negative potentials cause a rapid (<0.1 ms) FV current activation (Fig. 2c). The respective steady state

I_{ss}/V curve is N-shaped (Fig. 2d). A high-resistance region around −40 mV provides an estimate of the seal quality (see Note 22), as unspecific leak current, in contrast to FV, displays almost Ohmic (linear) I/V relation. In single channel mode (outside-out patches), FV channels may be distinguished by a very fast closed-open transitions (flickering), especially obvious at cytosol negative potentials (inset in Fig. 2d). Evaluate the FV activity at +100 or −100 mV and express as specific current (pA/pF), or FV channels density (mean number of open FV channels)/pF (see Fig. 2d).

3.6. Patch-Clamp Recording of the Currents Generated by H^+ Pumps

1. For protoplasts, due to a relatively low current density (0.1–1 pA/pF), pump-generated H^+ currents in the plasma membrane can be measured in the whole cell mode only. Register the pump current in a response to 1 s voltage ramps from −200 to +100 mV, from a holding potential of −90 mV. Minimize the passive and leak conductances, ensure that the whole cell current is mainly contributed by H^+ pumps, verifying that it reverses between −80 and −200 mV for given ionic conditions (22).
2. For vacuoles, pump-generated currents are also measured in whole vacuole configuration only. These relatively small (10–50 pA per vacuole) currents can be better resolved, minimizing passive and leak conductances. Clamp the voltage at 0 mV (for symmetric ionic conditions) and perform a stepwise increase in the concentration of Mg-ATP in the bath, resulting in a stepwise increase of H^+ pump-generated current (Fig. 3) (see Note 23).

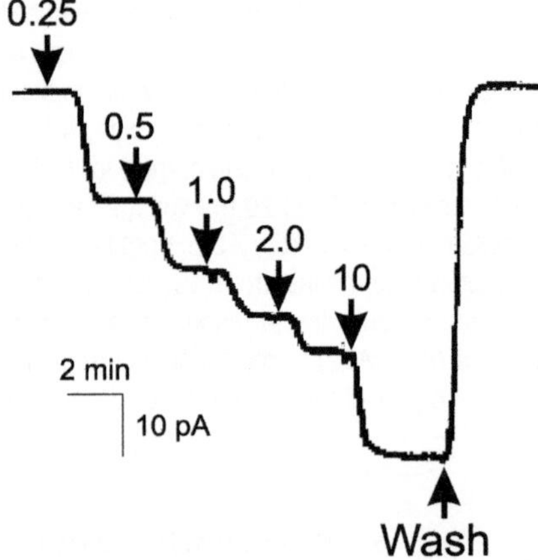

Fig. 3. Voltage-clamp recordings of ATP-induced pump currents in the whole vacuole mode. Membrane potential was held at 0 mV. Increasing the Mg-ATP concentration in steps (numbers above the *arrows* indicate Mg-ATP concentrations in mM) by bath perfusion caused a stepwise increase of the total inward current from 0 up to 65 pA at 10 mM Mg-ATP. When ATP was removed from the bath, the inward current declined to zero. (Reproduced from Hedrich et al. (29) with permission).

3. Verify the H^+ pump-specific current by bath application of dicylohexylcarbodiimid, DCCD (100 μM). Stimulation by fusicoccin (1 μM) allows the estimate of maximal activity of H^+ pumps present in the cell membrane.

4. Notes

1. Barley is monocotyledon and generally salt-resistant (yet some barley varieties such as Gairdner or Franklin are salt-sensitive). We also tested some salt-sensitive species as pea (*Pisum sativa* L. cv Greenfeast), a dicotyledon plant.
2. Mix 100 mM KCl and 3.5% agar in water for a 10 mL glass flask. Heat, but not boil, the mixture until the solution turns clear, then fill carefully in polyimide microtubing with suction, and leave at room temperature until agar solidifies. Some batches of commercially available agar remain liquid at room temperature even at 3.5%, thus the agar content needs to be increased in these cases.
3. Prepare Mg-ATP stock solution on the same day of the experiment and store it frozen in aliquots of 0.5 mL.
4. *N*-methylglucamine and Tris are impermeable monovalent cations and are added to reduce the background conductance via NSCCs. K^+ is excluded from all solutions (substituted for Mg^{2+} and *N*-methylglucamine). Gluconate and glutamate are considered to be relatively impermeable for most anion channels. However, the presence of several mM of Cl^- in bath and pipette is required for the operation of AgCl electrodes. Appropriate liquid junction potentials correction is required for solutions, containing low mobility cations and anions (consult JPCalc program by P. H. Barry, University of New South Wales, Sydney, NSW, Australia).
5. Take into account that each plant is different, therefore for every plant species/variety/tissue the solution osmolality needs to be adopted in accord with the actual cell sap osmolality. The degree of lignification will also affect the incubation times in the enzymatic solution.
6. It is more difficult to achieve high-resistance GOhm seals with protoplasts obtained from >6 days old plants.
7. Protoplasts from the mature root zone normally are released spontaneously into the measuring chamber. To release protoplasts from the elongation zone, use a longer exposure to the enzyme solution (1–2 h), plus a subsequent centrifugation step (3 min at 250 g).

8. This is more crucial for other plant preparations (e.g., pea root protoplasts).
9. Protoplast selection for patch-clamp experiments is based on the protoplast diameter (15 μm or less for the barley root mature zone), which is indicative of their epidermal origin; cortex protoplasts are twice as large on average.
10. Measure first the osmolality of the cell sap, which varies between 300 and 800 mOs, depending on the season, age, and root size. Set the osmolality of experimental solutions to isotonic or slightly (by <10%) hypertonic with respect to the beetroot sap.
11. Just make a few cuts to reduce tissue debris.
12. To release a vacuole, rupture the plasma membrane of a single selected protoplast by a strong suction pulse applied to a patch-microelectrode filled with the same solution as the bath (it could be the patch-pipette used in the previous experiment). Substitute the patch-electrode with a new (clean) one, wait until a large vacuole is released from a broken protoplast, and obtain a gigaseal contact between the patch-pipette tip and vacuolar membrane, to initiate patch-clamp measurements.
13. Protoplasts, running from the pipette or rotating, are not able to form tight seals. Select only those well attached to the bottom. Initiate the formation of a GOhm contact of the patch-pipette tip with a plasma membrane by a gentle suction and is often a slow (few minutes) process. Monitor sealing by applying 5 ms rectangular pulses of 5–10 mV amplitude. Sealing could be facilitated by a moderate negative (−50 mV) command potential applied to the interior of the patch-pipette in the voltage-clamp mode. At this moment, a background current reflects mainly leak current. Its decrease by absolute magnitude implies the seal resistance increase. Successful samples rapidly reach a seal resistance of >200 MOhm; reject those that do not approach this level within 1–2 min. Apply suction in a gradually increased and/or cyclic mode. Make a pause in suction every time, when the background current level decrement stops, or when it tends to increase. Normally, a transition from sub-GOhm to a final few GOhm seal resistance occurs spontaneously. After achieving a tight seal, apply brief and stronger suction pulse, until the patch is ruptured and low-resistance access to the protoplast interior is gained. Optionally, an extra short but potent voltage pulse (zap) could be delivered beforehand, destabilizing the membrane patch under the pipette tip; with some preparations, however, this often leads to a disruption of the whole protoplast. The transition to the whole cell configuration is manifested by large capacitance current artifacts in response to test voltage stimulation.

The magnitude of this artifact is inversely proportional to the access resistance (R_a) and characteristic decay time to R_aC_m, where C_m is the cell membrane capacitance. Flatter (low-magnitude and slowly decaying) artifacts imply a high access resistance. It is very important for a stable whole cell record to clean constantly the access into the cell via the pipette by periodic suction pulses in the first few minutes (the access to the cell tends to be re-plugged), monitoring the decrease of R_a (artifact shape) on the screen. Normally, after 5 min the R_a value stabilizes, and actual measurements may be started.

14. The whole cell capacitance is measured, using RC-compensation circuit of the patch-clamp amplifier. The capacitance varies according to the diameter of protoplast, depending on the tissue and plant species; e.g., in pea, typical epidermal protoplast's capacitance is in the range between 2 and 6 pF. Whole cell capacitance of a spherical protoplast can be approximated using the formula, $C_m = \pi d^2 10^{-2}$ pF/μm², where d is the protoplast diameter in μm.
15. Pea protoplasts, assayed under these conditions, mainly express KORC and NSCC. More than one individual current could be registered in a single root barley protoplast; some coupled currents expression as E-QUAC + KORC, E-IRAC + NORC-1, or E-IRAC + NSCC-2 are common. K^+-selective channels (KIRC and KORC) often display a rundown of activity, whereas anionic current E-QUAC on the contrary show a run-up, reaching a steady state level 20–30 min after achievement of a whole cell configuration.
16. Both anionic current, IRAC and QUAC, display high-resistance regions in their I/V relationships. These can be used to estimate the upper limit of the leak conductance. Unspecific leak current, in contrast to this generated by NSCC or VIC, is not very sensitive to gadolinium. Add 0.1 mM $GdCl_3$ into the bath at the end of experiment to reveal the leak contribution into the overall measured current.
17. In some cases whole vacuole SV currents are too large (>10 nA) to be correctly voltage-clamped. In this case, isolate a small (capacitance of 1–4 pF) right-oriented tonoplast vesicle from a large central vacuole. This is achieved by a slow lifting of the patch-pipette, initially (before start of the sealing procedure) poised approximately halfway from the edge to the center of vacuole.
18. The tonoplast is a much more elastic membrane than the plasma membrane. During withdrawal of the pipette to achieve the outside-out configuration, the tonoplast patch can spontaneously form a closed vesicle, accompanied by an apparent decrease of a temporal resolution and lower noise. Single

channel currents become distorted in shape: currents show rounded rising and falling time courses and progressively lower amplitudes. If this problem persists, reject the further record.

19. The threshold voltage for the SV channels activation increases with an increase in vacuolar Ca^{2+} or Mg^{2+} (in the virtual absence of these cations, SV channel openings may be observed at voltages as low as −100 mV, and the activation becomes much faster) and decreases with the increase of cytosolic Ca^{2+} or Mg^{2+} (26, 27).

20. This is an optimal duration to "trap" single channel in a certain state, closed or open. Longer ramps may result in a voltage-dependent channel closure or opening of multiple channels during the ramp. A shorter ramp will produce a progressively higher capacitance current artifact, interfering with a channel-mediated current.

21. FV channels, even maximally active, produce currents of few nA at the whole vacuole level. This could be correctly measured under patch-clamp conditions, providing the appropriate access resistance compensation is made (18). After achievement of the whole vacuole configuration, a run-up of the FV current is observed, until it reaches a steady state in some minutes. This probably reflects the relief of block by intravacuolar divalent cations due to their chelating with EGTA present in the pipette solution. No such lag exists or it is substantially shorter, when FV channels are assayed in tiny outside-out patches.

22. Presence of high (tens of mM) NH_4^+ at cytosolic side increases the FV channel mean open time and the FV channel open probability around −40 mV. This makes this minimum less pronounced and a steady state I_{ss}/V relation less non-linear; a position of minimum/rectification of I_{ss}/V relation is also modified by vacuolar and cytosolic K^+ (28). Vacuolar Ca^{2+} and Mg^{2+} abolish the FV-mediated inward (cytosol-directed) current and transform the N-shaped I_{ss}/V relation (Fig. 2d) into an outward-rectifying one (23).

23. The reported mean K_m for Mg-ATP is about 1 mM for H^+-ATPase, thus addition of 5 mM Mg-ATP is sufficient to produce 80–90% of maximal H^+-ATPase-generated current (29). H^+-PPase current could be assayed in the same mode, but instead of Mg-ATP, 0.1 mM sodium pyrophosphate, 0.2 mM ADP plus 1 mM potassium phosphate is added (30). The identity of the H^+-PPase and H^+-ATP can be resolved by the addition of 100 nm bafilomycin A1 or 1 μM tributyltin, H^+-PPase and H^+-ATP inhibitors, respectively (29, 30).

Acknowledgments

This work was supported by the CONACyT grant (CB-2007 82913) to IP, and CONACyT PhD fellowships to AMVB and RAEF.

References

1. Marschner H (1995) Mineral nutrition of higher plants. Academic, London
2. Maathuis FJM, Amtmann A (1999) K^+ nutrition and Na^+ toxicity: the basis of cellular K^+/Na^+ ratios. Ann Bot 84:123–133
3. Blumwald E, Aharon GS, Apse MP (2000) Sodium transport in plant cells. Biochim Biophys Acta 1465:140–151
4. Munns R (2002) Comparative physiology of salt and water stress. Plant Cell Environ 25:239–250
5. Tester M, Davenport R (2003) Na^+ tolerance and Na^+ transport in higher plants. Ann Bot 91:503–527
6. Apse MP, Blumwald E (2007) Na^+ transport in plants. FEBS Lett 581:2247–2254
7. Chen ZH et al (2007) Potassium and sodium relations in salinised barley tissues as a basis of differential salt tolerance. Funct Plant Biol 34:150–162
8. Sun J et al (2009) Ion flux profiles and plant ion homeostasis control under salt stress. Plant Signal Behav 4:261–264
9. Demidchik V, Tester M (2002) Sodium fluxes through nonselective cation channels in the plasma membrane of protoplasts from *Arabidopsis* roots. Plant Physiol 128:379–387
10. Silva P, Gerós H (2009) Regulation by salt of vacuolar H^+-ATPase and H^+-pyrophosphatase activities and Na^+/H^+ exchange. Plant Signal Behav 4:718–726
11. Chen ZH et al (2007) Root plasma membrane transporters controlling K^+/Na^+ homeostasis in salt stressed barley. Plant Physiol 145: 1714–1725
12. Zepeda-Jazo I et al (2007) Na-K transport in roots under salt stress. Plant Signal Behav 3: 401–403
13. Pottosin II et al (2003) Potassium-selective channel in the red beet vacuolar membrane. J Exp Bot 54:663–667
14. Maathuis FJM, Prins HB (1991) Inhibition of inward rectifying tonoplast channels by a vacuolar factor: physiological and kinetic implications. J Membr Biol 122:251–258
15. Maathuis FJM, Sanders D (2001) Sodium uptake in Arabidopsis roots is regulated by cyclic nucleotides. Plant Physiol 127:1617–1625
16. Volkov V, Amtmann A (2006) *Thellungiella halophila*, a salt-tolerant relative of *Arabidopsis thaliana*, has specific root ion-channel features supporting K^+/Na^+ homeostasis under salinity stress. Plant J 48:342–353
17. Demidchik V, Maathuis FJM (2007) Physiological roles of nonselective cation channels in plants: from salt stress to signalling and development. New Phytol 175:387–404
18. Hamill OP et al (1981) Improved patch-clamp techniques for high-resolution current recording from cells and cell-free membrane patches. Pflugers Arch 391:85–100
19. Hedrich R et al (1988) General mechanisms for solute transport across the tonoplast of plant vacuoles: a patch-clamp survey of ion channels and proton pumps. Bot Acta 101:7–13
20. Lohse G, Hedrich R (1992) Characterization of the plasma-membrane H^+-ATPase from *Vicia faba* guard cells. Modulation by extracellular factors and seasonal changes. Planta 188:206–214
21. Tyerman SD et al (2001) Oscillations in proton transport revealed from simultaneous measurements of net current and net proton fluxes from isolated root protoplasts: MIFE meets patch-clamp. Aust J Plant Physiol 28:591–604
22. Zhu J, Raschke K, Köhler B (2007) An electrogenic pump in the xylem parenchyma of barley roots. Physiol Plant 129:397–406
23. Tikhonova LI et al (1997) Fast-activating cation channel in barley mesophyll vacuoles. Inhibition by calcium. Plant J 11:1059–1070
24. Bertl A et al (1992) Electrical measurements on endomembranes. Science 258:873–874
25. Dobrovinskaya OR, Muñiz J, Pottosin II (1999) Asymmetric block of the plant vacuolar Ca^{2+}-permeable channel by organic cations. Eur Biophys J 28:552–563

26. Pottosin II et al (1997) Slowly activating vacuolar channels cannot mediate Ca^{2+} induced Ca^{2+} release. Plant J 12:1387–1398
27. Carpaneto A, Cantú AM, Gambale F (2001) Effects of cytoplasmic Mg^{2+} on slowly activating channels in isolated vacuoles of *Beta vulgaris*. Planta 213:457–468
28. Pottosin II, Martínez-Estévez M (2003) Regulation of the fast vacuolar channel by cytosolic and vacuolar potassium. Biophys J 84:977–986
29. Hedrich R, Flügge UI, Fernandez JF (1986) Patch-clamp studies of ion transport in isolated plant vacuoles. FEBS Lett 4:228–232
30. Hirata T et al (2000) Regulation and reversibility of vacuolar H^+-ATPase. J Biol Chem 275:386–389

Chapter 2

Studying Plant Salt Tolerance with the Voltage Clamp Technique

Zhong-Hua Chen, Dezhi Wu, Cornelia Eisenach, Adrian Hills, Guoping Zhang, and Michael R. Blatt

Abstract

Voltage clamp is one of the key techniques for the dissection, identification, and monitoring of ion transporters in plant cells. Voltage clamp-based research work on salinity stress in plants enables the characterization of many plant ATP-dependent pumps, ion channels, and ion-coupled carriers through heterologous expression in *Xenopus laevis* oocytes and *in vivo* measurements in salt-tolerant and salt-sensitive giant green algae such as *Chara* and many plant species. We have modified and developed a reliable set of procedures for voltage clamp analysis in intact guard cells and root epidermal cells from *Arabidopsis thaliana* with potentially broad applications in the salinity response of plants. These procedures greatly extend the duration of measurements and scope for analysis of the predominant K^+ and anion channels.

Key words: Voltage clamp, Microelectrode, Voltage-gated K^+ and anion channels, Guard cell, Root epidermal cell, *Arabidopsis thaliana*

1. Introduction

Our understanding of ion transport in living cells has been greatly improved by taking advantage of many technological breakthroughs including voltage- and patch clamp techniques (1–5). The voltage clamp method enables the dissection, identification, and monitoring of ionic currents carried by various ion transporters: ion channels, ATP-dependent pumps, and ion-coupled carriers across biological membranes in plants (4, 6, 7). Generally, this technique relies on impalements with two microelectrodes or a single microelectrode with two separate barrels to "clamp" the cell at sets of chosen membrane voltages. It is thus possible to measure

Sergey Shabala and Tracey Ann Cuin (eds.), *Plant Salt Tolerance: Methods and Protocols*, Methods in Molecular Biology, vol. 913, DOI 10.1007/978-1-61779-986-0_2,

ionic current across the membranes, to gather data for the calculation of membrane conductance of different ions, to monitor changes in membrane conductance over time in the order of *ms* or less, and to determine the equilibrium potential of the membrane to different ions (8, 9). Using the concept of voltage-operated gates for a formal description of conductance changes and conducting a series of voltage clamp experiments on squid giant axons, Hodgkin and Huxley (1952) formulated the modern concept of ion channels in cell membranes (10, 11). This was a landmark for giving insights into cellular function and formed the foundation for subsequent electrophysiological work (9). Later, the voltage clamp was employed for characterizing action potential in green algae and higher plants (12, 13). Since the 1980s, voltage clamp has been extensively used to monitor the activity of proton pumps (6, 14, 15), K^+-, and anion channels (16–18) in higher plant cells. Overall, the voltage clamp technique is one of the most powerful tools in connecting the molecular and physiological framework for the function of ionic channels in numerous organisms including walled cells, notably, fungi such as *Neurospora* (19), giant algal cells (20–22), stomatal guard cells (23–25), root hairs (16, 26), and root epidermal cells (27, 28).

The other major development in electrophysiology has been the introduction of the patch clamp technique, an advanced version of the voltage clamp, which increases the numbers of ion transporters discovered in the membranes of higher plant cells (29–31). The patch clamp technique resolves currents at both the whole cell and single channel level and allows the separate measurement of ionic currents through the plasma membrane and the vacuolar membrane (5, 7, 32). However, for patch clamp experiments, plant cell walls need to be removed by enzymatic digestion and protoplasts require stabilization against osmotic swelling in the absence of turgor. This affects the underlying homeostatic properties and physiological behavior of the cells (33, 34). Additionally, obtaining active and robust seals between the patch pipette and plant protoplast and retaining stable measurements over long time are quite challenging (31).

Plant salinity tolerance is a polygenic trait with contributions from genetic, developmental, and physiological interactions, in addition to interactions between the plant and its environment (35–37). Changes in plasma membrane potential and modulation of ion flux in plant cells are amongst the earliest cellular events in response to salinity (38–40). Therefore, the importance of membrane transport in plant adaptive responses to salt stress renders ion transporters as important targets in the research area of plant salinity tolerance. The combination of electrophysiological and molecular biological methods has greatly advanced our understanding of the plant salt tolerance in the past few decades (1, 4, 36, 37, 40). The voltage clamp technique, a key electrophysiological tool, is frequently used for characterizing plant ion transporters in response to salinity (1, 4, 41). The voltage clamp is heavily used for the identification and characterization of many plant ion channels and co-transporters heterologously expressed in oocytes of *Xenopus laevis* (42–44). Furthermore, voltage clamp studies of proton pumps, K^+ channels, Cl^- channels, and mechanosensory ion channels have been intensively deployed in salt-tolerant and salt-sensitive giant algae *Chara* (45–48).

It is well-known that intracellular K^+/Na^+ homeostasis is one of the key components of salinity tolerance in plants (40, 49, 50). First of all, high K^+/Na^+ is maintained by restricting unidirectional root Na^+ influx mediated by the large family of nonselective cation channels (NSCCs) (51–53) and high-affinity K^+ transporters (HKTs) (44, 54, 55). For instance, the mechanisms of salt response in salt-sensitive and salt-tolerant wheat genotypes are based on differences in the K^+/Na^+ selectivity of cation channels (56). Also, electrophysiological studies revealed that HKTs from many plant species mediate high-affinity Na^+–K^+ co-transport and Na^+-selective Na^+ transport in *X. laevis* oocytes and *saccharomyces cerevisiae* (42, 55, 57). Therefore, NSCCs- and HKTs-regulated Na^+ accumulation play a central role in plant salt tolerance (53, 58). Secondly, under salt stress, low cytosolic Na^+ concentrations can also be sustained by active transport of Na^+, which is usually mediated by plasma membrane Na^+/H^+ antiporters (59, 60) and tonoplast Na^+/H^+ antiporters (NHXs) (61, 62). For example, the salt overly sensitive (SOS) pathway involved in salt stress is critical for maintaining low cytosolic Na^+ concentration by virtue of Na^+ extruding plasma membrane Na^+/H^+ antiporters (SOS1) (60, 63). At the same time, overexpression of the *Arabidopsis* tonoplast membrane Na^+/H^+ antiporter, AtNHX1, results in salt-tolerant *Arabidopsis* (61), *Brassica napus* (64), and *Lycopersicon esculentum* (62). In addition, other salt stress-responsive ion transporters such as K^+ transporters from the KUP/HAK/KT and AKT families and anion channels have also been studied using voltage/patch clamp techniques, yielding profound insights into salinity tolerance (4, 58, 65).

Manipulation of ion transporters in order to decrease the entry of Na^+ through NSCCs and HKTs, to enhance H^+-pump-fuelled Na^+ extrusion via SOS/NHXs, and to reduce K^+ loss will all contribute to the improvement of plant salinity tolerance. Each of those approaches requires detailed investigation using electrophysiological techniques. Therefore, the reintroduction and refinement of the existing voltage clamp technique on higher plant cells will further extend our knowledge of the underlying mechanisms of membrane transport for plant salinity tolerance. Here we use mainly *Arabidopsis* guard cells to introduce the refined voltage clamp technique in details.

2. Materials

2.1. Laboratory Items

2.1.1. Chemicals and Reagents

1. KCl analytical grade.
2. $Ca(OH)_2$ analytical grade.
3. NaOH analytical grade.
4. HCl analytical grade.
5. $NaHCO_3$ analytical grade.
6. MES [2-(*N*-morpholino)ethanesulfonic acid] analytical grade.
7. CsCl analytical grade.
8. TEA-Cl (tetraethylammonium chloride) analytical grade.
9. K-Ac (potassium acetate) analytical grade.
10. MS Basal medium powder.
11. Triton-X100 detergent.
12. Double distilled water.

2.1.2. Glassware and Plasticware

1. Volumetric glassware: 50, 100, 250, 500, and 1,000 mL screw-topped bottles.
2. Glass pipettes.
3. Glass capillaries for double-barrelled microelectrodes.
4. 0.4 mm-thick glass for measuring chambers.
5. Cover slips.
6. Plastic measuring cylinders: 50, 100, 250, 500, and 1,000 mL.
7. Six-well sterile plastic plates.
8. 2-mL plastic pipettes.

2.1.3. Equipments

1. Autoclave system.
2. Refrigerator.
3. pH meter.
4. Analytical balance.
5. Calculator.
6. Narashige PD5 multipurpose microelectrode puller.
7. Conventional multichannel voltage clamp amplifiers and probes.
8. Desktop computer with Windows operating system (see Subheading 3.4.1 for minimal system requirements).
9. Data acquisition card connected to the computer.
10. Light microscope with a maximal total magnification at least ×400 or higher.

11. Car battery for DC power supply of microscope.
12. Three-dimensional Narashige micromanipulators.
13. Faraday cage.
14. Anti-vibration table.
15. Gravity-fed system for switching between experimental solutions.

2.1.4. Accessories

1. Optically clear and pressure-sensitive silicone adhesive.
2. Fine-tipped forceps and dressing forceps.
3. Razor blades.
4. Customized metal pieces for measuring chambers.
5. Parafilm.

2.2. Experimental Materials

2.2.1. Guard Cells

The vast amount of genetic resource available for *Arabidopsis* makes this species particularly attractive for molecular genetic studies of guard cell homeostasis, transport, and signaling processes and is not matched at the cellular level by accessible tools such as the voltage clamp technique for quantitative analysis of transport. Thus, we have conducted a series of technical refinements using guard cells of *Arabidopsis* wild type Columbia-0 (Col-0) and some mutants.

2.2.2. Root Epidermal Cells

The root epidermis is at the forefront of the response to salinity stress and is easily accessible by microelectrodes. Consequently, it is one of the most studied plant cell types for salinity stress (4, 37, 40). We have recently demonstrated that *Arabidopsis* root epidermal cells without root hairs are also accessible for the voltage clamp experiments (28). Some specific details on root epidermal cells are described in this article.

2.2.3. Other Types of Cells

Many studies characterizing plant membrane transporters and their response to salinity have been conducted in organisms such as giant algae *Chara* and *Xenopus* oocytes (27, 45, 66, 67) other than higher plant cells. Electrophysiological studies on these cell types are similar and often easier due to their size and membrane characteristics.

3. Methods

3.1. Plant Growth Conditions

Optimal growth conditions for cultivating *Arabidopsis* plants are critical for obtaining ideal guard cells and root epidermal cells (see Note 1) for the voltage clamp experiments.

3.1.1. Arabidopsis Plants for Voltage Clamp in Guard Cells

1. Uniform seeds are sown onto the nutrient-rich Levington F2+S 3 compost (Coulders, Glasgow, UK).
2. Compost is pre-treated with Intercept 70WG (Scotts, Ipswich, UK), a systemic insecticide.
3. After sowing, seeds are stratified at 4°C for 48 h and germinated under a plastic lid to give >90% relative humidity for 1 week.
4. All the plants are cultivated in a controlled environment growth chamber under long-day conditions (photosynthetically active radiation at 100 μmol/m^2/s^1, light/dark cycle at 16 h/8 h, temperature of 22/18°C, and 55/70% relative humidity).
5. Plants are evenly and regularly watered from below to reduce mechanical damage.
6. To avoid soil contact of the abaxial leaf surface and soil-borne stress factors, plants are grown in individual 60-mm pots covered with polyester mesh.
7. Plants are grown under propagators with an NITEX 200-μm mesh fabric (Sefar, Heiden, Switzerland) at all four sides, allowing the maintenance of relative humidity, CO_2 level, and temperature, while keeping out any insects.

3.1.2. Arabidopsis Seedlings for Voltage Clamp in Root Epidermal Cells

1. Seeds of *Arabidopsis* Col-0 are sterilized for 10 min in 10% $NaHClO_3$ with 1% Triton-X100.
2. Seeds are stratified in the dark for at least 2 days at 4°C.
3. Seeds are then incubated in 6-well plates containing 3 mL 0.5× autoclaved MS medium.
4. Seedlings are germinated and grown for 6–8 days under constant light at 80 mmol/s^1/m^2 light and 22°C.
5. All the preparations are conducted to minimize contamination to the seedlings.

3.2. Microelectrodes

3.2.1. Microelectrodes for Arabidopsis Guard Cells

1. All microelectrodes are pulled to give tip resistances of 300–500 MΩ for easy impalement through *Arabidopsis* guard cell wall and plasma membrane (see Note 2).
2. The typical characteristics for the time of pulling and the shape of the double-barrelled microelectrodes based on the PD5 Narashige horizontal puller are initial heating at 3.5, second heating at 3.0, and magnetic pulling force at 9.0.
3. After 60 s of initial heating, the microelectrodes are twisted for 360°, followed by 30 s of cooling before magnetic pulling for around 25 s with heating.
4. Leave the two double-barrelled microelectrodes and heating coil to cool off for 30 s, and then remove them from the puller.

5. Microelectrodes are stored in a small glass desiccator and are coated with paraffin before impalement to reduce capacitance.

3.2.2. Microelectrodes for Arabidopsis Root Epidermal Cells

Similar approaches are suitable to pull microelectrodes for the voltage clamp experiments on root epidermal cells. However, due to the thick wall and large size of root epidermal cells, the microelectrodes should have much stronger tips and shorter shanks to avoid bending during impalement. It also requires a reasonably fine tip with resistance of 100–200 MΩ.

3.3. Impalement

3.3.1. Impaling Guard Cells on Epidermal Strips

1. The fifth and sixth true leaves of approximately 20-day-old young plants are excised as close to the hypocotyls as possible and peels are taken by wrapping each excised leaf over a finger, adaxial side down, and cutting into the mesophyll near the base of the mid-vein with fine forceps.
2. Leaf epidermis is then stripped off the mesophyll from the middle vein towards the leaf margin with a section larger than 10 mm^2 in order to gain enough guard cells for the experiments.
3. The epidermis attached to the leaf is gently placed back onto the mesophyll and a fine cut is made near the leaf margin using a sharp razor blade.
4. The abaxial side of the leaf is then gently pressed onto the glass of a measuring chamber covered with an optically clear, pressure-sensitive adhesive without delay. The remaining leaf tissue is removed and the peel is immediately covered with an opening buffer solution (see Note 3) to avoid drying of the strip.
5. Switch on all the relevant electronic equipment 1 h before experiments to "warm up" the voltage clamp system.
6. Fix the measuring chamber containing the opening buffer-pretreated leaf epidermal strips under a microscope in a Faraday cage.
7. Firmly connect a 1 M KCl|Ag-AgCl halfcell (68) to a reference electrode (one inner diameter 1.5-mm rubber tube containing 3% Agar and 1 M KCl) using dental impression compound.
8. The reference electrode is then connected to the buffer solution, mounted with the measuring chamber and grounded.
9. Firmly connect another halfcell with the voltage barrel of the double-barrelled microelectrode.
10. Link halfcells with two voltage clamp headstages. These are mounted on the 3-D Narashige micromanipulator with a 45° and are connected to the voltage clamp amplifiers.
11. Adjust micromanipulator and microscope to position the microelectrode right on top of the adjacent guard cell, where

the tip of microelectrode is positioned 1–2 μm away from the target guard cell.

12. The initial movement of the tip of the microelectrode towards the guard cell requires very gentle manipulation.
13. Successful penetration is marked by a "snapping" of the tip through the cell wall and into the guard cell (see Note 4) and, by an increase in input resistance and decrease (more negative) in membrane voltage over 2–3 min.
14. After the impalement, wait 2–3 min for a seal to stabilize before running the first voltage clamp protocol.
15. Run the voltage clamp protocol 2–3 times within the first 10 min and use the second or third recording as control (see Note 5), followed by switching to different treatments via the gravity-fed system.
16. Leave a few minutes between each run for the guard cell to rest and readjust.

3.3.2. Penetrating Root Epidermal Cells

Similar approaches are used to impale root epidermal cells, except the specific conditions mentioned in the above sections.

3.4. Software

Although other commercial software is available, Henry EP Suite (Y-Science, University of Glasgow) is our primary tool in recording and controlling voltage clamp experiments. We present some details of the requirements, installation, and procedure of this software for voltage clamp experiments. For more details, please refer to the Henry's EP Suite website: http://www.psrg.org.uk/henrys-ep-suite.htm.

3.4.1. System Requirements

1. CPU: Intel Pentium II equivalent or higher.
2. RAM: 256 MB or more.
3. HDD: 100 MB or more space for installation and data storage.
4. Microsoft Windows 2000, XP, Vista, or 7.

3.4.2. Software Installation

1. Gain administrator's rights in order to install the software package.
2. Download the Henry's EP Suite setup program for 32- or 64-bit windows from the above website.
3. Run the package file ep-win32.msi (or ep-win64.msi) and follow the instructions in the setup wizard.
4. After installation, the program can be found in Henry's EP Suite folder in the Programs from the Start menu. Henry III (the main program) and Vicar V3 (the virtual chart recorder) are both in the Henry's EP Suite folder.

3.4.3. Voltage Clamp Protocol Design

A protocol defines the way in which a particular voltage- or current-clamp experiment is performed. The basic parameters for the protocol include: type of clamp, number of cycles, time for each cycle, data-saving options, etc.

1. Open Henry III and choose the *File* menu and *New* command: a default protocol will show on the main screen.
2. Choose the *Edit* menu and *Protocol Option*, and *Protocol Editor*, to further edit appropriate protocols suitable for specific ion channels.
3. Save the protocol after editing as a Henry's Protocol File with the "hpr" extension in a defined or user-specified folder.

3.4.4. Running a Protocol for Voltage Clamp

1. Choose the *File* menu and *Open* command to select a specific protocol file for the voltage clamp experiment.
2. Choose the *File* menu and *Run* command when you have impaled and obtained a cell with a reasonable seal.
3. Depending on the Trigger Mode you use, the program will either run immediately or wait for a Keyboard or External event before starting the run.
4. A run in progress can be aborted using the *Close* command from *File* menu or closing the Run Monitor window.
5. A run usually stops automatically after completing all the voltage clamp cycles in a protocol and data will be saved as a Henry's Raw-Data file "hrd."
6. It is convenient to monitor voltage clamp traces while the protocol is running. All the Run-Time Windows can be arranged through the commands in the *Window* menu.
7. To create a new sketchpad image (see Note 6), use the *Copy* command from the *Sketchpad* menu; to overlay the active view onto the existing sketchpad image, use the *Add* command from the *Sketchpad* menu.

3.4.5. Running the Virtual Chart Recorder (Vicar V3)

The Vicar-V3 Virtual Chart Recorder is useful for monitoring the input of sophisticated recording of multichannel data over long periods of time without having to use a standard paper-based recording device. It can run simultaneously with Henry III to collect data from up to 16 channels such as membrane potential, clamp voltages, ion channel current, etc. The data collected by Vicar is essential for analyzing voltage clamp data at a later stage.

1. Open the Start menu and choose Vicar V3 from Henry's EP Suite from Programs menu.
2. Choose the *Start* command in the *File* menu to record all the measurements for the day.

3. Select the channels, usually one voltage and one current channel, from which the data are to be collected.
4. Choose *Define Strings* command in the *Edit* menu to define the actions, treatments, and other important information while collecting data.
5. Files can be saved during a run or after finishing all the experiments at the end of the day.

3.5. Data Recording and Analysis for K^+ and Anion Channels

Electrophysiological recordings using double-barrelled microelectrodes on guard cells and root epidermal cells are described elsewhere (28, 69). For $I_{K,in}$ and $I_{K,out}$ measurements, current- and voltage-recording barrels are filled with 200 mM K-Ac at pH 7.5 to minimize the interference of anion current, while a standard MES buffer was used as the bathing solution. When it comes to measuring I_{anion}, both microelectrode barrels are filled with 200 mM CsCl at pH 7.5 and a second standard buffer (15 mM CsCl, 15 mM TEA-Cl, 5 mM Ca^{2+}-MES at pH 6.1). Moreover, all recordings are analyzed and leak currents subtracted using standard methods (7) with Henry's EP Suite software. Surface areas and volumes of impaled guard cells are estimated and current density is then calculated according to the cell surface area for all the measurements using Henry's EP Suite. Data analysis and curve fittings can be carried out using SigmaPlot 11 (Systat Software, Inc., USA). Here, we use three guard cell ion channels as examples to emphasize some important issues when it comes to measuring specific membrane conductance.

3.5.1. $I_{K,in}$ and $I_{K,out}$ of Guard Cells

$I_{K,in}$ and $I_{K,out}$ channels are typical Shaker-like inward- and outward-rectifiers, judging by the current traces and reversible inhibition of Cs^+ and TEA^+, and are comparable with early publications (14, 70).

1. A 40% increase of membrane potential (*Em*) by pre-incubation indicates that opening buffer is critical to maintain guard cell K^+ channel activity over long periods of time.
2. Also, guard cells pre-treated with opening buffer show significantly longer seal lasting time and steadier $I_{K,in}$ and $I_{K,out}$ current, regardless whether they are from wild type, null, or knockout mutants.
3. $I_{K,in}$ and $I_{K,out}$ currents in guard cells of Col-0 and mutants are faster in "rundown" under control conditions compared with guard cells pre-treated with opening buffer.
4. Curve fitting and statistical analysis show maximal conductance (g_{max}) in guard cells of Col-0 and mutants significantly increases in response to opening buffer pre-treatment, while half maximal voltage ($V_{1/2}$) and apparent gating charge (δ) of $I_{K,in}$ and $I_{K,out}$ remain unchanged.

5. There is a causal link between K^+ channel current and stomatal aperture; stomata of mutants in relatively closed condition exhibit up to twofold increase in $I_{K,in}$ and $I_{K,out}$ currents in response to opening buffer, whereas $I_{K,in}$ and $I_{K,out}$ currents of stomatal guard cells with open stomata are relatively unaffected.

3.5.2. I_{anion} of Guard Cells

In *Arabidopsis*, the SLAC/I_{anion} was identified with a greatly impaired current in guard cell protoplasts of the *slac1* mutant, demonstrating the importance of SLAC1 and of I_{anion} in the regulation of stomatal closure (71–73). So far, there has been no equivalent report on I_{anion} in intact *Arabidopsis* stomatal guard cells.

1. In *Arabidopsis* guard cells, the average *Em* of all the I_{anion} measurements is around −10 mV, and the majority of guard cells show significant steady-state anion currents in Col-0. Both of these are comparable to our previous work on *Nicotiana tobacum* and *Vicia faba* (17, 18).
2. The seal lasting time of I_{anion} experiments in *Arabidopsis* guard cells increases by over twofold in response to pre-incubation, with steadier I_{anion} current for wild type, null, or knockout mutants.

4. Notes

1. Selection of electrophysiologically active plant cells.
Growth history and preparation of leaves, epidermal strips, and roots has a strong impact on stable recordings of the voltage clamp from *Arabidopsis* guard cells and root epidermal cells. The ideal leaves display a more elliptical shape and are more serrated than other leaves, and their abaxial epidermis is stretchy and flexible. Most active guard cells are found on leaves with a high stomatal density, a large number of stomatal primordial, and a small size of epidermal pavement cells. We favor plants grown under long-day conditions because growth under a short-day gives lower stomatal density that leads to less viable guard cells per unit leaf area. Ideal epidermal peels are free from any wrinkles, folds, dirt, and air bubbles. In general, successful impalements are commonly obtained from open stomata with young guard cells, judged by the thickness of the stomatal lip and squat shape of the guard cells. For the selection of root epidermal cell, uniform *Arabidopsis* seedlings with root length at 1–2 cm are selected. Seedlings with a large number of root hair are used for impalements, while any hairless roots are discarded. Seedlings are mounted with soft transparent rubber pieces on glass chamber coated with optically clear, pressure-sensitive adhesive, and then covered with a standard MES buffer

(10 mM KCl, 5 mM Ca^{2+}-MES, pH 6.1). The seedlings in the chamber are incubated with the standard MES buffer solution for 2 h to remove the residual MS liquid medium. This makes the root epidermal cells easier to impale. Successful impalements are commonly obtained from mature root epidermal cells without long root hairs. Epidermal cells with long root hairs, at the elongation region and at root tips, are not suitable for voltage clamp using our current technique.

2. Special microelectrodes for *Arabidopsis* guard cells.
First of all, we observed that 50–100 MΩ tips, which are ideal for *V. faba* guard cells, easily destroy the *Arabidopsis* guard cells, resulting in extremely low seal succeeding rate and reductions in the length of time that the seal will last. Secondly, fine adjustments should be made for an appropriate setting because microelectrode pullers vary from one to another, even for the same model. Thirdly, magnetic pulling time is critical for shaping the tip of the microelectrodes. Shorter magnetic pulling time normally results in microelectrodes with overly long shanks and tips that are easily bent against the guard cell wall and plasma. Longer time always leads to microelectrodes with abnormal shapes and sometime damage. Finally, a higher coil heat gives correspondingly finer-tipped microelectrodes with 1.8–2.0 cm-long shanks and tips that tapered with a 1.5–2°.

3. Pre-treating the stomata with opening buffer.
Experiments have been intensively conducted in guard cells of other species, with some measurement on guard cells with open stomata lasting for over 2 h (18, 21, 24, 25). We also found that pre-treating stomata with the opening buffer [50 mM KCl and 5 mM MES, titrated to a pK_a of 6.1 with NaOH (74)) is crucial for long-term (>30 min) voltage clamp experiments in *Arabidopsis* guard cells. Impalements are more successful on stomata with a bigger aperture for resisting the impaling microelectrode and forming a better seal. Stomata pre-treated with opening buffer usually show larger aperture than those without pre-treatment for all the *Arabidopsis* lines. Moreover, the opening buffer pre-treatment extends the seal lasting time much longer than those guard cells impaled in control condition.

4. Impalement of stomatal guard cells.
Given the guard cell and microelectrode are both in good conditions (i.e., guard cell has high turgor pressure and thin wall, microelectrode shows the right shape of tip, and tip potential is in the right range), the microelectrode should pierce the cell wall and the plasma membrane without any obvious bending of the microelectrode or notable curling of the plasma membrane.

5. Voltage clamp recordings.
   The first run of voltage clamp protocol is usually not steady since the guard cells are not used to extreme clamp voltages (e.g., -250 and +50 mV) in normal growth conditions. The second and third runs, around 5–10 min, are normally quite steady and serve as the standard recordings for the control condition. This is then followed by various treatments or time course experiments for different ion channels.
6. The *Sketchpad* in the Henry EP suite software.
   The *Sketchpad* is a special type of utility window that allows users to store and compare the data from different files of the same type. For instance, it is easy to compare the voltage clamp I/V curves from the same cell before and after a specific treatment, while the experiment is still in progress. The *Sketchpad* simplifies judgement on the quality of data and effect of treatment without formal data analysis.

References

1. Tester M (1990) Plant ion channels: whole-cell and single channel studies. New Phytol 114:305–340
2. Hille B (1992) Ionic channels of excitable membranes, 2nd edn. Sinauer, Sunderland, MA
3. Neher E (1992) Ion channels for communication between and within cells. Biosci Rep 12:1–14
4. Tyerman SD, Skerrett IM (1999) Root ion channels and salinity. Sci Hort 78:175–235
5. Sucher NJ, Deitcher DL, Baro DJ et al (2000) Genes and channels: patch/voltage-clamp analysis and single-cell RT-PCR. Cell Tissue Res 302:295–307
6. Hedrich R, Schroeder JI (1989) The physiology of ion channels and electrogenic pumps in higher plants. Annu Rev Plant Phys 40: 539–569
7. Sokolovski S, Blatt MR (2004) Nitric oxide block of outward-rectifying K^+ channels indicates direct control by protein nitrosylation in guard cells. Plant Physiol 136:4275–4284
8. Kandel ER, Schwartz JH, Jessell TM (2000) Principles of neural science, 4th eds. McGraw-Hill, New York, pp 152–153
9. Huxley AF (2002) From overshoot to voltage clamp. Trends Neurosci 25:553–558
10. Hodgkin AL, Huxley AF (1952) A quantitative description of membrane current and its application to conduction and excitation in nerve. J Physiol 117:500–544
11. Liem LK, Simard JM, Song YM et al (1995) The patch-clamp technique. Neurosurgery 36:382–392
12. Higinbotham N (1973) Electropotentials of plant cells. Annu Rev Plant Phys 24:25–46
13. Hope AB, Walker NA (1975) The physiology of giant algal cells. Cambridge University Press, London, pp 181–197
14. Blatt MR (1987) Electrical characteristics of stomatal guard cells: the contribution of ATP-dependent, "electrogenic" transport revealed by current–voltage and difference-current–voltage analysis. J Membr Biol 98:257–274
15. Blatt MR (1988) Potassium-dependent, bipolar gating of K^+ channels in guard-cells. J Membr Biol 102:235–246
16. Meharg AA, Blatt MR (1995) NO_3^- transport across the plasma-membrane of *Arabidopsis Thaliana* root hairs: kinetic control by PH and membrane voltage. J Membr Biol 145:49–66
17. Grabov A, Blatt MR (1997) Parallel control of the inward-rectifier K^+ channel by cytosolic free Ca^{2+} and pH in *Vicia* guard cells. Planta 201:84–95
18. Chen ZH, Hills A, Lim CK et al (2010) Dynamic regulation of guard cell anion channels by cytosolic-free Ca^{2+} concentration and protein phosphorylation. Plant J 61:816–825
19. Blatt MR, Rodriguez-Navarro A, Slayman CL (1987) Potassium-proton symport in *Neurospora*: kinetic control by pH and membrane potential. J Membr Biol 98:169–189
20. Beilby MJ, Walker NA (1981) Chloride transport in *chara*. 1. Kinetics and current–voltage curves for a probable proton symport. J Exp Bot 32:43–54

21. Blatt MR, Thiel G, Trentham DR (1990) Reversible inactivation of K^+ channels of *Vicia* stomatal guard-cells following the photolysis of caged inositol 1,4,5-trisphosphate. Nature 346:766–769
22. Beilby MJ, Cherry CA, Shepherd VA (1999) Dual turgor regulation response to hypotonic stress in *Lamprothamnium papulosum*. Plant Cell Environ 22:347–359
23. Thiel G, MacRobbie AC, Blatt MR (1992) Membrane transport in stomatal guard cells: the importance of voltage control. J Membrane Biol 126:1–18
24. Garcia-Mata C, Gay R, Sokolovski S et al (2003) Nitric oxide regulates K^+ and Cl^- channels in guard cells through a subset of abscisic acid-evoked signalling pathways. Proc Natl Acad Sci USA 100:11116–11121
25. Levchenko V, Konrad KR, Dietrich P et al (2005) Cytosolic abscisic acid activates guard cell anion channels without preceding Ca^{2+} signals. Proc Natl Acad Sci USA 102:4203–4208
26. Lew RR (1991) Electrogenic transport properties of growing *Arabidopsis* root hairs: the plasma membrane proton pump and potassium channels. Plant Physiol 97:1527–1534
27. Grefen C, Chen ZH, Honsbein A et al (2010) A novel motif essential for SNARE interaction with the K^+ channel KC1 and channel gating in *Arabidopsis*. Plant Cell 22:3076–3092
28. Chen ZH, Grefen C, Donald N et al (2011) A bicistronic, Ubiquitin-10 promoter-based vector cassette for transient transformation and functional analysis of membrane transport demonstrates the utility of quantitative voltage clamp studies on intact *Arabidopsis* root epidermis. Plant Cell Environ 34:554–564
29. Neher E, Sakmann B (1976) Single-channel currents recorded from membrane of denervated frog muscle fibres. Nature 260:799–802
30. Hamill OP, Marty A, Neher E et al (1981) Improved patch-clamp techniques for high-resolution current recording from cells and cell-free membrane patches. Eur J Physiol 391:85–100
31. White SH, Wimley WC (1999) Membrane protein folding and stability: physical principles. Annu Rev Biophys Biomol Struct 28:319–365
32. Sakmann B, Neher E (1984) Patch clamp techniques for studying ionic channels in excitable membranes. Annu Rev Physiol 46: 455–472
33. Meckel T, Hurst AC, Thiel G et al (2005) Guard cells undergo constitutive and pressure-driven membrane turnover. Protoplasma 226:23–29
34. Sutter JU, Sieben C, Hartel A et al (2007) Abscisic acid triggers the endocytosis of the *Arabidopsis* KAT1 K^+ channel and its recycling to the plasma membrane. Curr Biol 17: 1396–1402
35. Hasegawa PM, Bressan RA, Zhu JK et al (2000) Plant cellular and molecular responses to high salinity. Annu Rev Plant Biol 51: 463–499
36. Tester M, Davenport R (2003) Na^+ tolerance and Na^+ transport in higher plants. Ann Bot-London 91:503–527
37. Munns R, Tester M (2008) Mechanisms of salinity tolerance. Annu Rev Plant Biol 59:651–681
38. Chen ZH, Newman I, Zhou M et al (2005) Screening plants for salt tolerance by measuring K^+ flux: a case study for barley. Plant Cell Environ 28:1230–1246
39. Chen ZH, Zhou M, Newman I et al (2007) Potassium and sodium relations in salinised barley tissues as a basis of differential salt tolerance. Funct Plant Biol 34:150–162
40. Shabala S, Cuin TA (2008) Potassium transport and plant salt tolerance. Physiol Plantarum 133:651–669
41. Beilby MJ (2007) Action potential in charophytes. Inte Rev Cytol 257:43–82
42. Rubio F, Gassmann W, Schroeder JI (1995) Sodium-driven potassium uptake by the plant potassium transporter HKT1 and mutations conferring salt tolerance. Science 270:1660–1663
43. Dreyer I, Horeau C, Lemaillet G et al (1999) Identification and characterization of plant transporters using heterologous expression systems. J Exp Bot 50:1073–1087
44. Ren ZH, Gao JP, Li LG et al (2005) A rice quantitative trait locus for salt tolerance encodes a sodium transporter. Nat Genet 37:1141–1146
45. Beilby MJ, Shepherd VA (2001) Modeling the current–voltage characteristics of charophyte membranes. III. K^+ state of *Lamprothamnium*. Aust J Plant Physiol 28:541–550
46. Beilby MJ, Shepherd VA (2006) The characteristics of Ca^{2+} activated Cl^- channels of the salt-tolerant charophyte *Lamprothamnium*. Plant Cell Environ 29:764–777
47. Shepherd VA, Beilby MJ, Shimmen T (2002) Mechanosensory ion channels in charophyte cells: the response to touch and salinity stress. Eur Biophys J 31:341–355

48. Shepherd VA, Beilby MJ, Al Khazaaly SA et al (2008) Mechano-perception in Chara cells: the influence of salinity and calcium on touch-activated receptor potentials, action potentials and ion transport. Plant Cell Environ 31:1575–1591
49. Niu X, Bressan RA, Hasegawa PM et al (1995) Ion homeostasis in NaCl stress environments. Plant Physiol 109:735–742
50. Maathuis FJ, Amtmann A (1999) K^+ nutrition and Na^+ toxicity: the basis of cellular K^+/Na^+ ratios. Ann Bot 84:123–133
51. Demidchik V, Tester M (2002) Sodium fluxes through nonselective cation channels in the plant plasma membrane of protoplasts from *Arabidopsis* roots. Plant Physiol 128:379–387
52. Essah PA, Davenport R, Tester M (2003) Sodium influx and accumulation in *Arabidopsis*. Plant Physiol 133:307–318
53. Demidchik V, Maathuis FJM (2007) Physiological roles of nonselective cation channels in plants: from salt stress to signalling and development. New Phytol 175:387–404
54. Rus A, Yokoi S, Sharkhuu A et al (2001) *AtHKT1* is a salt tolerance determinant that controls Na^+ entry into plant roots. Proc Natl Acad Sci USA 98:14150–14155
55. Horie T, Hauser F, Schroeder J (2009) HKT transporter-mediated salinity resistance mechanisms in *Arabidopsis* and monocot crop plants. Trends Plant Sci 14:660–668
56. Schachtman DP, Tyerman SD, Terry BR (1991) The K^+/Na^+ selectivity of a cation channel in the plasma membrane of root cells does not differ in salt-tolerant and salt-sensitive wheat species. Plant Physiol 97:598–605
57. Uozumi N, Kim EK, Rubio F et al (2000) The *Arabidopsis HKT1* gene homolog mediates inward Na^+ currents in *Xenopus laevis* oocytes and Na^+ uptake in *Saccharomyces cerevisiae*. Plant Physiol 122:1249–1259
58. Kronzucker HJ, Britto DT (2011) Sodium transport in plants: a critical Review. New Phytol 189:54–81
59. Shi HZ, Quintero FJ, Pardo JM et al (2002) The putative plasma membrane Na^+/H^+ antiporter SOS1 controls long-distance Na^+ transport in plants. Plant Cell 14:465–477
60. Zhu JK (2002) Salt and drought stress signal transduction in plants. Annu Rev Plant Biol 53:247–273
61. Apse MP, Aharon GS, Sneddon WA et al (1999) Salt tolerance conferred by overexpression of a vacuolar Na^+/H^+ antiport in *Arabidopsis*. Science 285:1256–1258
62. Zhang HX, Blumwald E (2001) Transgenic salt-tolerant tomato plants accumulate salt in foliage but not in fruit. Nat Biotechnol 19:765–768
63. Shi HZ, Ishitani M, Kim C et al (2000) The *Arabidopsis thaliana* salt tolerance gene *SOS1* encodes a putative Na^+/H^+ antiporter. Proc Natl Acad Sci USA 97:6896–6901
64. Zhang HX, Hodson JN, Williams JP et al (2001) Engineering salt-tolerant *Brassica* plants: characterization of yield and seed oil quality in transgenic plants with increased vacuolar sodium accumulation. Proc Natl Acad Sci USA 98:12832–12836
65. Felle HH (1994) The H^+/Cl^- symporter in root-hair cells of *Sinapis alba*. An electrophysiological study using ion-selective microelectrodes. Plant Physiol 106:1131–1136
66. Berthomieu P, Conejero G, Nublat A et al (2003) Functional analysis of *AtHKT1* in *Arabidopsis* shows that Na^+ recirculation by the phloem is crucial for salt tolerance. EMBO J 22:2004–2014
67. Honsbein A, Sokolovski S, Grefen C et al (2009) A tripartite SNARE-K^+ channel complex mediates in channel-dependent K^+ nutrition in *Arabidopsis*. Plant Cell 21:2859–2877
68. Blatt MR (1991) Ion channel gating in plants: physiological implications and integration for stomatal function. J Membr Biol 124: 95–112
69. Blatt MR, Armstrong F (1993) K^+ channels of stomatal guard cells-abscisic-acid-evoked control of the outward rectifier mediated by cytoplasmic pH. Planta 191:330–341
70. Roelfsema MRG, Prins HBA (1997) Ion channels in guard cells of *Arabidopsis thaliana* (L) Heynh. Planta 202:18–27
71. Schroeder JI, Allen GJ, Hugouvieux V et al (2001) Guard cell signal transduction. Annu Rev Plant Phys 52:627–658
72. Negi J, Matsuda O, Nagasawa T et al (2008) CO_2 regulator SLAC1 and its homologues are essential for anion homeostasis in plant cells. Nature 452:483–486
73. Vahisalu T, Kollist H, Wang YF et al (2008) SLAC1 is required for plant guard cell S-type anion channel function in stomatal signaling. Nature 452:487–491
74. Allen GJ, Chu SP, Harrington CL (2001) A defined range of guard cell calcium oscillation parameters encodes stomatal movements. Nature 411:1053–1057

Chapter 3

Using the Multifunctional Xylem Probe for in situ Studies of Plant Water and Ion Relations Under Saline Conditions

Lars H. Wegner

Abstract

By insertion into an individual xylem vessel at the root base, the multifunctional xylem probe allows the monitoring of the xylem pressure, the radial electrical gradients in the root (the so-called trans-root potential, TRP), as well as the activity of a particular ion such as K^+ in the xylem sap of intact, transpiring plants. The biophysical and physiological significance of these parameters with respect to salt stress is briefly explained, and the assembly of the probe, the setup used for these measurements, and the experimental procedure are outlined in detail.

Key words: Xylem pressure, Trans-root (electrical) potential, Xylem sap, Xylem K^+ activity, Radial reflection coefficient, Xylem pH

1. Introduction

When plant roots are exposed to an acute salt shock, their physiological response is complex and multi-facetted with processes being elicited at time scales ranging from seconds to days. Biophysical parameters related either to plant hydraulics (i.e., turgor pressure and xylem pressure) or to ion transport (cellular membrane potentials, ion fluxes at the root surface or into the xylem vessels) are among the first to be affected and are therefore well suited for "speed diagnosis," with respect to salt tolerance of a particular cultivar, for example (see also Chap. 7). Much effort has been invested over the last 60 years to design refined experimental strategies for monitoring these effects in an exact and reliable way. Sophisticated techniques have been developed to unravel the physiological processes elicited by NaCl treatment and to identify

Sergey Shabala and Tracey Ann Cuin (eds.), *Plant Salt Tolerance: Methods and Protocols*, Methods in Molecular Biology, vol. 913, DOI 10.1007/978-1-61779-986-0_3, © Springer Science+Business Media, LLC 2012

multiple traits to cope with salt stress. QTL analysis provides a powerful strategy to find the genetic basis of traits that confer salinity tolerance to the plant, albeit the method relies, among other things, on an exact and reproducible quantification of these traits. Much information can be obtained from work on excised roots and on protoplasts derived from root tissue (Pottosin et al. Chap 1). However, isolated protoplasts and excised roots may not always be representative of cells and roots functioning as a part of an intact organism (e.g. (1)), hence, experiments on intact plants should also be performed.

A range of biophysical parameters can simultaneously be assessed on intact plants by making use of *xylem probes*. Originally, these probes were designed to measure the hydrostatic pressure in the xylem vessels (2), in analogy to the recording of turgor in single cells with the turgor pressure probe (3, 4). This is achieved by inserting a fine-tipped glass capillary attached to a microbaric chamber made from Perspex, into a cell or an individual xylem vessel. One major difference is that turgor pressure probes are filled with silicone oil, forming, once inserted into the cell, a meniscus with the cell sap that is kept at a constant position close to the cell surface to eliminate artifacts related to the elasticity of the probe on pressure recording. In contrast, xylem probes are filled with degassed water (or electrolyte solution) for reasons that will be explained in more detail below.

Xylem probes have been extended to functions beyond mere pressure recording. By introducing an Ag/AgCl electrode into the body of the probe and filling it with electrolyte solution instead of pure water, the electrical potential in a xylem vessel with respect to an external electrode can be measured (5). This device was termed a "xylem pressure-potential probe." A still further advanced version, the "multifunctional xylem probe," also allows the monitoring of the activity of a particular ion (K^+, H^+, NO_3^-) in a xylem vessel. This is achieved by using double-barreled electrodes. One barrel is attached to the Perspex body of a xylem pressure-potential probe, whereas the other barrel is designed as an ion-selective electrode. The tip of this ion-sensing barrel is filled with a resin that selectively binds a certain ion at its surface, thus leading to a corresponding shift in the voltage drop across the tip upon changes of the activity of this particular ion in the solution in which the tip is immersed (for more detailed information on ion-selective electrodes, see Felle (6)). With this tool, xylem pressure, xylem electrical potential, and the activity of a particular ion (see Note 1) can be measured on-line in an intact, transpiring plant during the imposition of a stress such as salinity (7).

For convenience, I will discuss technical challenges as well as the solutions to overcome them separately for measurement of xylem pressure, xylem electrical potential, and xylem ion activities. After reviewing technical aspects, the most important experimental results relevant to salinity stress are briefly summarized and discussed. Note that some aspects covered here have been treated in more detail in previous review articles (8–10).

1.1. Xylem Pressure

When the root system of higher plants is exposed to salt shock, a rapid decrease in xylem pressure is induced, starting within seconds. Two different tools, the root pressure probe and the xylem pressure probe, are available to quantify these changes in real time. The root pressure probe is designed to monitor changes in root pressure (i.e., above-atmospheric pressure) on excised roots, whereas the xylem pressure probe allows measurements of the hydrostatic pressure in the vessels of intact, transpiring plants. These pressure values are usually sub-atmospheric when tension develops in the xylem (see Note 2). Frequently, the tension exceeds 0.1 MPa, i.e., the pressure in xylem vessels drops below vacuum. Sub-vacuum pressures imply the absence of any gas phase in these vessels (with the exception of extremely small gas bubbles in the sub-micron range, stabilized by surface tension). Coexistence of liquid and gas phase will lead to an equilibrium pressure that obeys Clausius–Clapeyron's law (2 kPa above vacuum at room temperature). A drop of xylem pressure below this value implies that the liquid phase is in a metastable state that is continuously threatened by the spontaneous formation of a gas phase. This type of event, known as "cavitation," is associated with a net release of energy and a jump of hydrostatic pressure to the equilibrium value (for more details on the related physics of water, the reader is referred to Zimmermann et al. (9)). Cavitation can lead to an embolism of a particular vessel and, in turn, to a dramatic loss in conductivity that may eventually result in vessel dysfunction. Plants have developed strategies to protect against cavitation and to re-fill embolized vessels; this has been a field of vibrant research over the last 20 years ((11); for a review, see ref. (12)).

The metastable state of xylem water is also the most challenging aspect of measuring xylem pressure. This is because the insertion of the probe into a vessel may itself cause cavitation, either by a leakage induced at the insertion point of the glass micropipette, or by the presence of a gas phase in the interior of the probe (which can, however, be avoided when the probe is prepared properly before use; see below). It has to be kept in mind that a pressure equilibrium between the probe and vessel is reached once continuity between both water-filled compartments is established. Hence pre-existing gas bubbles or a cavitation event within the probe following the insertion into a vessel will artificially shift local xylem pressure to a value above vacuum. Contact between hydrophilic and hydrophobic fluids anywhere in the probe increases the danger of cavitation events and is avoided by filling it with degassed water instead of silicon oil; a meniscus is of no use in xylem recording anyway because any attempt to adjust its position by moving a metal piston (as usually performed in turgor recording) is by itself likely to induce cavitation. Water is much less compressible than oil, thus making it more suitable for use in this system. Nonetheless, some residual compressibility has to be tolerated. However, since the xylem is (compared to a single cell) a large, hydraulically highly coupled compartment, a shift of small volumes of water from the probe into the impaled vessel (and *vice versa*) has only a minor effect on xylem pressure, as demonstrated by the insertion of two probes into one xylem conduit (13) and by osmotic experiments (see below). For the filling of the probe, only thoroughly degassed water is used to prevent the formation of gas bubbles.

In spite of these precautions, measurement of xylem pressure remains particularly challenging from a technical point of view. Consequently, several authors have questioned the feasibility of this approach (14, 15). To meet their concerns, a broad range of meticulously planned control experiments have been performed. An example is the "Hepp-type bio-osmometer," especially tailored for this purpose (for experimental details and theory, see ref. (2)). Briefly, a *Plantago* leaf was cut under water and fitted to a small reservoir in such a way that the xylem sap and the reservoir were separated by a semi-permeable membrane that tightly sealed the cut surface of the leaf petiole. By adjusting the osmotic pressure in the reservoir with PEG, negative hydrostatic pressures (tensions) of different magnitudes could be established in the leaf xylem. Tensions measured experimentally by the insertion of a xylem pressure probe into an individual vessel exactly matched those calculated according to the theory of the Hepp-type osmometer. These, as well as other experiments, revealed that xylem walls form a tight seal at the insertion point of the probe once the xylem sap is under tension, thus usually preventing the formation of leaks. In vessels damaged by probing (or when impaled with a pressure probe that contained gas bubbles), xylem pressure would not drop below vacuum ((13); see Note 3). Clogging of the tip during advancement of the probe through the root tissue or once a vessel has been impaled can also easily be detected, since the pressure trace becomes noisy due to the susceptibility to minor fluctuations in temperature. Moreover, pressure ceases to respond to changes in light intensity or to osmotic challenge when the hydraulic continuity between the xylem and probe interior is impaired. Experiments with dyes also confirmed that the pressure recording accurately reflects the positioning of the probe in a xylem vessel (16). Therefore, it is no surprise that the technique originally developed by Zimmermann and co-workers was later also adopted by other laboratories (with some modifications (17, 18)) and is now generally accepted.

In more than 20 years of research, detailed studies have been undertaken on the impact of environmental factors such as light, humidity, nutritional status, water availability, etc. on xylem pressure ((9, 19), and quotations therein). In this technical review, the focus is on the effects of an acute increase in the sodium chloride concentration in the external medium in which the root is immersed. This salt shock exerts an unspecific osmotic force on the root that leads to an efflux of water. As a consequence, xylem pressure decreases with an exponential time course, as registered by the xylem pressure probe (Fig. 1 (5, 13, 20–22)). This tension surplus is transmitted to the shoot and has a rapid effect on growth rate (23) and increases the probability of cavitation events. Cavitation may induce vessel dysfunction and, in turn, impair the supply of water and nutrients to the shoot.

The pressure response to an osmotic shock can be quantified by dividing the maximum amplitude of the xylem pressure response by the imposed change in the external osmotic pressure, $\Delta\pi$ (either calculated from the change in concentration Δc by using the van't Hoff equation $\Delta\pi = RT\Delta c$ or, more accurately, determined with an osmometer). This ratio is called the radial reflection coefficient of the root (σ_r). In terms of the thermodynamics of irreversible processes, it indicates the extent of coupling between radial water and solute flow in the root (for a detailed account of the theory, see (8, 9)). For a perfect osmometer, σ_r attains a value of 1. A value below 1 indicates that the impact of the osmotic gradient is reduced by a back-flow of water coupled to solute flux across the membrane or transport barrier (in the present situation, the tissue separating the lumen of the xylem and the external medium). Root σ_r obtained with the xylem pressure probe on intact plants should be considered an "operational parameter" since the root xylem is not a closed compartment and no true steady state is established. In contrast, these criteria are met when σ_r is measured with the root pressure probe (24). However, it can be shown that transpiration remains invariant with osmotic treatments at the time scale considered here, i.e., transpiration and osmotic challenge act independently on P_x (and J_v) with the exception of very high transpiration rates in some species. Therefore, determination of σ_r by making use of the xylem pressure probe is feasible.

Detailed studies using the xylem pressure probe revealed that the σ_r value determined experimentally varies greatly, ranging from 0.1 to 1, depending on the species, the concentration of the osmolyte, and the transpiration rate of the plant. The reflection coefficient decreases with the osmolyte concentration (in wheat and barley, but less so in maize) and is lower in non-transpiring plants than in plants transpiring at a low rate (20, 21), indicating that unstirred layer effects (concentration-polarization) have a strong impact on the xylem pressure response to an osmotic challenge. The interpretation of σ_r values considerably lower than 1 with respect to the mechanisms of radial water and solute transport in roots is still under debate; the "composite transport model" advocated by Steudle and co-workers (24, 25) is not generally accepted and has

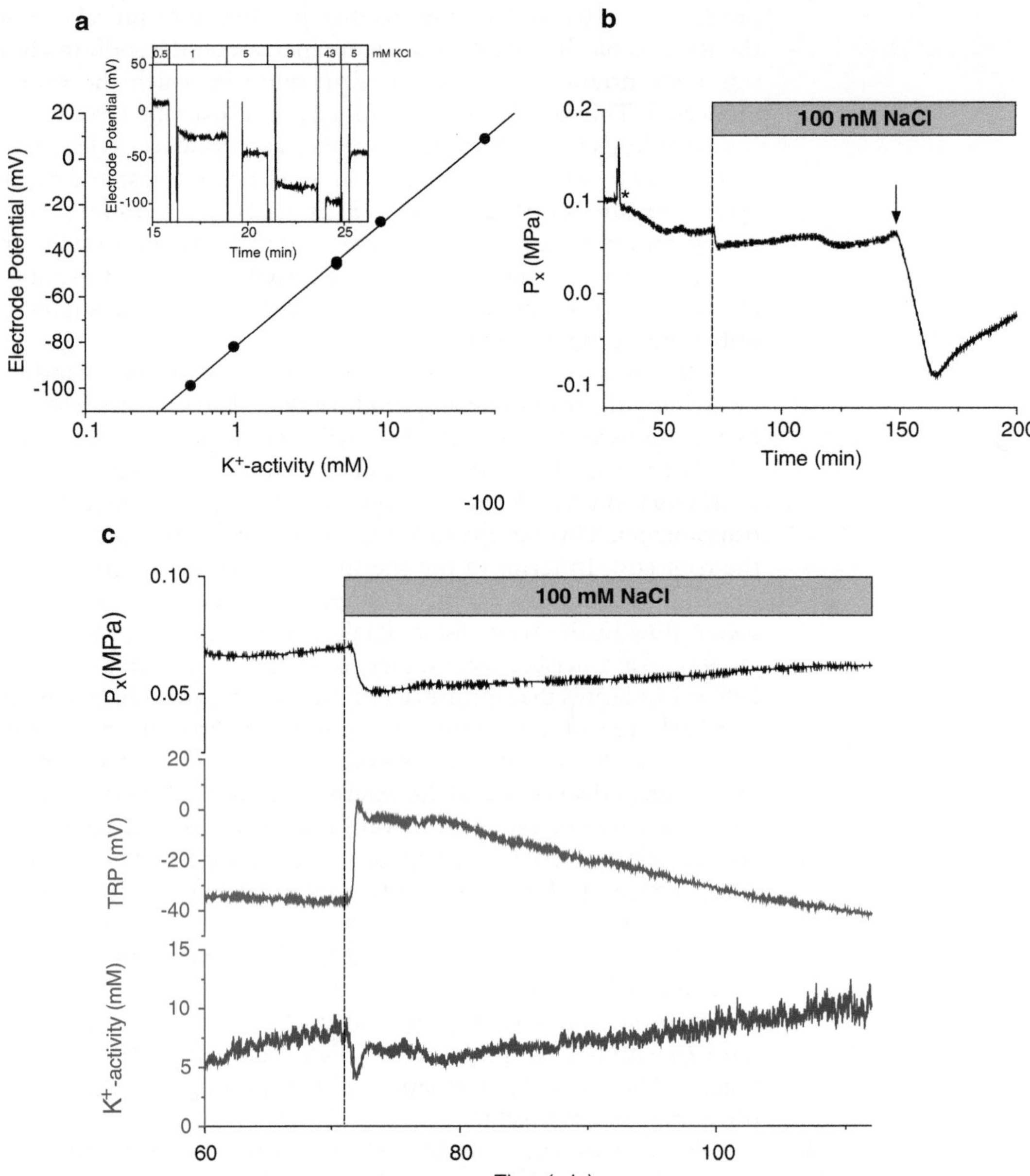

Fig. 1. Typical experiment showing the response of a maize root exposed to acute salt stress, as measured with a K^+ selective multifunctional probe. (**a**) The pre-calibration of the K^+ selective barrel used in this experiment is documented. The *slope* of the semi-logarithmic plot of the ion-selective electrode potential (see also *inset*; *numbers* in the *top bar* denote KCl activities in mM) against the K^+ activity of various calibration media was 56 mV. (**b**) The time course of the xylem pressure is depicted on an extended time scale. Upon impalement in a dilute medium (0.1 mM $CaCl_2$, 1 mM KCl, 2 mM $MgCl_2$, 10 mM MES/BTP, pH 5.5) at low light irradiation (~10 μmol/m²s), the pressure dropped to 0.092 MPa. Subsequently pressure decreased slowly further until a steady state of about 0.069 MPa was established. Upon salt exposure (by adding an aliquot of a 3 M NaCl stock solution to the bath, *dotted horizontal line*), the xylem pressure (P_x) rapidly decreased by −0.019 MPa (within about 3 min). In order to test whether above-vacuum xylem pressures, despite high external salt concentration, were due to a failure of the probe, irradiation was increased to 250 μmol/m²s. As a consequence, pressure dropped rapidly below zero and passed through a pressure minimum of almost −0.1 MPa in accordance with previous reports. This indicates that the experiment was properly conducted. (**c**) The time course of P_x, trans-root potential (TRP), and xylem K^+

recently been challenged by a range of authors for methodological and other reasons (13, 21, 26, 27). From a physiological point of view, it is interesting to note that structural features of the root can considerably attenuate the effect of external osmotic fluctuations on xylem pressure, so minimizing the effects on plant hydraulics and transpiration and protecting the vessels against cavitation. However, the impact of this buffering effect is apparently limited, especially at a high transpiration rate (13, 21).

1.2. Trans-Root Potential

Besides affecting osmotic and hydraulic balance in the root, exposure to a sudden salt stress also has an almost immediate impact on the electrical gradients in the root. This is due to either diffusion potentials that result from differences in the mobility of the cation and the anion (Na^+ and Cl^-), or indirectly, by an effect of Na^+ and/or Cl^- on other membrane transport processes. The interaction of both ions with transport proteins in the plasma membrane of cortical cells, as well as feedback of altered transport properties on the membrane potential, has been analyzed in detail previously ((28); see also below). Much less is known about transport processes elicited in tissues located deeper inside the root, notably in the stele. This is mainly due to the poor accessibility of cells of the inner cortex and the stele. Radial profiles in membrane potential can be obtained by slowly advancing a microelectrode through the root tissue (29). However, since mechanical access requires penetration of the neighboring cell(s) and root cells are electrically coupled via plasmodesmata, the physiological relevance of voltages measured in this way is at least questionable (30). In some species, the stele can be isolated using a mechanical procedure, making the stelar cells accessible for membrane potential measurements in the usual way (7, 31), albeit again under artificial experimental conditions. Another technical approach to monitor electrical changes in the root of both cortical and stelar cells simultaneously is provided by the measurement of the (extracellular) electrical potential in the xylem with respect to an external electrode, the so-called trans-root (electrical) potential (TRP; to my knowledge the name was first introduced by Dunlop and Bowling (32)). These as well as other researchers (33, 34) have developed a theoretical framework for the interpretation of the trans-root potential that is supported by many experimental observations: A non-zero potential difference across the root tissue separating the xylem from the outside

Fig. 1. (continued) activity is shown for the same experiment at an enlarged scale immediately before and after salt administration. Note that the TRP depolarized rapidly by about 38 mV and passed through a minimum after about 40 s, after which repolarization occurred. Note also that the time course of repolarization was much slower (*slope* −0.019 mV/s) than in the experiment documented by Wegner et al. ((7); Fig. 1 of that publication; slope −0.23 mV/s) giving evidence of the variability among individual plants in that respect. Xylem K^+ activity responded by a rapid transient drop from 7.6 to 4.3 mM. The activity subsequently recovered; a rapid and a slow component are clearly discernible.

reflects the polarity of the tissue for ion (and water) transport. The TRP mainly results from membrane potentials generated at the plasma membrane of cortical cells (MP_c) and stelar cells (MP_{xpc}), the latter contributing with an opposite sign to the overall potential difference. The symplast extending between both membranes is supposed to be largely isopotential. Moreover, the apoplast contributes little to the overall conductance due to the presence of a Casparian strip or band in the cell walls of the endodermis separating the cortical and the stelar apoplast. Hence, the TRP can be estimated as

$$TRP \approx MP_c - MP_{xpc}$$

Originally, the TRP was measured on excised roots by placing one electrode into the exudate secreted at the cut surface, whereas the reference electrode was positioned in the medium in which the root was immersed (for a detailed description of the method see ref. (35)). This method shares disadvantages with other techniques that rely on measurements with excised roots. Since radial ion gradients as well as osmotic/hydrostatic pressure gradients are known to be strongly affected by root excision, concomitant changes in the electrical gradients that are likely to affect the TRP have to be taken into account.

An alternative to these "traditional" techniques is provided by inserting an electrode into the Perspex body of the xylem pressure probe, thus transforming it to a xylem pressure-potential probe (5, 36). Again, this extension of the pressure probe is associated with a range of technical challenges to be solved:

1. The pressure transducer can strongly interfere electrostatically with voltage recording. Initially, this was circumvented by grounding the electrode in the probe and connecting the bath electrode to a high-impedance amplifier (5). In an advanced version of the probe, the pressure transducer and the electrode are separated by a polyester foil that transmitted the pressure but acted as an electrical isolator. Both versions of the probe rendered identical results.
2. With the integration of an Ag/AgCl electrode into the body of the probe, another surface that may seed cavitation is introduced, especially since AgCl crystals tend to create a rough surface that may harbor tiny gas bubbles. This can be overcome by coating the electrode with Ag-alginate.
3. The body of the probe and the capillary to be inserted into the xylem both have to be filled with electrolyte solution in order to establish a stable electrical contact between the electrode and the xylem sap. Since the compressibility of the probe cannot be neglected as discussed above, pressure changes are necessarily associated with volume flow across the tip in both

directions. A pressure increase is associated with a volume transfer from the capillary into the vessel, whereas a pressure decrease comes with flow in the opposite direction. This could have a direct effect on TRP measurements since local concentrations in the xylem as well as the electrical resistance of the tip are affected. However, injection of large volumes into the xylem or volume extraction by applying pressure (volume) pulses has no effect on the measured electrical potential, or only minor, short-lived effects have been measured, very different from TRP responses to changes in light intensity, or to an osmotic challenge (5).

4. As a result of tiny leaks along the insertion canal of the microcapillary, the "original" TRP could be partly short-circuited. However, results obtained with the probe indicated that this is not the case (see below). Moreover, the presence of tiny leaks precludes the existence of stable negative pressures that were frequently measured with the xylem pressure-potential probe.

So far, TRP recordings obtained on intact plants have rendered results that are qualitatively in accordance with measurements previously performed on excised roots. TRP recordings with the multifunctional probe on low-salt maize seedlings exposed to nitrate or ammonium fully reflected characteristic membrane potential (MP_c) changes of root cortical cells to both N-forms (37). This is in accordance with the model proposed previously, whereas an attenuated TRP response would be taken as evidence for an apoplastic shunt. Hence, the data also indicate that the apoplast is a high-resistance pathway that contributes little to radial salt transport into the stele. The difference between TRP and MP_c is equal to $-MP_{xpc}$, provided that the symplast can be considered as an isopotential compartment. This is definitely a simplifying assumption that needs further experimental testing. It should also be emphasized that the interpretation of the TRP is restricted to low volume flow rates across the root; when transpiration-driven volume flow across the root tissue increases, streaming potentials will be superimposed and start to shape the TRP patterns (36).

Recently, the effect of a sudden salt shock on maize and barley seedlings was studied with the xylem probe technique (7) under a low light regime (see also Fig. 1). Salt treatment (100 mM NaCl) elicited a rapid depolarization, within about a minute. The maximum amplitude was ~35 and ~14.4 mV in the presence of 0.1 or 2 mM Ca^{2+} in the bath. This depolarization was followed by a slow recovery of the TRP almost to its original value. The transient excursion of the TRP lasted up to about 30 min. Subsequently, it remained constant for about 1 h, until the light irradiation was increased. Based on a previous study on salt-induced shifts in cortical membrane potential in maize roots (38), the rapid depolarization could be assigned to the response of the cortical cells, whereas the subsequent repolarization was apparently due to a delayed depolarization of stelar parenchyma (note again that MP_{xpc} contributes with opposite sign to the TRP). The characteristic delay of electrical events in the stele correlated quite well with the time course of stelar Na^+ accumulation, as measured in separate experiments with sodium green, a Na^+-sensitive dye (7).

1.3. Ion Concentrations in the Xylem Sap

Measuring ion concentrations in the xylem sap is a particularly challenging task, especially if one wants to determine ion activities in absolute numbers. The possible pitfalls of conventional techniques such as those based on the analysis of (root) exudate or the use of xylem-feeding insects have been discussed in detail elsewhere (10, 39). In summary, it can be stated that traditional techniques tend to over-estimate the actual salt concentrations in the xylem sap (for further details, the reader is referred to the cited review articles). Multifunctional xylem probes provide a viable alternative to record ion concentrations continuously in real time on intact, transpiring plants, with the limitation that, at the moment, only one ion can be measured at a time (see also Note 1). Multifunctional probes make use of double-barreled micropipettes, with one barrel serving as an ion-selective electrode and the other barrel, for the measurement of xylem pressure and TRP. Since both barrels form a single tip, the pressure recording gives evidence of the probe tip being located in the lumen of a xylem vessel. To ensure that leakage of salts from the pressure/potential sensing barrel does not interfere with concentration measurements, the capillary is filled with an organic salt solution (50 mM NMG-Mes). Even though these precautions were taken, various tests had to be performed again to make sure that the probe is actually reading the correct ion concentration/activity values. An experiment designed to validate the technique was described by Wegner and Zimmermann (36, 40): A multifunctional probe was inserted into a late metaxylem vessel at the root base of a transpiring maize seedling. Subsequently, the root was cut about 1 cm away from the site of impalement in apical direction. As expected, this led to a rapid increase of the pressure in the open vessel from a value below vacuum to atmospheric. The lumen of the open vessels equilibrated rapidly with the bath upon cutting; this was reflected by an increase in the K^+ signal or pH (depending on the type of probe that was used) to the level adjusted in the bath, indicating that the probe was reading the correct local concentration values. For a pH-recording multifunctional xylem probe, this experiment was extended to in situ calibration of the probe by successively replacing the bath by media with different pH (40).

Unfortunately, it is not yet possible to measure the Na^+ concentration in the xylem sap in situ with this method, due to the lack of Na^+ selective electrodes with sufficient selectivity against K^+ and/or Ca^{2+}. However, it is possible to monitor the effects of Na^+ on xylem K^+ using a K^+ selective multifunctional probe (7). Interestingly, a transient drop in xylem K^+ activity (A_{K+}) by 2–3 mM was observed to start within ~1 min after the onset of stress (see Fig. 1). After several minutes, the value returned more or less to its original level. The initial drop in xylem K^+ was explained by the following scenario (7). Exposure of the root to high Na^+ concentrations leads to a massive efflux of K^+ from the root cortex, and in turn, to a K^+ depletion of the symplast as shown previously for *Arabidopsis* (41). As long as the plasma membrane of stelar parenchyma cells remains in a hyperpolarized state (note that these cells respond with a few minutes delay as evidenced by the TRP measurements), a decrease in symplastic K^+ will transiently shift the electrochemical potential gradient towards K^+ resorption from the stelar apoplast (including xylem vessels) by xylem parenchyma cells. The gradient is reversed again with the onset of depolarization of these cells, corresponding to the second phase of the TRP response to an acute salt shock. This correlates with the arrival of Na^+ in the root centre, i.e., from 1 min after the salt application onwards. As a consequence, K^+ efflux from xylem parenchyma is initiated (as observed on isolated steles), and the K^+ activity in the xylem gradually recovers.

1.4. Potential for the Application of the Multifunctional Xylem Probes in Research on Salinity Stress

Recent work on short-term effects of NaCl treatment of maize and barley roots has demonstrated that multifunctional probes are very efficient and useful tools in research on salinity stress and tolerance (7). When this tool is combined with simultaneous recordings of water uptake (either gravimetrically or by monitoring gas exchange), even more detailed information can be provided (42), including quantification of the radial net flow of K^+ into the xylem or the hydraulic conductivity of the root. Hydraulic conductivity provides information on the activity of aquaporins in root cell membranes. It will be of particular interest to study the effect of NaCl on aquaporin activity in roots of intact plants.

The probe is superior to alternative "whole plant physiology" techniques in that it provides simultaneous data on both water and solute transport in the intact plant. "Realistic" measurements of this kind are urgently needed to feed models that can make reliable predictions on, for example, the physiological status and yield of stands of crop species irrigated with mildly saline waste water. So far, only experiments with hydroponically grown crop plants have been performed with the probe technique, but it is also suitable for experiments on potted plants when xylem vessels at the root base just above the ground are impaled; these are not likely to differ much from conditions in the root. Moreover, long-term experiments on the effect of salinity should also be conducted in the near future. Research is under way to develop Na^+ sensors with an improved selectivity that may soon be available.

Admittedly, employing the multifunctional xylem probe requires some skill and the technique is not (yet) suitable for screening salinity tolerance of a large collection of cultivars. Nevertheless, it can be used to calibrate other techniques that are more suitable for this purpose. For example, it was recently shown that K^+ concentrations in xylem sap obtained by rapid pressurization of cut barley leaves in a pressure bomb were very similar to values measured with the multifunctional probe. This gives credit to data obtained with the former method (31).

The author of this review article is happy to support groups that want to establish multifunctional xylem probes as a routine lab tool, both with technical support and with advice. For the time being, the reader is referred to the detailed experimental protocol given below.

2. Materials

Below, all materials required for the use of multifunctional xylem probes are listed, with the exception of facilities, tools, or chemicals that are likely to be part of every laboratory.

2.1. The Probe

1. A custom-made microbaric chamber fabricated from Perspex with an integrated pressure transducer (Siemens KPY-16, Erlangen, Germany), Ag/AgCl electrode (for the chloridation procedure, see Subheading 4.1, step 6), and a stainless steel piston attached to a micrometer screw. The precise composition of the probe is shown in Fig. 2. The assembly is described in detail below.
2. Two types of microcapillaries made from borosilicate glass: MTW 100F-6 and M1B100-6 (World Precision Instruments (WPI) Sarasota, FL, USA). Length 150 mm.
3. An amplifier to record the voltage output of the pressure transducer. We use a custom-made apparatus fabricated at the Department of Biotechnology, University of Würzburg, Germany.
4. A differential amplifier with a sufficiently high input resistance to operate ion-selective electrodes (e.g., FD 223, MPI, Sarasota, FL, USA). The amplifier should have an input resistance of about 10^{15} Ohm to make sure that the voltage drop across the ion-selective fluid membrane is properly measured.
5. Black plastic foil that is impermeable to light.
6. Polyester foil (thickness 0.1 mm).
7. Silicone grease.
8. Silicone rubber blocks, thickness 0.8 mm.
9. Blu-tack.
10. A set of O-rings.

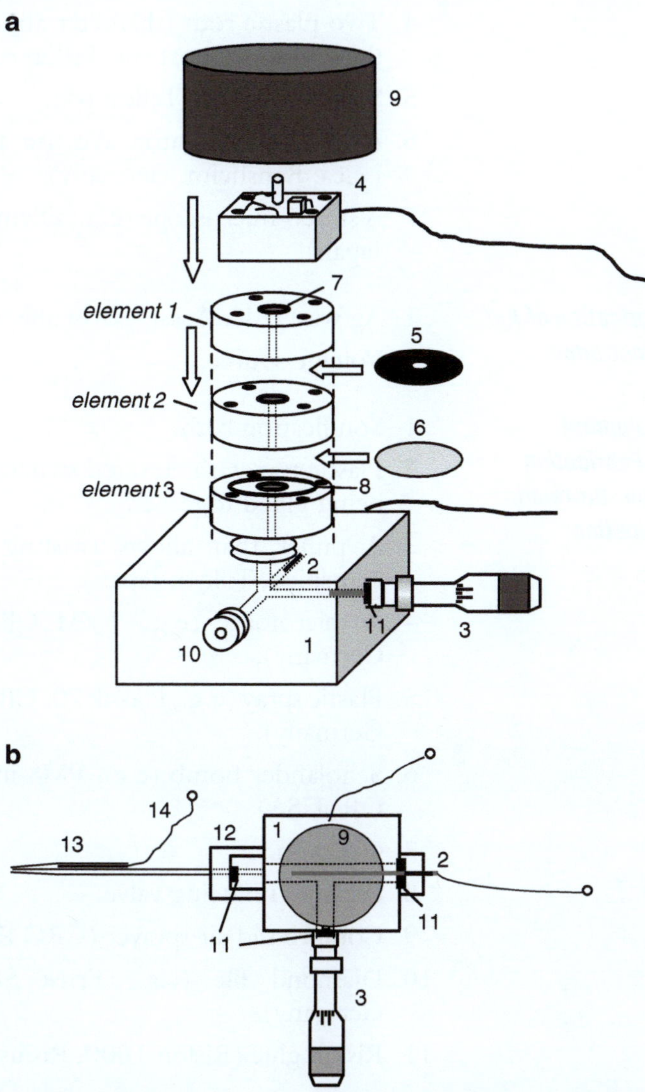

Fig. 2. Schematic drawing of the Perspex body of the xylem pressure-potential probe (**a**; side view) and the fully assembled multifunctional xylem probe (**b**; *top view*). The numbers denote: 1, main Perspex body; 2, Ag/AgCl electrode, 3, piston attached to a micrometer screw; 4, pressure transducer; 5, dark foil with central borehole; 6, polyester foil; 7, rubber O-ring; 8, circular notch filled with grease; 9, dark plastic lid; 10, nozzle at which the micropipette is attached; 11, rubber seals; 12, Perspex screw receiving the double-barreled microcapillary; 13, double-barreled microcapillary; 14, Ag/AgCl electrode to measure the potential of the ion-selective electrode.

2.2. Further Experimental Setup

1. A bath electrode (see Note 4).
2. An elongate plastic box, sufficiently long to receive the root (approximate dimensions: height 50 mm, width 70 mm, length 200 mm).
3. A Teflon rod used as a support for the root during measurements (diameter about 10 mm).

4. Two plastic rods (diameter about 15 mm) with a borehole at the end to support the Teflon rod.
5. Stands to fix the Teflon rod.
6. A micromanipulator. We use a mechanical one (Type "R," Leica, Bensheim, Germany).
7. A stereo-microscope (e.g., Olympus SZX 10, Olympus, Tokyo, Japan).

2.3. Fabrication of Ag/AgCl Electrodes

1. Ag wire (Alfa-Aesar, Karlsruhe, Germany).
2. Voltage source.

2.4. Equipment for the Fabrication of Double-Barreled Micropipettes

1. Sonification bath.
2. Plastic tray with elevated structures to support glass capillaries being glued together.
3. A puller that allows twisting the capillaries (e.g., PE21, Narishige, Tokyo, Japan).
4. A microforge (e.g., L/M CPZ101, Luigs and Neumann, Germany).
5. Plastic spray (e.g., Plastik 70, CRC Kontakt Chemie, Iffezheim, Germany).
6. Scholander bomb (e.g., PMS instrument company, Corvallis, OR, USA).
7. Gas cylinder.
8. Pressure reducing valve.
9. Compressed air sprayer (CRC Kontakt Chemie).
10. Diamond file (e.g., Ernst Stamm Stahlwaren, Solingen, Germany).
11. Rislon glue (Rislon 1000, Risius GmbH, Pullheim, Germany).

2.5. Data Recording

1. Personal computer.
2. A/D converter (e.g., DAS 1601, Keithley, Taunton, MA, USA).
3. Software for data recording and evaluation. We use home-made software that was developed with a Testpoint platform (Testpoint, Sydney, Australia).

2.6. Further Equipment

1. Precise digital manometer (e.g., GMSD 10br, GMH3150, Greisinger electronic GmbH, Regenstauf, Germany).
2. Exsiccator.
3. Vacuum pump (e.g., CPS-5B Vacuum pump, US vacuum pumps LLC, Dallas, TX, USA).

3. Chemicals

1. Chemicals for producing ion-selective fluid membranes. Chemicals are available from Sigma-Aldrich (München, Germany) except for tetrahydrofuran (J.T. Bakker, Deventer, the Netherlands) and for Vinnolit S1565™ (Vinnolit Kunststoff, Burghausen, Germany).
2. Na-alginate (Sigma-Aldrich).

4. Methods

4.1. The Assembly of the Probe

This description refers to the advanced version of the pressure-potential probe that was used by Wegner and Zimmermann (36) and in several follow-up studies (see Note 5). For assembly of the probe, the following steps are undertaken:

1. Position the pressure transducer, mounted on a commercially available Perspex support, upside-down in a beaker and fill the cavity below the pressure-sensitive membrane with degassed, double-distilled water (see Note 6). Form a small water reservoir on top of the Perspex support with a ring of Blu-tack that can easily be removed afterwards. Place this construction inside an exsiccator. Vacuum is established by means of a vacuum pump (see Note 7). The procedure should be continued until all gas bubbles attached to surfaces in the water-filled cavity have been removed.
2. Remove the Blu-tack ring and attach element 1 to the Perspex support of the pressure transducer using plastic screws. Fill the central borehole with degassed water and remove air bubbles in the same way as described above (see Subheading 4.1, step 1). A pressure-tight connection between the elements is insured by rubber O-rings surrounding the central borehole at each junction.
3. Mount the light protecting foil, and then element 2. The filling and degassing procedure is repeated as described in steps 1 and 2.
4. When water in the borehole extending from the pressure transducer to the outer surface of element 2 is air-bubble free, close it with a circular polyester foil (thickness 0.1 mm). Fill the ring-shaped notch engraved in the surface of element 3 with grease (see Note 8) and attach element 3 in such a way that the grease-filled notch is oriented towards the foil. For a detailed explanation of this design and its significance for the functioning of the probe, see Note 9. The borehole of element 3 is again filled with water and degassed by the usual procedure.

5. Attach the pressure transducer + periphery to the main body of the probe that receives the electrode for TRP recording and the microcapillary to be inserted into the tissue. The main body is initially also filled with double-distilled water (see Note 10).
6. Fabrication of the Ag/AgCl electrode (see Note 11):
   (a) Dip an Ag wire (diameter 0.1 mm) into a 0.1 M KCl solution and attach it to the positive pole (anodic connection) of a voltage source.
   (b) A second Ag wire, serving as the cathode, is immersed in the same solution and attached to the negative pole.
   (c) Start chloridation of the anode at a voltage of 0.5–1 V. Clean both Ag wires thoroughly before the procedure starts. Make sure that residual AgCl or hydrophobic layers on the metal surface have been removed. The chloridation process takes about 20 min (see Note 12).
   (d) The electrode prepared for the pressure-potential probe is additionally covered with a layer of Ag-alginate to create a smooth surface, thus minimizing the danger of cavitation. For Ag-alginate coating, the process described above (see step 6a–c) is repeated with the chlorided electrode, but now KCl is replaced by a 0.1 M Na-alginate solution. During the galvanic process, the electrode is covered by a brownish gel-like layer.
7. After this procedure, push the end of the chlorided wire through a silicone seal with a central borehole and insert it at the designated port of the main body of the probe. The seal is pressed to the cavity in the main Perspex body by a custom-made plastic screw with a metal core to make electrical contact with the electrode.
8. Contact the screw at the rear with a cable connecting it to a differential amplifier.
9. Another port receives a stainless steel piston attached to a micrometer screw designated to manipulate the pressure inside the probe (see Note 13). Cover the pressure transducer with a solid black plastic lid to shield it against light.

The body of the probe is fully assembled now and is ready for use. At this stage, the interior of the probe must be free of air bubbles (see Note 14).

4.2. Fabrication of a Bath Electrode

For TRP recording, an external reference electrode is required. It is fabricated as follows:

1. Drill a small hole into the wall of a yellow plastic tip as used with automatic pipettes.
2. Insert a chlorided Ag wire (diameter 1 mm; chlorided as described in Subheading 4.1, step 6a–c) partly through the pre-formed borehole.

3. Fill the tip with a 50 mM KCl solution gelled with 2% agarose.
4. Plug the basal end of the plastic tip with flexible silicone mass to prevent drying of the electrode.
5. Contact the extending part of the Ag wire to the signal ground of the differential amplifier used for measuring TRP and the potential of the ion-selective electrode.

4.3. The Experimental Setup

The setup is depicted in Fig. 3.

1. Fill an elongate plastic box, sufficiently long to receive the root (approximate dimensions: height 50 mm, width 70 mm, length 200 mm), with the bath medium in which the root is to be immersed during the experiment (termed "cuvette" hereon).
2. The intact seedling is supported by a Teflon rod that can be lowered into the cuvette in a controlled way by two rotary arms made from plastic. This construction must be firmly attached to a stable stand for maximum stability, as required during the measurements. A longitudinal notch along the rod receives the root (notches with various diameter and depth can be engraved among which can be selected for perfect fitting of an individual root). At a distance of 5 mm along the notch, small holes (diameter 0.8 mm) are drilled into the Teflon rod (see Notes 15 and 16).
3. Mount a seedling to this setup by fixing the root with several small hair clips of an appropriate size to the Teflon rod. Stabilize the shoot at the stand.

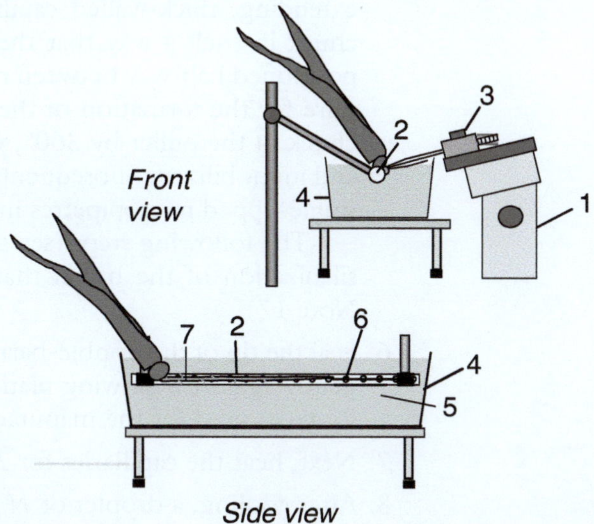

Fig. 3. Experimental setup. Numbers denote: 1, micromanipulator; 2, Teflon rod; 3, multi-functional xylem probe; 4, cuvette; 5, nutrient solution; 6, air-filled borehole; 7, root. After (42), with modifications. For more details see text.

4.4. Fabrication of Double-Barreled Micropipettes

A cartoon summarizing the fabrication process is shown in Figs. 4 and 5.

1. Double-barreled micropipettes are fabricated from two types of borosilicate glass capillaries (MTW 100F-6, M1B100-6). Clean the capillaries by sonification in a beaker filled with 50% w/v ethanol for 10 min (see Fig. 4, step 1). Before drying the capillaries in an oven, remove both the cleaning solution inside the capillaries using compressed air and the solution attached externally, by rolling the capillaries on tissue paper.
2. The ion-selective electrode is produced from the thin-walled capillary type MTW 100F-6 (outer diameter 1 mm; inner diameter 0.75 mm; with a filament; length 15 cm) that is shortened by 1.4 cm (by cutting the glass with a diamond file and subsequently breaking the capillary back).
3. Take one capillary of both types and place them alongside each other in such a way that the thick-walled capillary M1B100-6 (outer diameter: 1 mm; inner diameter: 0.58 mm; no filament; length 15 cm) extends at both ends by 0.7 cm. The capillaries are positioned on a plastic tray with elevated supports at both ends.
4. Glue capillaries together by placing two droplets of Rislon glue about 3 cm away from the centre in each direction. The glue is specific for glass and sufficiently inert to heat. Allow the glue to dry for at least 1 h in air.
5. Pull double-barreled single-tipped micropipettes by using a puller that allows twisting the capillaries. The pulling procedure is illustrated in Fig. 5. Fix the twinned capillaries at the extending, thick-walled capillary using the upper and lower chuck in such a way that the coiled wire used for heating is positioned half-way between the chucks. Use a two-step procedure for the formation of the capillary: First rotate the upper chuck of the puller by 360°, while the lower one is supported and immobilized. Subsequently, remove support and draw two single-tipped micropipettes in the usual way.

   The following steps (see steps 6–11) describe the selective silanization of the barrel that will serve for ion sensing (see Note 17).
6. Seal the tip of the double-barreled microcapillary by pushing it gently against a glowing platinum/tungsten wire on a microforge, as used for the manufacture of patch clamp electrodes.
7. Next, heat the capillaries for 2 h at 150°C in an oven.
8. After cooling, a droplet of *N,N*-dimethyltrimethylsilylamine is back-filled at the blunt ends of the thin-walled capillary shank with a syringe, and immediately removed. The rest of the silane that remains attached to the walls is sufficient for successful silanization of the barrel (see Note 18).

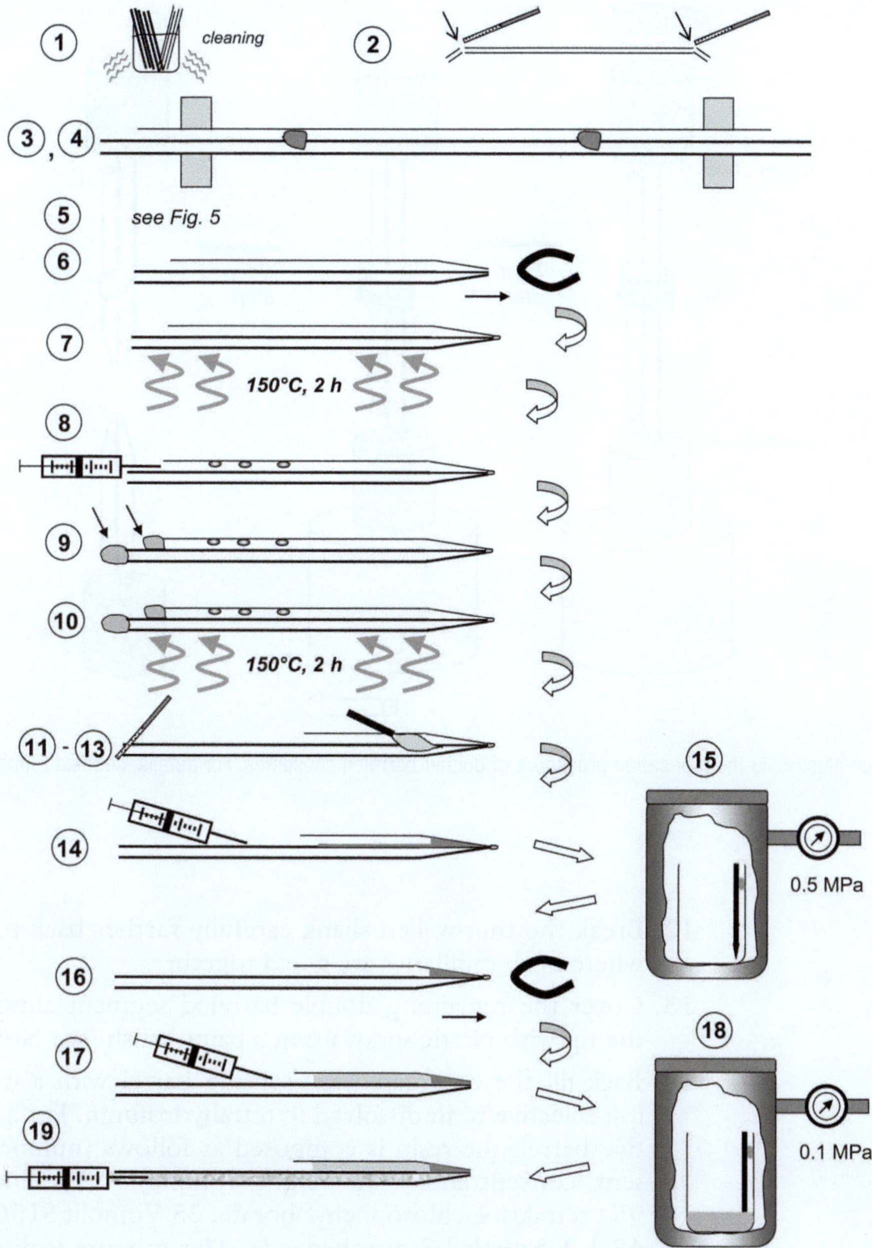

Fig. 4. Illustration of the procedure for electrode manufacture (compare text, Subheading 4.4).

9. Close the blunt ends of both shanks with a drop of Silgard™ 182 (Dow Corning, Midland, MI, USA). This is immediately hardened by brief exposure to a bunsen burner flame.
10. Next, the heating procedure as described under Subheading 4.4, step 7 is repeated.
11. Remove the Sylgard droplets by cutting the ends back with the diamond file.

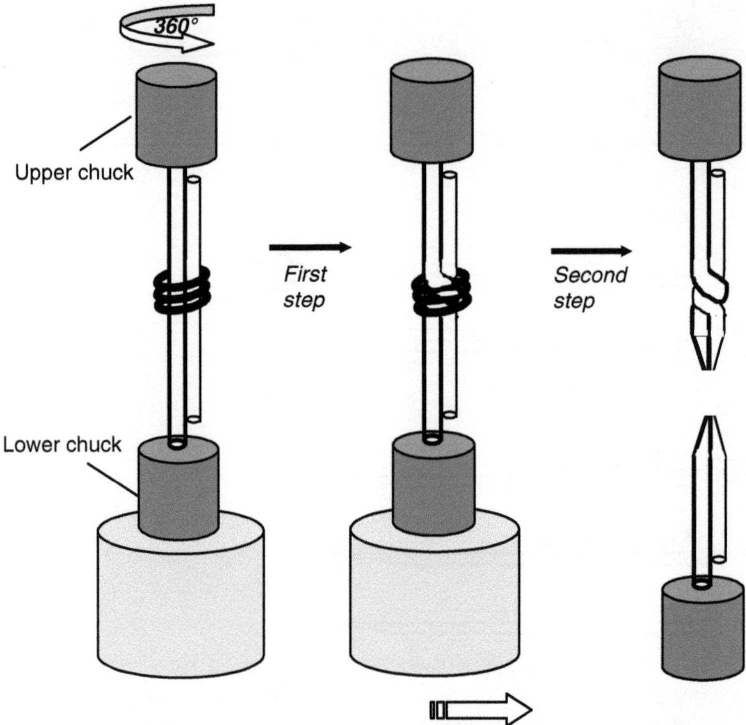

Fig. 5. Cartoon illustrating the fabrication procedure of double-barreled capillaries. For details, see text (Subheading 4.4, step 5).

12. Break the thin-walled shank carefully further back to the site where both capillaries are glued together.
13. Cover the remaining double-barreled segment almost up to the tip with plastic spray, using a paint brush (see Note 19).
14. Back-fill the designated ion-sensing barrel with a droplet of ion-selective resin dissolved in tetrahydrofuran. For a K^+ selective barrel, the resin is composed as follows (numbers represent concentrations in weight/volume): 18 valinomycine, 9K^+-tetrakis(4-chlorophenyl)borate, 25 Vinnolit S1565™, and 48 1,2-dimethyl-3-nitrobenzene. This mixture is dissolved in 4 volumes of tetrahydrofuran. Vinnolit, a derivative of polyvinylchloride, forms hydrophilic surfaces, so is most suitable for use in multifunctional xylem probes. Xylem pH is recorded with a resin containing (in %w/v) 12 4-Nonadecyl pyridine (Fluka trade name: H-Ionophore II); 6K-tetrakis-(4-chlorophenyl)borat; 27 Vinnolit™ S1565; 5 Nitrocel S; 50 2-Nitrophenyloctylether.

    After dissolution in tetrahydrofuran, the ion-selective resins will flow down into the tip without further assistance (see Note 20).

15. A pressure-tight plug is formed by evaporation of the tetrahydrofuran. Drying of the resin is best performed at room temperature and under an overpressure of about 0.5 MPa in a pressure bomb of the Scholander type for at least 2 h. Pressurization prevents the formation of gas-filled cavities in the resin and favors a more rigid texture of the resin. Take a beaker that fits into the bomb and plug a ring of Blu-tack to the inner wall. The double-barreled micropipettes are plugged to the Blu-tack ring in tip-down position and can easily be removed again at the end of the procedure.
16. After pressure-drying the ion-selective plug, the sealed tips of the micropipettes have to be broken back again. This is achieved by gently pushing the tip once again against the platinum/tungsten wire of the microforge (which is not heated this time). With this procedure, sharp edges are generated that will facilitate penetration of the tissue and the vessel walls. The final diameter of the tip should be about 7 μm.
17. Back-fill the ion-selective shank with 50 mM KCl solution. For pH sensing, the back-filling solution should additionally contain 10 mM Mes/BTP, adjusted to a pH value of 5.5.
18. Pre-condition the ion-selective electrode by dipping the tip in the solution used for back-filling and applying a pressure of 0.1 MPa for 30 min. Use the setup as described under step 15.
19. The pressure/voltage sensing barrel should be filled with a thoroughly degassed solution (see above for the procedure) either 50 mM KCl or NMG-Mes.
20. The double-barreled micropipette is now ready for use.

4.5. Experimental Procedure

1. After fixing a plant to the setup as described above (see Fig. 3), allow it to equilibrate for at least 30 min.
2. Concomitantly, the probe is prepared for an experiment:
   (a) *Calibration of the pressure transducer:* Attach the nozzle of the probe that is designated to receive the microcapillary to a commercially available digital manometer via a system of pressure-tight connections. A further connection makes contact to a gas cylinder via a pressure reducing valve. Determine the constant calibration factor by increasing and decreasing the pressure stepwise and taking notes of the exact pressure from the manometer and the output voltage signal coming from the pressure transducer. Make sure that no hysteresis occurs with decreasing and increasing pressure. With the pressure-potential probe, this rapid procedure should be routinely repeated every second day to detect possible changes in the sensitivity to pressure.

(b) Fix the pressure-potential sensing barrel of the double-barreled capillary to the main Perspex body via a rubber seal. The rubber seal is embedded in a Perspex screw and screwed to the chamber in a pressure-tight manner. At the inside of the screw, a thread is engraved that fits to its designated counterpart at the nozzle of the microbaric chamber. The blunt end of the double-barreled microcapillary is pushed through the seal via a central borehole so that after mounting, the end of the microcapillary extends into the interior of the probe (see Fig. 2b). Surfaces of the screw and rubber seal should be wetted prior to use. This is achieved by immersing the screw in a beaker filled with degassed water that is placed in an exsiccator (compare Subheading 4.1, step 1). Vacuum is applied with a vacuum pump until gas bubbles attached to the inner surfaces are removed. The screws are then stored in degassed water until they are used for an experiment. When the screw has received the micropipette and the blunt end of the capillary has been pushed about 0.5 cm beyond the rubber seal, the tip of the micropipette is downward-directed and the cavity of the screw is filled with degassed electrolyte solution (either 50 mM KCl or NMG-Mes). Surfaces are checked again for attached tiny air bubbles (see Note 21). Upon attachment of the capillary to the Perspex body, wait until the overpressure generated in the probe has relaxed (see Note 22). The voltage reading of the probe should be at infinity (beyond the range of the amplifier) when the tip is exposed to air and should immediately read a value close to 0 once the tip is located in the bath (see Note 23).

3. Insert a chlorided silver wire (diameter 0.3 mm; for the chloridation procedure, see Subheading 4.1, step 6a–c)) into the blunt end of the ion-selective electrode that makes contact with a differential amplifier. Close the blunt end with a droplet of glue that also fixes the position of the wire.

4. Calibrate the ion-selective barrel of the probe. Dip the tip into a series of solutions with known activities (in the case of K^+) or pH (for pH probes). For this purpose, a number of small Petri dishes (diameter about 2 cm) are glued to a large Petri dish lid (diameter about 15 cm). The small Petri dishes are pre-filled with a series of solutions, and the microelectrode can be moved from one to the other by turning the large Petri dish (see Note 24).

5. Fix the probe to a micromanipulator and lower the tip into the bath. As soon as the tip makes contact with the bath solution, pressure should rapidly relax to atmospheric level.

6. Wait for at least for 2 min before the impalement is started to verify that the voltage recording is drift-free. Success of the experiment depends much on the site at which the probe tip enters the root tissue (see Note 25). Advance the probe tip to the root surface under visual control using a stereo-microscope. Once the tip makes contact with the tissue, the voltage registered by the probe will immediately drop to negative values.
7. Slowly penetrate the root tissue at a rate of about 10 μm/s until a vessel is hit (see Note 26).
8. Once a vessel has been hit, this is almost instantaneously registered by a drop in pressure below the atmospheric level (see Fig. 1b). Stop moving the probe as soon as the pressure starts to decrease (see Note 27).
9. The "quality" of the impalement can be critically assessed from the smoothness of the pressure trace (large fluctuations indicate tip clogging) and the TRP trace that should relax to a new steady state value within a few minutes (see Note 28). The experiment should be terminated once a cavitation event has occurred in the probe and/or in the impaled vessel (see Note 29).
10. At the end of the experiment, withdraw the probe carefully from the xylem until the tip is situated in the bath solution again. The potential should be stable and identical to the one measured before impalement (within about ±4 mV).
11. Re-calibrate the ion-selective electrode as described in Subheading 4.5, step 4.
12. After the experiment, take notes on details of the plant specimen: age, fresh weight, and length of root and shoot.

4.6. Data Acquisition and Evaluation

The handling of data for multifunctional xylem probes is relatively straightforward. Three parameters, xylem pressure, xylem electrical potential (here: TRP), and the signal registered by the ion-selective electrode (see Fig. 6), are recorded in units of volt or millivolt and stored on a personal computer. Conventional a/d converters can be used to digitize the data. Further details are given below.

1. *Evaluation of xylem pressure data*. The output of the pressure transducer (after suitable amplification) is sampled at a low frequency, typically 0.1 Hz. Calculate the pressure by dividing the output of the pressure sensor (in volt or millivolt) by the slope factor obtained as a result of the pre-calibration of the probe (see Subheading 4.5, step 2.1). Note again that according to the definition, xylem pressure is zero at vacuum (see Note 2).
2. *Evaluation of TRP data*. TRP data, usually being more noisy than the pressure data, are sampled at a rate of 10 Hz. After recording, they are digitally low-pass filtered at a rate of 0.1 Hz to facilitate the identification of long-term trends. The offset

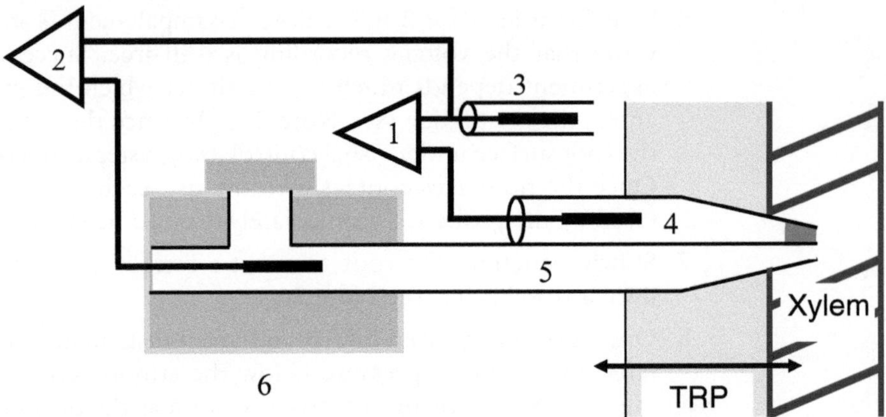

Fig. 6. Schematic drawing of the electrical periphery for recording the TRP and the potential of the ion-selective electrode. Numbers denote: 1, amplifier 1; 2, amplifier 2; 3, bath (reference) electrode; 4, ion-selective barrel; 5, TRP sensing barrel; 6, Perspex chamber of the probe.

between the pressure/potential sensing barrel and the reference electrode is measured, while the electrode tip is immersed in the bath, before and after impalement. Subtract this value from the data recorded, while the probe remains inserted in the xylem. Note that values obtained during penetration of the tissue do not reflect the actual membrane potential of the penetrated cells.

3. *Calculation of xylem sap ion activities from the output of the ion-selective electrode.* The voltage that is relevant for the calculation of the activity of the ion of interest is the voltage drop across the ion-selective resin filling the tip (E_I). This voltage drop is equal to the voltage difference between values recorded by the ion-sensing barrel (E_{ISE}) and the pressure-potential sensing barrel (TRP), both measured with respect to the bath:

$$E_I = E_{ISE} - \text{TRP}$$

4. The electrical arrangement is shown in Fig. 6. As with the TRP, E_{ISE} is sampled and digitally filtered at rates of 10 and 0.1 Hz, respectively. E_I is equivalent to the voltage drop across the tip of the ion-selective electrode measured during calibration, so ion activities can be re-calculated from E_I by making use of the calibration curve (see Fig. 1a). For this purpose, values obtained during the calibration procedure are plotted against the corresponding known ion activities. Pool the data obtained before and after insertion of the probe into the root. From the semi-logarithmic plot, the individual slope of the ion-selective electrode, S, and the E_I value at an activity

of 1 mM, $E_{I,1\,mM}$, are obtained. Ion activities in the xylem, A_x, can be calculated according to:

$$A_x = 10((E_{ISE} - TRP - E_{I,1mM}) / S)$$

See also Note 30.

5. Notes

1. Concentrations versus activities. The parameter measured by the ion-selective electrode is the ion activity (A) that reflects the "effective" concentration, taking non-ideal properties of solutions into account. The ion activity is also the parameter that is relevant for thermodynamic calculations, i.e., the driving force across a membrane. In physiology we usually deal with "concentrations" (c) rather than activities. Both parameters are related by the simple equation $A = \gamma c$, with γ being the dimensionless activity coefficient. γ decreases with the concentration, but is close to 1 in very dilute solutions such as the xylem sap (maximum K^+ activity ~10 mM, with K^+ being the main cation). Hence, for the xylem it is justified to postulate $A \approx c$.
2. According to convention, *xylem pressure* is given in absolute numbers, i.e., zero pressure corresponds to vacuum. Consequently, atmospheric pressure is at +0.1 MPa. Note that this is different from *turgor (pressure)* that is usually defined as an overpressure (i.e., zero turgor is equivalent to atmospheric pressure). When pressure drops below atmospheric level, *tension* builds up. Tension and pressure are opposite in sign and tension at vacuum attains a value of +0.1 MPa.
3. Sub-atmospheric xylem pressure values above vacuum occur frequently but are somewhat ambiguous and should be interpreted with care. They could be an experimental artifact, for example, due to an improperly prepared probe or to a leak at the site of insertion. Pressure values can be verified by making sure that the pressure drops below vacuum at least once during an experiment, e.g., when light intensity is increased or an osmoticum is added to the bath.
4. A home-made bath electrode is required as a reference for measuring TRP and xylem ion activities. For the fabrication procedure, see Subheading 4.2. This electrode is grounded at the amplifier (see Fig. 6).
5. A schematic representation of the body of the probe is shown in Fig. 2. In this version, the main Perspex body that hosts the potential sensing electrode is separated from the pressure

transducer by three circular Perspex elements, each with a central borehole (diameter 0.8 mm). A black foil with a central hole, shielding the pressure transducer against light, and a 12-μm-thin polyester foil are inserted between element 1 and 2 and between element 2 and 3, respectively. The polyester foil functions to separate the pressure transducer and the internal electrode galvanically while providing (almost) no resistance to pressure transmission. Below and above this foil, the central borehole is extended to circular cavities with a diameter of about 3 mm.

6. Assembly of the probe starts with the pressure transducer. This has to be perfectly wettable at the surface exposed to the interior of the probe and should have linear properties both in the positive and the negative pressure range. Preferably, a Siemens KPY 16 sensor (Siemens, Erlangen, Germany) should be used as this meets these requirements (2). A suitable amplifier is also required to register the voltage output of the probe; we use home-made amplifiers built at the Department of Biotechnology, University of Würzburg, Germany.
7. Tiny air bubbles attached to the surface of the pressure transducer expand under these conditions and can easily be discerned by inspection through the transparent lid. They will coalesce and finally rise to the surface of the water reservoir.
8. This notch is filled with grease to prevent formation of an electrical leak by a thin film of electrolyte solution that might spread between polyester foil and the surface of element 3.
9. An overpressure of about 0.1 MPa will be generated in the closed compartment that is formed between polyester foil and pressure transducer when element 3 is tightly fixed. However, this overpressure will dissipate within about 24 h. Subsequently, pressure remains sufficiently stable so that the probe is ready to be used the following day. Small pressure offsets can be easily detected before each experiment prior to mounting the capillary when the interior of the probe is at atmospheric pressure. These offsets are taken into account during data evaluation (see Note 1). The calibration factor will be stable for about 4 weeks, but should be checked regularly. After this time the sensitivity of the foil to increments in pressure decreases dramatically. In order to replace the foil, the probe has to be re-assembled.
10. When the Perspex body of the probe is assembled, it is completely filled with degassed double-distilled water. Only later, before attachment, is the water in the central borehole hosting the electrode replaced by 50 mM KCl. KCl will diffuse into the other boreholes, including the one contacting the pressure transducer, up to the polyester foil that serves as an efficient diffusion barrier.

11. An Ag/AgCl electrode is used to avoid drift of the electrode potential during the experiment.
12. During this galvanic process, the Ag wire to be chlorided is covered with a brownish-grey layer of AgCl. Small H_2 gas bubbles are formed at the cathode.
13. The piston should be electrically isolated from the metal parts of the micrometer screw. If it is not, the screw that picks up electrical noise from the environment transduces it to the interior of the probe.
14. One major issue is the elimination of air bubbles entrapped in the interior of the probe after assembly or the local inclusion of vapor after the occurrence of a cavitation event during an experiment. The probe is most "vulnerable" at the piston for applying volume/pressure pulses or at the Ag/AgCl electrode. Attached bubbles (easily identified by inspection since the Perspex body is transparent) are eliminated by removing these elements from the probe and immersing them in a beaker filled with degassed, double-distilled water and placing them in an exsiccator. When vacuum is established in the exsiccator, bubbles detach from the surface. The Perspex body of the probe is also put into the exsiccator; the nozzle for attaching the glass capillary is closed with a tight screw (without a central borehole) and the remaining open port receiving the piston or electrode is also filled with the degassed water. When all air bubbles have been removed, the probe is re-assembled. Air can also be dragged into the probe at the ports when the rubber seals are not fitting tightly. In these cases, the rubber seals have to be replaced and/or the screws have to be fastened more tightly. Another way of eliminating small air bubbles inside the Perspex body of the probe is to establish high pressures for a short time when it is completely assembled. This can be done by closing the opening usually receiving the capillary with the tight screw. This should be done carefully in order not to damage the pressure transducer by exceeding its pressure limit. If overpressure is dissipated rapidly under these conditions, the probe is apparently leaky and has to be re-assembled to eliminate the leak after replacing the rubber parts that cause the leak.
15. Due to the hydrophobicity of the material, these holes remain air-filled when the rod carrying the seedling is submersed, serving as an oxygen reservoir for the attached root tissue to prevent hypoxic conditions. Regular cleaning of the borehole is required to remove algae.
16. It should be noted that the rate of water uptake by the root can additionally be measured with this setup when the cuvette is placed on a digital balance and the loss of weight is continuously monitored, as described previously (42). This aspect will not be covered in more detail here.

17. Glass surfaces that are usually strongly hydrophilic can be made hydrophobic by silanization. The barrel designated for K^+ sensing is silanized to avoid short-circuiting of the fluid ion-selective membrane along the hydrophilic glass surface. However, silanization of the pressure/potential sensing barrel must be prevented, since cavitation is favored by a hydrophobic surface.
18. If the silane does not evaporate completely during the subsequent heating process, the residual fluid tends to accumulate in the very tip, taking the place later to be filled with ion-selective resin. In this case, the capillary has to be discarded.
19. This is required to prevent inadvertent filling of the gap between both capillaries with electrolyte solution when the microelectrode is in use, leading to a short-circuiting of the ion-selective barrel.
20. Void volume enclosed by resin during the filling process of the very tip will vanish as drying proceeds.
21. Only after gas inclusions attached to glass, Perspex, or rubber surfaces have completely been removed (mechanically with a toothpick for example) is the probe ready for assembly. Also, check the glass capillary and the probe for the presence of any gas phase by inspection. Air bubbles can be removed by washing the central borehole or the pressure/potential sensing barrel of the micropipette with degassed electrolyte solution as required for the particular experiment (50 mM KCl or NMG-Mes, see Subheadings 1 and 4.4, step 19). This is done by injecting the solution via a Hamilton syringe. The solution in the central borehole should regularly be replaced, especially when the pressure/potential sensing barrel is filled with a solution that is different from the one inside the Perspex chamber.
22. When the double-barreled micropipette supported by the screw is attached to the body of the probe, an overpressure will be generated in the probe that will dissipate with an exponential time course, depending on the hydraulic conductance of the tip.

    When the capillary is attached to the main Perspex body, be aware that the microcapillary should be inserted into a xylem vessel by keeping the K^+ sensing barrel in upstream direction in order to minimize artifacts arising from the release of small amounts of solution from the pressure-potential sensing barrel into a xylem vessel. This can be controlled by turning the ion-selective electrode after insertion at a certain position with respect to the thread of the Perspex screw that should be marked on the outside of the screw with a text-marker. Orientation of the probe is controlled visually via the stereo-microscope.

23. When the electrical signal does not respond in the described way, no closed electrical circuit can be established, even though the tip has been lowered into the bath. This could be due to an inclusion of an air bubble either at the very tip of the pressure/potential sensing electrode (easily removed by a pressure pulse) or in the bath electrode. Alternatively, an air bubble could have been enclosed at the blunt end of the glass capillary when the micropipette was attached to the Perspex body of the probe. The electrical signal will also not be as expected if the Perspex body of the probe has not been assembled properly, so as a consequence, the pressure sensor interferes with the potential recording electrode.
24. Ensure that the electrodes respond within less than a minute to a change in ion activity (see Fig. 1a). While the tip is transferred from one solution to another, voltage recordings of both microelectrodes should rapidly move away from 0. Make sure that increasing and decreasing series of ion activities render the same results. The slope of the ion-selective electrode (i.e., the voltage shift associated with a tenfold increase in activity) should at least be 45 mV for monovalent cations such as K^+ and H^+. If this criterion is not met, the electrode should be discarded.
25. The impalement is done "blindly," but the chance of hitting a vessel depends to a high degree on the site at which the organ (here, the root) is impaled. A detailed knowledge of the anatomy of the plant organ is a prerequisite for successful measurements. Based on this knowledge and using simple geometric considerations, an optimum position and angle for the probe to make contact with the tissue can be calculated.
26. During passage through the tissue, a slight overpressure will be registered due to the turgor of the root cells (this is not identical to the original turgor of the cells); the voltage will be negative with respect to the bath due to the negative membrane potential of the cells.
27. The time course of pressure relaxation depends on the hydraulic conductance of the tip (and is much slower than the speed of pressure transmission in a water column that equals the speed of sound), since pressure equilibration between the probe and the vessel is associated with mass flow between the compartments due to the compressibility of the probe (see above). When a tip of a well-prepared probe is inserted into a vessel, the pressure should drop to a new steady value within ~30 s. A slower time course indicates that the tip is partially blocked. Sometimes an obstruction can be removed by a pressure pulse applied by moving the metal piston designed for this purpose.
28. Critical assessment of an experiment as a whole is based on the pressure trace that discloses problems such as tip clogging or a

leak induced by the impalement of a vessel. Criteria for a "good" experiment are: (i) rapid response of the pressure signal to impalement and retraction of the sensor, (ii) response to a stepwise increase in light intensity and osmotic pressure in the bath according to the "typical" patterns (see Fig. 1), and (iii) a low noise level.

29. Detection of a cavitation event and significance of cavitation events for the experiment: Cavitation can only occur at negative pressures when the sap is in a metastable state. A cavitation event is detected by an instantaneous increase in pressure to a value slightly above vacuum that remains invariant to changes in light or humidity. Application of a volume pulse by moving the position of the piston will also not affect the pressure signal. If a gas bubble is formed in the capillary or (more frequently) in the body of the probe, a permanent water flow from the probe into the xylem is induced that will lead to a gradual volume increase of the gas phase in the probe. The experiment should be stopped and the data discarded since under these conditions both TRP and ion activity recordings are strongly affected. Furthermore, a cavitation event will render data for all three parameters obsolete.

30. For the ion activity trace, the noise level increases strongly with increasing activities. This is due to the fact that the signal is recorded at a logarithmical scale but presented at a linear scale (see Fig. 1).

References

1. Shabala S, Pang J, Zhou M et al (2009) Electrical signalling and cytokinins mediate effects of light and root cutting on ion uptake in intact plants. Plant Cell Environ 32:194–207
2. Balling A, Zimmermann U (1990) Comparative measurements of the xylem pressure of *Nicotiana* plants by means of the pressure bomb and pressure probe. Planta 182:325–338
3. Zimmermann U, Räde H, Steudle E (1969) Kontinuierliche Druckmessung in Pflanzenzellen Naturwissenschaften 56:634
4. Zimmermann U (1989) Water relations of plant cells: Pressure probe techniques. Methods Enzymol 174:338–366
5. Wegner LH, Zimmermann U (1998) Simultaneous recording of xylem pressure and trans-root potential in roots of intact glycophytes using a novel xylem pressure probe technique. Plant Cell Environ 21:849–865
6. Felle HH (1993) Ion-selective microelectrodes: their use and importance in modern plant cell biology. Bot Acta 106:5–12
7. Wegner LH, Stefano G, Shabala L et al (2011) Sequential depolarization of root cortical and stelar cells induced by an acute salt shock – implications for Na^+ and K^+ transport into xylem vessels. Plant Cell Environ 34:859–869
8. Zimmermann U, Schneider H, Thürmer F, Wegner LH (2001) Pressure probe measurements of the driving forces for water transport in intact higher plants: Effects of transpiration and salinity. In: Läuchli A, Lüttge U (eds) Salinity: environment-plants-molecules. Kluwer Academic, Dordrecht
9. Zimmermann U, Schneider H, Wegner LH, Haase A (2004) Water ascent in tall trees: does evolution of land plants rely on a highly metastable state? New Phytol 162:575–615 (Tansley Review)
10. Wegner LH, Schneider H, Zimmermann U (2007) On-line measurements of ion relations in the xylem sap of intact plants. In: Sattelmacher B, Horst WJ (eds) The apoplast: compartment of transport, storage and reaction. Kluwer Academic, Dordrecht, pp 221–234

11. Kaufmann I, Schulze-Till T, Schneider HU et al (2009) Functional repair of embolized vessels in maize roots after temporal drought stress, as demonstrated by magnetic resonance imaging. New Phytol 184:245–256
12. Zwieniecki MA, Holbrook NM (2009) Confronting Maxwell's demon: biophysics of xylem embolism repair. Trends Plant Sci 14:530–534
13. Schneider H, Wistuba N, Miller B et al (1997) Diurnal variation in the radial reflection coefficient of intact maize roots determined with the xylem pressure probe. J Exp Bot 48: 2045–2053
14. Milburn JA (1996) Sap ascent in vascular plants: challengers to the cohesion theory ignore the significance of immature xylem and the recycling of Münch water. Ann Bot 78:399–407
15. Sperry JS, Saliendra NZ, Pockman WT et al (1996) New evidence for large negative xylem pressures and their measurement by the pressure chamber method. Plant Cell Environ 19:427–436
16. Benkert R, Balling A, Zimmermann U (1991) Direct measurements of the pressure and flow in the xylem vessels of *Nicotiana tabacum* and their dependence on flow resistance and transpiration rate. Bot Acta 104:405–464
17. Wei C, Tyree MT, Steudle E (1999) Direct measurement of xylem pressure in leaves of intact maize plants. A test of the Cohesion-Tension theory taking hydraulic architecture into consideration Plant Physiol 121:1191–1205
18. Wei C, Steudle E, Tyree MT, Lintilhac PM (2001) The essentials of direct xylem pressure measurement. Plant Cell Environ 24:549–555
19. Schneider H, Wegner LH, Haase A, Zimmermann U (2007) Long-distance water transport under controlled transpirational conditions: minimal-invasive investigations by means of pressure probes and NMR imaging. In: Sattelmacher B, Horst WJ (eds) The apoplast: compartment of transport, storage and reaction. Kluwer Academic, Dordrecht
20. Zhu JJ, Zimmermann U, Thürmer F, Haase A (1995) Xylem pressure response in maize roots subjected to osmotic stress: determination of radial reflection coefficients by use of the xylem pressure probe. Plant Cell Environ 18:906–912
21. Schneider H, Zhu JJ, Zimmermann U (1997) Xylem and cell turgor pressure probe measurements in intact roots of glycophytes: transpiration induces a change in the radial and cellular reflection coefficients. Plant Cell Environ 20:221–229
22. Bai XF, Zhu JJ, Zhang P et al (2007) Na^+ and water uptake in relation to the radial reflection coefficient of root in arrow leaf saltbush under salt stress. J Int Plant Biol 49:1334–1340
23. Tang A-C, Boyer JS (2008) Xylem tension affects growth-induced water potential and daily elongation of maize leaves. J Exp Bot 59:753–764
24. Steudle E (2001) The cohesion-tension mechanism and the acquisition of water by plant roots. Ann Rev Plant Physiol Plant Mol Biol 52:847–875
25. Steudle E, Peterson CA (1998) How does water get through roots? J Exp Bot 49: 775–788
26. Bramley H, Turner NC, Turner DW, Tyerman SD (2007) Comparison between gradient-dependent hydraulic conductivities of roots using the root pressure probe: the role of pressure propagations and implications for the relative roles of parallel radial pathways. Plant Cell Environ 30:861–874
27. Knipfer T, Fricke W (2010) Root pressure and a solute reflection coefficient close to unity exclude a purely apoplastic pathway of radial water transport in barley. New Phytol 187(1): 159–170
28. Shabala S, Cuin TA (2008) Potassium transport and plant salt tolerance. Physiol Plant 133:651–669
29. Dunlop J (1982) Membrane potentials in the xylem in roots of intact plants. J Exp Bot 33:910–918
30. Anderson WP, Higinbotham N (1975) A cautionary note on plant root electrophysiology. J Exp Bot 26:533–535
31. Shabala S, Shabala L, Cuin TA et al (2010) Xylem ionic relations and salinity tolerance in barley. Plant J 61:839–853
32. Dunlop J, Bowling DJF (1971) The movement of ions to the xylem exudate of maize roots. III The location of the electrical and electrochemical potential differences between the exudate and the medium J Exp Bot 22:453–464
33. Okamoto H, Ichinao K, Katou K (1978) Radial electrogenic activity in the stem of *Vigna sesquipedalis*: involvement of spatially separate pumps. Plant Cell Environ 1:279–284
34. De Boer AH, Prins HBA, Zanstra PE (1983) Biphasic composition of trans-root electrical potential in roots of *Plantago* species: involvement of spatially separated electrogenic pumps. Planta 157:259–266
35. De Boer AH (1989) Xylem transport. Methods Enzymol 174:277–287
36. Wegner LH, Zimmermann U (2002) On-line measurements of K^+ activity in the tensile water of the xylem conduit of higher plants. Plant J 32:409–417

37. Wegner LH, Sattelmacher B, Läuchli A, Zimmermann U (1999) Trans-root potential, xylem pressure, and root cortical membrane potential of 'low-salt' maize plants as influenced by nitrate and ammonium. Plant Cell Environ 22:1549–1558
38. Hua JM, Wang XL, Zhai FQ, Yan F, Feng K (2008) Effects of NaCl and Ca^{2+} on membrane potential of epidermal cells of maize roots. Agricult Sci China 7:291–296
39. Schurr U (1998) Xylem sap sampling - new approaches to an old topic. Trends Plant Sci 3:293–298
40. Wegner LH, Zimmermann U (2004) Bicarbonate-induce alkalinization of the xylem sap in intact maize seedlings as measured in situ with a novel xylem pH probe. Plant Physiol 136:3469–3477
41. Shabala S, Demidchik V, Shabala L et al (2006) Extracellular Ca^{2+} ameliorates NaCl-induced K^+ loss from Arabidopsis root and leaf cells by controlling plasma membrane K^+-permeable channels. Plant Physiol 141: 1653–1665
42. Wegner LH, Zimmermann U (2009) Hydraulic conductance and K^+ transport into the xylem depend on radial volume flow, rather than on xylem pressure, in roots of intact, transpiring maize seedlings. New Phytol 181:361–373

Chapter 4

Measuring Intracellular Ion Concentrations with Multi-Barrelled Microelectrodes

Anthony J. Miller and Susan Smith

Abstract

Ion-selective microelectrodes can be used to measure intracellular ion concentrations. The use of multi-barrelled electrodes enables the identification of the cellular compartment. For example, the inclusion of a pH-selective electrode enables the cytoplasm and vacuole to be distinguished. The ion-selective barrels of microelectrodes are filled with a sensor cocktail containing several different components.

- An ion-selective molecule, sensor or exchanger.
- Membrane solvent or plasticizer.
- Additives, e.g., lipophilic cation/anion.
- Membrane matrix to solidify the ion-selective membrane; essential for measurements in plant cells with a cell wall and turgor.

For many ions, the ready-made membrane cocktail can be purchased, but the individual chemical components can be bought from suppliers and mixing the cocktail oneself is cheaper. For commercially available liquid membrane cocktails, the membrane matrix is not normally included. A matrix is needed if the microelectrodes are to be used in plants because cell turgor will displace a liquid membrane from the electrode tip, thereby changing or eliminating the sensitivity to the measuring ion. The matrix used is usually a high molecular weight poly(vinyl chloride), but can include other polymers, such as nitrocellulose for additional strength.

Key words: Microelectrodes, Intracellular ion concentrations, Cytoplasm, Vacuole

1. Introduction

Intracellular ion concentrations are important indicators of cellular health and nutrient status. In addition, changes in these concentrations can provide signals that can translate an environmental signal into altered gene expression and post-translational modifications.

Sergey Shabala and Tracey Ann Cuin (eds.), *Plant Salt Tolerance: Methods and Protocols*, Methods in Molecular Biology, vol. 913, DOI 10.1007/978-1-61779-986-0_4, © Springer Science+Business Media, LLC 2012

Methods for measuring intracellular ion concentrations suffer from the difficulty that the technique itself can alter the parameter that is being measured. Reporter molecules such as fluorescent dyes or proteins function by binding the ion being sensed and therefore will change the concentration of the ion being measured. Inserting glass microelectrodes into a cell measures at a single location and gradients may occur within a cell. No method is perfect and the best approach is to use more than one method to check any intracellular ion measurement.

The term *microelectrode* is used to describe a glass micropipette that is pulled to a fine tip at one end and filled with an aqueous salt solution that provides an electrical bridge to a metal wire contact. The junction between the salt solution inside the microelectrode and the input to the electrometer amplifier is provided by a *half-cell.* There are different types of *half-cell*, but usually the metal contact is AgCl-coated silver wire and the salt solution is 0.1 M KCl. The simplest microelectrodes measure voltage, and when inserted into cells, measure the membrane potential, in mV, between the inside and outside of the cell. An ion-selective microelectrode contains an ion-selective membrane in the tip of the glass micropipette and is responsive both to the membrane potential and the activity (not actually concentration) of the ion sensed by the selective membrane. Ion-selective membranes are composed of specific types of chemicals and the roles played by each component are described in detail by Ammann (1). Good electrodes have a low detection limit, a near ideal slope, and a small selectivity coefficient for physiologically important interfering ions.

Ion-selective microelectrodes are used to measure ion gradients across membranes. These measurements can be made outside and inside cells. For example, ion fluxes at the surface of roots can be measured by using either directly ion-selective microelectrodes or by using an ion-selective vibrating probe (2). Microelectrodes have been used to report the cellular compartmentation of nutrients, dynamics of changes in ion activities (e.g., in intracellular signaling), and transport mechanisms, particularly the energy gradients for cotransported ions (3). For intracellular ion measurement, the simultaneous recording of membrane potential is required. This is done by insertion of a second electrode or, for small cells, by combining the ion-selective and voltage-measuring electrodes into a double-barrelled microelectrode (see Fig. 1). Five distinct stages for the manufacture of ion-selective electrodes can be identified:

1. Pulling glass micropipettes.
2. Silanization of the inside surface of the designated ion-selective micropipette.
3. Back-filling with ion-selective membrane cocktail.

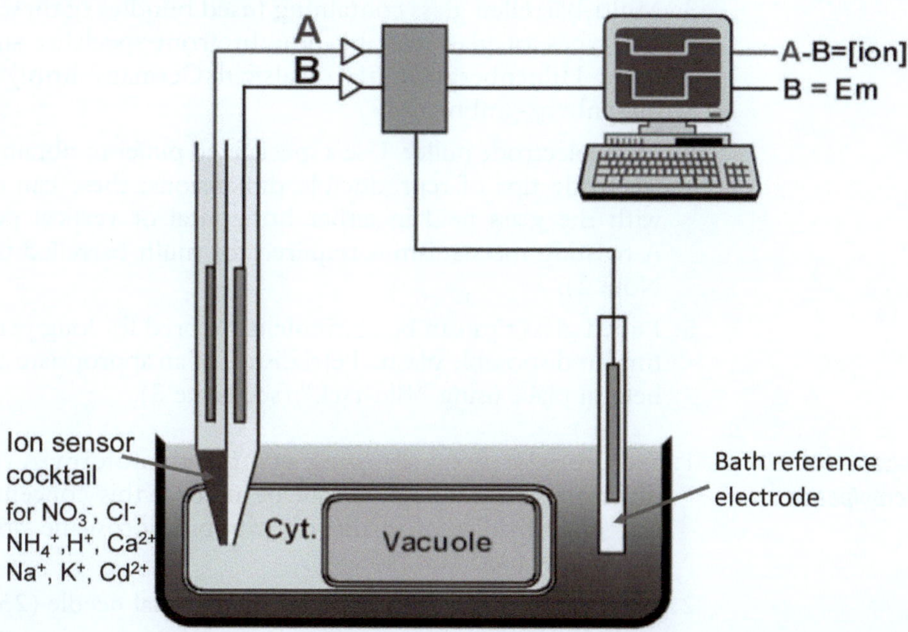

Fig. 1. Diagram showing the electrical arrangement for double-barrelled ion-selective microelectrode measurements. The bath reference electrode provides electrical contact to complete the recording circuit. These reference electrodes can be obtained from electrometer suppliers or made using a half-cell holder filled with 0.2 M KCl and connected to the solution via a short piece of plastic tubing (1 mm outer diameter) plugged with 1% agar (w:v) made using the same KCl solution.

4. Back-filling with salt solution.
5. Calibration.

These stages are easily defined, do not need to be done at the same time and are therefore convenient for heading the manufacturing sections within this chapter.

2. Materials

Prepare and store all reagents at room temperature unless specified otherwise. Wear eye protection at all times when pulling and breaking glass. For pulling micropipettes, glass tubing can be purchased of varying diameter and wall thickness according to requirements. An outer diameter of 1 mm is convenient and fits many commercially available half-cell holders.

2.1. Glass Micropipettes

1. Filamented borosilicate glass (see Note 1) is used for membrane potential recording electrodes that are filled with 0.1 M KCl solution to make filling the fine tips easier.
2. Non-filamented glass for ion-selective glass electrodes because this is easier to keep dry and treat with a silanizing agent.

3. Multi-barrelled glass containing fused bundles of these different types of glass can be bought from specialist suppliers (e.g., Hilgenberg GmbH, Malsfeld Germany http://www.hilgenberg-gmbh.com/).
4. Microelectrode puller. Use a specialized puller to obtain microelectrode tips of reproducible dimensions; these can operate with the glass held in either horizontal or vertical position. A twisting mechanism is required for multi-barrelled tips (see Note 2).
5. Pulled glass tips can be conveniently stored for long periods of time in disposable plastic Petri dishes of an appropriate size and held in place using "Blu-tack" (see Note 3).

2.2. Silanization of Glass Micropipettes

1. Prepare a 2% (w/v) silanizing agent in chloroform. A range of different silanizing agents can be used at this concentration, but dimethyldichlorosilane or tributylchlorosilane are most commonly used (see Note 4).
2. Disposable plastic syringe (1 mL) and metal needle (25 G).
3. Heating lamp to give 140°C on the electrode tips.

2.3. Back-Filling with Ion-Selective Cocktail

1. Ion-selective cocktail dissolved in THF. This can be purchased direct from a supplier or mixed from the components chemicals (see Subheading 3.3 below). A cocktail for measuring nitrate (see Subheading 3.3), first reported in 1991 (4), is now sold commercially (Sigma-Aldrich product number 72549).
2. Glass 1 mL syringe and metal needle (30 G is a convenient size).
3. Glass beaker covered with parafilm.
4. After about 48 h, the THF has evaporated to leave the sensor cocktail in a plastic membrane embedded in the tip of the glass microelectrode. The microelectrodes are now ready for use or they can be conveniently stored at this stage (see Note 5).

2.4. Back-Filling with Salt Solution, Calibration, and Measurements

1. Like back-filling with cocktail (see Subheading 2.3, item 2 above), a syringe and needle can be conveniently used to back-fill behind the ion-selective membrane.
2. Back-filling solutions. Typically for plant cell measurements, 0.2 M KCl solution is used for back-filling reference barrels. For an ion-selective barrel, the choice of back-filling solution depends on the sensor, but it is best to use a high concentration of the ion that is being sensed. For example, pH 4 for proton-selective microelectrodes or 0.1 M KNO_3 for nitrate electrodes.
3. An amplifier voltmeter (or electrometer) with a high input impedance that is at least 1,000 times higher than the

ion-selective electrode, e.g., 10^{15} Ω. Furthermore, the input leakage current from the electrometer must be low so that no significant offset voltage (>1 mV) is produced across the ion-selective electrode. Other useful electrometer facilities include GΩ range resistance tester and a difference-voltage output so that a direct output equivalent to cell ion activity can be obtained.

4. The back-filled microelectrode can be conveniently connected to a commercially available holder (or "half-cell"). A range of holder sizes are available, choose one that fits the outer diameter of the microelectrode glass. Also the holder pin connection size and type (male or female) can be chosen as appropriate for the input on the electrometer. Microelectrode holders can be bought from electrometer suppliers and they are often supplied with the equipment (see Note 6).
5. Standard solutions of known fixed concentrations are required for the calibration of the ion-selective electrode tip. Calibration solutions can be stored at 4°C for several weeks (see Note 7). For these reasons, the calibration of microelectrodes generally uses solutions that resemble the intracellular environment in terms of interfering ions and ionic strength (see Note 8). A standard bath reference electrode is used to complete the circuit (see Fig. 1).

3. Methods

3.1. Pulling Glass Micropipettes

1. Break pieces of glass micropipettes to convenient lengths (10 cm).
2. Pull tips using a puller and following the manufacturers' instructions. Twisting multi-barrelled glass before the final stage of the pull helps to ensure that the tips are aligned; see Walker et al. (5) for details.
3. Check that all the microelectrode tips can be used to obtain stable intracellular recordings of membrane potential (see Note 9).
4. For multi-barrelled glass electrodes, after pulling the open glass ends can be broken back to different lengths to enable treatment with a silanizing agent at the end of specific barrels.
5. Tips can then be stored dry until they are ready to be used.

3.2. Silanizing Specific Glass Barrels of Micropipettes

1. Pulled glass tips are placed under a heating lamp at 140°C for at least 30 min.
2. Under the lamp, the open blunt end of the glass barrel designated to receive ion-selective electrode sensor is treated with a

few drops of silanizing agent, dispensed using a syringe and needle.

3. The silanizing solution should quickly vaporize giving the ion-selective barrel a hydrophobic coating and leaving no liquid remaining within the glass (see Note 10).
4. After silanizing, the microelectrode should remain under the heat for 30 min before directly back-filling with the ion-selective cocktail (see Subheading 3.3 below).

3.3. Preparation of a Nitrate-Selective Cocktail

1. Weigh the MTDDA.NO_3 (3 mg), nitrocellulose (2.5 mg), poly(vinyl chloride) (11.5 mg), and a lipophilic cation—methyltriphenyl phosphonium bromide (0.5 mg) into a 1 mL glass screw-topped vial using a balance accurate to 0.1 mg (see Note 11).
2. Add the liquid nitrophenyl octyl ether (32.5 mg) to the vial using a glass microcapillary on the balance pan.
3. Dissolve the cocktail in approximately 4 volumes of THF (see Note 12). The cocktail takes at least 30 min to dissolve completely.
4. This cocktail can then be stored at 4°C for several weeks and is enough to make about 70 nitrate-selective microelectrodes.

3.4. Back-Filling with Ion-Selective Cocktail into Silanized Glass

1. Ion-selective cocktail mix dissolved in THF is back-filled into the blunt end of a glass microelectrode using a 1 mL glass syringe and metal needle. The cocktail composition depends on the type of ion-selective microelectrode (1, 6).
2. The filled microelectrode is placed with the tip down in a glass beaker covered with parafilm.
3. After about 48 h, the THF has evaporated to leave the sensor cocktail in a plastic membrane embedded in the tip of the glass microelectrode. The microelectrodes are now ready for calibration and use or they can be conveniently stored at this stage (see Note 13).

3.5. Back-filling with Salt Solution, Calibration, and Use for Intracellular Measurements

1. Back-fill the electrode barrels with salt solutions. Care must be taken to avoid air bubbles in the fine tip and at the surface of the ion-selective membrane.
2. Newly back-filled ion-selective microelectrodes require "conditioning" for a minimum of 30 min before they become stable and responsive. This process involves immersing the tip in a solution containing a high concentration (e.g., 0.1 M) of the ion to be measured.
3. Calibration of microelectrodes can be performed in the chamber built to take the plant tissue, or using a U-shaped funnel (see Note 14). Good electrodes should have a low detection

limit, a near ideal slope, and a small selectivity coefficient for physiologically important interfering ions (see Note 15).

4. The microelectrodes are now ready for intracellular recordings using a standard electrophysiological equipment rig that includes a high input impedance electrometer and off-the-shelf data-logging software. If the dynamics of changes in ion activity are particularly of interest, then the response time of the electrode must be carefully recorded (see Note 16). Usual electrophysiological preparation is needed, ensuring that the plant tissue is well anchored in position under a microscope for the electrode impalement and wet contact is retained between the ion-selective electrode and reference (Fig. 2). Problem solving can be also followed in the routine way for these types of electrical measurements (see Note 17). Specialized electrodes can be used to identify the intracellular location of the tip (see Note 18).

5. Several criteria for acceptable intracellular measurements can be defined. Firstly, after impalement the ion-selective microelectrode should be recalibrated and should give a very similar response to that shown before the cell impalement particularly at activities similar to those measured in vivo. Sometimes the recalibration shows a displacement up or down the *Υ* mV output axis. More often the detection limit of the ion-selective

Fig. 2. Impaling Arabidopsis leaf cells with an ion-selective microelectrode (3).

microelectrode has changed, but provided the measurement was on the linear response range of the electrode calibration curve, this is not usually a reason to disregard the result. Sometimes the performance of the ion-selective microelectrode can even improve with the detection limit actually becoming lower (see Note 19). A comparison between the electrical resistance of the ion-selective microelectrode before and after impalement provides a good indicator of whether the tip will recalibrate. If the resistance decreases below 1 GΩ, the ion-selective membrane has probably been displaced during impalement and the electrode will not recalibrate. Throughout the recording, the state of the cell can be assessed by monitoring the membrane potential (which should remain stable unless deliberately perturbed) or processes like cytoplasmic streaming.

4. Notes

1. Glass can be cleaned before pulling by washing in a solvent like ethanol or acetone, but this is not essential. Various types of glass can be used, but check the melting temperature. For example, aluminosilicate glass is tougher than borosilicate glass, but it requires a hotter heating element temperature. Electrode puller heating elements may need changing to reach a higher temperature.
2. Avoid positioning the microelectrode puller in an air draught and this can strongly influence the tip dimensions.
3. “Blu-tack” may not be available, but equivalent products are available in most countries, try a web search. The glass electrode should be lightly pressed into the Blu-tack, if attached too firmly the glass can break during removal.
4. Silanizing agents can generate a very toxic vapor (see Note 10) and must be prepared and used in a fume hood. Only small volumes are required typically a syringe needle drop is sufficient for each tip and a stock solution of just 10 mL is prepared.
5. For long-term storage, the ion-selective microelectrodes should be stored without back-filling, in a silica-gel dried sealed container in the dark. This can be done in a screw-cap glass jar or parafilm-covered beaker containing dry silica gel, with the microelectrodes attached to the inner wall using plasticine or Blu-tack. Ion-selective microelectrodes stored for several years in this way can still give a reasonable performance when back-filled.
6. Commercially available holders have a limited life of a few months, but they can be regenerated by replacing the metal

silver/silver chloride pellet electrode in the base of the holder. Microelectrode holders can be made using cut down glass pipettes of an appropriate size and 2 mm jack pins with a melted wax seal on to the electrode glass. The advantage of this method is that a silver wire can be directly soldered to the jack pin and easily re-coated with chloride. The symptoms of deterioration in the silver/silver chloride coating are unstable recordings.

7. Note that the slope of the calibration curve is temperature-sensitive and both calibrations and intracellular measurements should be done at the same temperature. If the temperature of the calibration solutions is 4°C and the cell is at 20°C, the slope of the electrode calibration for a monovalent ion will be 55 mV per decade change in activity, not the 58 mV expected at 20°C. Ion-selective microelectrodes can be calibrated using concentration or activity, but the electrodes actually respond to changes in *activity*. Therefore, calibrating with ion activity gives a microelectrode output which can be used directly without any assumptions of the intracellular activity coefficient for the ion. Furthermore, activity is actually the important parameter for all biochemical reactions.

8. The calibration of some microelectrodes for intracellular measurements (e.g., Ca^{2+} and H^+) requires the use of buffering agents such as EGTA for Ca^{2+} because of the very low concentrations being measured.

9. An estimate of the tip geometry of the microelectrode is provided by measuring its electrical resistance when filled with KCl, larger tips having lower resistances. For tips of 2–0.1 μm diameter the electrical resistances of ion-selective microelectrodes are usually in the GΩ range, while microelectrodes filled with 0.1 M KCl have 10^3 smaller resistances in the MΩ range. The dimensions of the microelectrodes are usually a compromise between obtaining a stable membrane potential and a good calibration response (detection limit).

10. Care must be taken to ensure that the reagent does not enter the membrane potential-measuring barrel of multi-barrelled microelectrodes. Silanizing agents are corrosive and toxic, protective glasses and gloves must be worn, and glass must be treated in a fume hood.

11. These additions are conveniently made one by one on a balance. The cocktail composition is altered to obtain different types of ion-selective electrode. Ion-selectivity is provided by a specific chemical sensor, but in addition most cocktails contain a lipophilic cation or anion, a plasticiser, and plastic polymer (often PVC). Some ready-mixed cocktails can be obtained from commercial suppliers.

12. No plastic can be used in contact with THF. Glass syringes and metal needles can be conveniently used to dispense small

volumes up to 1 mL. Excess THF is required to dissolve the cocktail components for mixing, but as it evaporates to carry the cocktail into the glass tip, the volume used need not be measured very precisely.

13. After back-filling and THF evaporation, the lifetime of the ion-selective microelectrodes is type-dependent. For example, nitrate-selective electrodes can be stored for more than a year in a dry atmosphere without any detrimental decline in their calibration and use, but proton-selective tips are best used within 4–6 weeks after manufacture.
14. When calibrating, it is important to make sure that the electrode tip is directly exposed to the complete calibration solution. Flow calibration systems can allow some mixing of the solutions resulting in poor calibration curves. Also, it is best to run through a complete set of calibration solutions returning again to the first one in the series to check for the stability of the electrode.
15. For statistical analysis and the calculation of means, these should be calculated using the data which is distributed normally, that is using the log activity or output voltages not the actual activities (7). Therefore, when mean activity value is used, it can only be expressed with 95% confidence limits, whereas log [activity] can be given standard errors or standard deviations.
16. When measuring changes in intracellular ion concentrations, artifacts can be caused by the differential response times of the two barrels; the ion-selective barrel usually has a slower response time than the membrane potential-sensing barrel. This can be corrected for when the response time of the electrode is known (8). The electrode response time can limit detection of rapid changes in ion activity.
17. The best approach is to try to solve problems by a process of elimination. Firstly, establish whether a problem occurs in the circuitry or is specific to the ion-selective microelectrodes. The circuitry can be tested by putting a broken-tipped KCl-filled microelectrode in place of the ion-selective microelectrode. A broken-tipped electrode should give a stable zero output. It may be necessary to re-coat AgCl-silver contact in the half-cell or there may be a wiring problem. Noisy recordings can be caused by poor earthing or air bubbles in back-filling solutions. If the circuitry has no problems, then the ion-selective microelectrode must be the cause. When the ion-selective microelectrode does not respond to the calibration solutions, then the membrane can be checked by deliberately breaking the tip to expose a larger area of ion-selective membrane. Breaking the tip can displace the ion-selective membrane from the tip, so it

is important to measure the resistance to check it is still in the GΩ range. If the broken tip gives a good response to changes in ion activity, then the problem is independent of the composition of the membrane. When the microelectrode tip diameter becomes too fine, the output from the ion-selective electrode will no longer respond to changes in ion activity.

18. In plant cells, identifying in which internal cell compartment (cytoplasm or vacuole) the microelectrode tip is located can be a problem for some ions and it may be necessary to grow the plant under conditions in which two populations of measurements can be identified. Alternatively, a triple-barrelled microelectrode can be used where one barrel is pH or Ca^{2+}-selective (5). Large gradients of these two ions are known to exist across the tonoplast, with the cytoplasm maintained at relatively constant values (pH 7.2, Ca^{2+} 100 nM) so compartment identification is possible. Another approach is to use tissues where the two major cell compartments can be identified under the microscope, e.g., root hairs, or cell cultures that have no large vacuole. However, identifying the compartmental location of the electrode can still be problematic, particularly if the tip indents the tonoplast, but does not penetrate it.
19. For this reason, it can be best to quickly impale a cell with a new tip before calibrating prior to measuring the activity in a cell.

Acknowledgments

John Innes Centre and Rothamsted Research are grant-aided by the Biotechnology and Biological Sciences Research Council (BBSRC) of the UK.

References

1. Ammann D (1986) Ion-selective microelectrodes, principles, design and application. Springer, Berlin, Heidelberg, Germany
2. Henriksen GH, Bloom AJ, Spanswick RM (1990) Measurement of net fluxes of ammonium and nitrate at the surface of barley roots using ion-selective microelectrodes. Plant Physiol 93:271–280
3. Miller AJ, Cookson SJ, Smith SJ, Wells DM (2001) The use of microelectrodes to investigate compartmentation and the transport of metabolized inorganic ions in plants. J Exp Bot 52:541–549
4. Miller AJ, Zhen R-G (1991) Measurement of intracellular nitrate concentrations in Chara using nitrate-selective microelectrodes. Planta 184:47–52
5. Walker DJ, Smith SJ, Miller AJ (1995) Simultaneous measurement of intracellular pH and K^+ or NO_3^- in barley root cells using triple-barreled ion-selective microelectrodes. Plant Physiol 108:743–751
6. Miller AJ (1995) Ion-selective microelectrodes for measurement of intracellular ion concentrations. Methods Cell Biol 49:275–292
7. Fry CH, Hall SK, Blatter LA, McGuigan JAS (1990) Analysis and presentation of intracellular measurements obtained with ion- selective microelectrodes. Exp Physiol 75:187–198
8. Sanders D, Slayman CL (1982) Control of intracellular pH. Predominant role of oxidative metabolism, not proton transport, in the eukaryotic microorganism *Neurospora*. J Gen Physiol 80:377–402

Chapter 5

Single-Cell Sampling and Analysis (SiCSA)

Wieland Fricke

Abstract

Single-cell sampling and analysis allows the determination of solute concentrations in individual cells and tissues. This is particularly important when studying a stress such as salinity, where the cell- and tissue-specific distribution of sodium and chloride may decide a plant's fate. In this chapter, some selected SiCSA methods are described in detail, and their advantages and possible pitfalls discussed. These methods include pressure-driven extraction of cell contents (cell sap sampling) and the analysis of extracted cell sap through picolitre osmometry (osmolality), energy-dispersive X-ray analysis (concentrations of Na, K, P, S, Cl, Ca), and microfluorometry (concentrations of, for example, nitrate and total amino acids).

Key words: Microcapillary, Picolitre-sized samples, EDX-analysis, Microfluorometry, Picolitre osmometry, Cell contents, Sodium, Chloride, Turgor pressure

1. Introduction

Salinity affects plants through three major stress components: an osmotic stress, which dominates the stress response in the very short term (minutes to days), and ion toxicity and nutrient imbalance, both of which increase in severity the longer the plant is exposed to a saline root environment (1). Each stress component has solute relations at its core: (1) to osmotically adjust; plant cells and tissues need to increase the solute load, (2) an ever-increasing load of sodium and chloride has to be dealt with, and (3) solutes such as potassium that become increasingly difficult to acquire have to be distributed efficiently. Some plants that have adapted to live on saline substrates have evolved specialized structures such as glandular hairs, salt glands, or bladder cells through which the surplus of Na^+ and Cl^- can be disposed. However, the majority of plants lack these options; the remaining option left to these plants

Sergey Shabala and Tracey Ann Cuin (eds.), *Plant Salt Tolerance: Methods and Protocols*, Methods in Molecular Biology, vol. 913, DOI 10.1007/978-1-61779-986-0_5, © Springer Science+Business Media, LLC 2012

is to compartmentalize the accumulating Na^+ and Cl^- and decreasing K^+ in a way most beneficial to the plant's performance. Solute compartmentation can occur between organs (root, shoot, flower), growing and non-growing regions, tissues (e.g., epidermis, mesophyll, vascular bundle), and cell compartments (cytosol, organelles, vacuole) (e.g., see refs. (2–9)). In this chapter, we will be looking at methods that make it possible to study compartmentation of solutes between cells and tissues.

A number of different ways of sampling contents of individual cells, so gaining information about compartmentation of solutes between tissues, are available. Protoplasts can be isolated from different tissues (10, 11), or leaf tissues can be fractionated according to their partitioning in non-aqueous media (12). Outlaw and colleagues developed techniques for manual micro-*dissection*, weighing and analyzing individual cells, more than 40 years ago (13). More recently, manual dissection of cells has been replaced by laser micro-dissection of tissues, followed by chemical or molecular (PCR) analyses (14, 15). An alternative to dissection of tissue is sampling of single-cell contents in vivo, by inserting a silicon-oil-filled microcapillary directly into cells (16). Since living plant cells have turgor pressure and an exponential relation between cell turgor and volume (compare inflating a bicycle tube and tyre, where the pressure increases rapidly as the final tube and tyre volume is reached), between 10 and 20% of cell contents shoot rapidly into the capillary when the cell is punctured. This single-cell sampling technique, which will be described here, is suited ideally for the analysis of highly vacuolated cells (too high a proportion of cytoplasm can clog the capillary) and surface tissues such as the leaf or root epidermis, since these can easily be accessed (17–19). Deeper-lying tissues such as mesophyll and bundle sheath or root cortex and stele can also be sampled, albeit with some modification of the sampling approach, i.e., inserting the microcapillary through a stomata (20). The extracted cell sap consists of almost 100% vacuolar sap in the case of an epidermal cell, but can contain significant portions of cytoplasm in the case of mesophyll or bundle sheath cells (20). The volume of extracted sap is in the lower (mesophyll) or higher (epidermis) picolitre range. Due to the small volume, the extracted sap must be handled under conditions that suppress evaporative loss. All sample manipulations have to be carried out under a stereomicroscope with the aid of micromanipulators.

The extracted cell sap can be analyzed for osmolality using picoliter osmometry (for details, see below). In short, a droplet (5–20 pL) of sap is placed under a drop of liquid paraffin onto a small (1 cm^2) copper stage on which a cover slip with a black and white background is attached using heat-conducting paste. Standards (NaCl) of known osmolality are placed nearby, and the stage is cooled to −40°C. Thereafter, the stage is reheated slowly and melting of ice crystals is observed under the stereomicroscope.

The temperature at which the last ice crystal in a particular droplet melts is recorded and used, together with values of standards, to calculate the osmolality of the sample. Up to 25 droplets can be analyzed in parallel.

Solutes in extracted cell sap can be analyzed using three types of techniques: energy-dispersive X-ray (EDX)-analysis, microfluorometry, and capillary-zone electrophoresis. The first two techniques will be described here in detail. In short, during EDX-analysis, sample droplets (10–20 pL) are pipetted onto an electro-microscope copper folding grid (100 and 200 mesh), which is coated with a film, ideally Pioloform, and placed under liquid paraffin. The same glass constriction pipette is used for samples and the internal standard (typically $RbNO_3$), which is placed in a 1:1 ratio together with sample droplets. Following pipetting, the liquid paraffin is removed by successive washes in hexane and isopentane, the latter having also a freeze-drying effect on droplets. Grids can then be analyzed with a scanning electron microscope (SEM) equipped with an X-ray analyzer (see Note 1). Solutes such as nitrate, sugars, or malic acid can be analyzed by microfluorometry. The enzymatic assays employed in this method are essentially the same as used in conventional spectro- or fluorimetric assays of metabolites, except that a set of constriction pipettes (10 pL–5 nL) is used, that all pipetting is done under liquid paraffin, and that sample fluorescence is determined under an inverted fluorescence microscope. This technique can also be used to measure enzyme activities in extracted saps (21).

Capillary-zone electrophoresis offers potentially the most extensive range of solutes that can be analyzed in single-cell extracts. However, it has been used by only a few laboratories (22, 23), possibly because of difficulties in injecting very small (pL) and reproducible volumes into the capillaries used for this technique.

2. Materials

Apart from a picolitre osmometer and possibly a micromanipulator and capillary puller, no other specialist equipment is required. An inverted fluorescence microscope and a SEM-EDX facility are available in most universities. Unless stated otherwise, all solutions are prepared in distilled water. Most reagents are not toxic, yet general precautions should be taken when dealing with laboratory chemicals. If not stated otherwise, chemicals are supplied by VWR or Sigma. The chemicals that require handling under a fume cupboard are:

1. Isopentane.
2. Hexane.

3. 2-Mercaptoethanol (also referred to as "β-mercaptoethanol").
4. Dimethyl-dichlorosilane.

2.1. Cell Sap Sampling and Pipetting Components

1. Glass capillaries (inner diameter 0.56 mm; e.g., from Clark Electromedical Instruments, Harvard Apparatus).
2. Horizontal or vertical capillary puller (e.g., Harvard Apparatus).
3. Microforge (e.g., Narashige).
4. Oven (80–200°C).
5. Fume cupboard.
6. Custom-built aluminum capillary holder.
7. Glass beaker (0.5 L; tall version).
8. Dimethyl-dichlorosilane.
9. Silicon oil (AS4 Wacker).
10. Liquid paraffin ("Paraffin oil").
11. Hypodermic needle attached to a 5-mL disposable plastic syringe.
12. Micromanipulator (e.g., Leitz, Narashige) or micropositioner (e.g., Prior, Narahsige, World Precision Instruments).
13. Capillary holder (commercially available or custom-built).
14. Silicone tubing.
15. T-connector (small, as available from any pet shop for connecting water tank tubing).
16. Disposable 50-mL plastic syringe.
17. Aluminum rings (custom-made; 3–4 mm deep; diameter just less than width of standard microscope slide).
18. Two-component glue (e.g., Araldite).
19. Stereomicroscope (standard type, e.g., Leica, Mejri, Olympus, Nikon), with magnifications ideally in the range ×50–×200.
20. Cold-light source with swan necks (standard type; such as from Volpi or Leica).
21. Petri dish (standard size).
22. Leaf holder (custom-built) to fix leaf.
23. Blu-Tack.

2.2. Picolitre Osmometry Components

1. Picolitre osmometre (has been available commercially from Bangor University; a Clifton nano-liter osmometer will also do).
2. Cover slips (standard size).
3. Plaster of Paris (also referred to as "Gypsum plaster"). Mix water with $CaSO_4 \cdot 1/2H_2O$ (calcium sulphate semihydrate) to get a slurry. The slurry should not be too fluid and too dense either.

4. Permanent-marker pen (black).
5. Acetone.
6. Heat-conducting paste (e.g., RS components).
7. Source of running (tap) water.
8. Wash bottle with silica gel.
9. Water tank pump (standard type as available from pet shops; only smaller versions required).
10. Silicon tubing.
11. NaCl (sodium chloride) standards. Dissolve the following quantities (g) of NaCl in 1 L each of distilled water to reach NaCl concentrations of 0, 100, 200, 300, 400, and 500 mM: 0 mM, 0 g; 100 mM, 5.84 g; 200 mM, 11.69 g; 300 mM, 17.53 g; 400 mM, 23.38 g; 500 mM, 29.22 g. The osmolality of these solutions is 0, 188, 373, 553, 738, and 924 mOsmol/kg, respectively.

2.3. EDX-Analysis Components

1. Electron microscopy folding grids (100/200 mesh) coated with Formvar (available commercially; Agar Scientific). It is even better if the grids are bought non-coated and are coated "in-house" with a film of Pioloform (1%; Agar Scientific), following standard procedures. The grids can be in copper, which makes them easier to use for pipetting. Since the Cu signals of grids interfere with the Na signal of the EDX analyser, nickel grids are recommended for analyses which focus on Na (salinity).
2. Internal standard solution containing 0.7 M mannitol and 200 mM rubidium nitrate ($RbNO_3$). Dissolve 12.75 g of mannitol in 90 mL distilled water and make up to 100 mL final volume. Take 2 mL of this and dissolve 59 mg of $RbNO_3$ in it. Store 1-mL aliquots in microcentrifuge tubes at 4° for up to 1 month.
3. Elemental standard series A. Pour 400 mL of distilled water into a 1-L glass beaker and add 5.84 g of NaCl and 23.62 g of $Ca(NO_3)_2 \cdot 4H_2O$ (calcium sulphate tetrahydrate). Dissolve the chemicals with a magnetic stirrer (takes just a few minutes). Make up solution to 500 mL. This solution has concentrations of 200 mM NaCl and 200 mM $Ca(NO_3)_2$ (200 mM Na, 200 mM Cl, and 200 mM Ca, which are relevant for EDX). Take X mL of this stock solution and add Y mL distilled water to reach final concentrations of each element of 150 mM (X/Y; 150/50), 100 mM (100/100), 50 mM (50/150), and 25 mM (25/175). Store the stock and dilutions in closed glass containers at 4° for up to 6 months. Check periodically for precipitates. If these form, discard the solution.

4. Elemental standard series B. Pour 400 mL of distilled water into a 1-L glass beaker, add 5.1 g of KH_2PO_4 (dipotassium phosphate, also known as "potassium dihydrogen phosphate"), dissolve it, and then add to the solution 6.53 g of K_2HPO_4 (monopotassium phosphate, also known as "potassium monohydrogen phosphate"). Dissolve it with a magnetic stirrer (takes just a few minutes). Make up solution to 500 mL. This stock solution contains 75 mM KH_2PO_4 and 75 mM K_2HPO_4. Together this amounts to EDX-relevant elemental concentrations of 225 mM K and 150 mM P. Make the following dilutions: (1) 150 mL stock solution plus 50 mL distilled water (final concentrations of 169 mM K and 112.5 mM P), (2) 100 mL stock solution plus 100 mL distilled water (122.5 mM K and 75 mM P), and (3) 50 mL stock solution plus 150 mL water (61.3 mM K and 37.5 mM P). Store the stock and dilutions in closed glass containers at 4°C for up to 6 months. Check periodically for precipitates. If these form, discard the solution.
5. Fume cupboard.
6. Tweezers (fine, curved, for electron microscopy; Agar Scientific).
7. Glass beakers (5-mL); three beakers.
8. Isopentane.
9. Hexane.
10. Gelatine capsules to store grids (Agar Scientific).
11. Silica gel (dried) in small (50 mL) plastic container.
12. Carbon stub (Agar Scientific) to mount grid in SEM.
13. SEM with EDX analyser.

2.4. Microfluorometry Components

1. Micromanipulation, pipetting, and viewing equipment as for single-cell sampling.
2. Constriction capillaries of approximately 10 pL, 0.5 nL, and 5 nL constriction volume. Construct these from glass microcapillaries using a microforge.
3. Fluorescence microscope (inverted), ideally with photomultiplier (e.g., MPV Leitz), filter combination to detect NAD(P)H-dependent fluorescence (e.g., Leitz filter block A, excitation 340–380 nm, dichromatic mirror 400 nm, and emission filter 430 nm) and, if required, standard image analysis software such as ImageJ (free download) (see Note 2).
4. Triethanolamine (TEA) hydrochloride buffer (480 mM; pH 7.6). Dissolve 8.91 g of TEA hydrochloride in 90 mL distilled water in a tall 200-mL glass beaker. On a magnetic stirrer, adjust pH to 7.0 using 0.1 M or 1 M KOH. Make up solution to a final volume of 100 mL. Store in a glass bottle at 4°C for up to 2 months.

5. NADPH solution. Prepare a 1% (w/v) $NaHCO_3$ solution by dissolving 1 g of $NaHCO_3$ in 90 mL of distilled water and make up to a final volume of 100 mL. This solution can be stored at 4°C for 6 months. On the day of analyses, take 0.5 mL of the $NaHCO_3$ solution and transfer this into a 1.5-mL microcentrifuge tube ("Eppendorf tube"). Depending whether you use NADH or NADPH, weigh out either 4.45 mg NADH (dipotassium salt; Sigma) or 6.85 mg NADPH (tetracyclohexylammonium salt; Sigma), add to the 0.5 mL $NaHCO_3$ solution, and dissolve quickly by inverting the tube. The final concentration of either NADH or NADPH is 12 mM. Wrap tube in tin foil and store on ice. Prepare fresh on the day of analyses.
6. FAD (flavin adenine dinucleotide) solution. Take 100 mL distilled water and dissolve 3.32 mg of FAD (FAD disodium salt; Sigma). The final concentration of FAD is 40 μM. Transfer a 1-mL aliquot into a microcentrifuge tube, wrap in tin foil, and store on ice on the day of analyses. The remaining FAD solution can be kept at 4°C for 1 week.
7. BSA (Bovin serum albumin, fraction V; Sigma) solution. Make a 1% (w/v) BSA solution by adding 100 mg BSA to 10 mL of distilled water contained in a 20-mL Erlenmeyer. Stir gently with a magnetic stirrer while avoiding foam formation. Store for up to 1 week at 4°C.
8. Nitrate reductase (from *Aspergillus niger*; Sigma; 300 units per gram lyophilized powder). Dissolve 1 mg of nitrate reductase lyophilized powder in 0.6 mL TEA buffer (see Subheading 2.4, item 4) in a 1-mL microcentrifuge tube. Keep on ice, wrapped in tin foil, on the day of analyses and store at 4°C for up to 1 week.
9. Nitrate standards. Prepare a nitrate standard stock solution containing 100 mM nitrate by dissolving 5.06 g KNO_3 (potassium nitrate) in 500 mL distilled water. From this stock make dilutions of (1) 75 mL stock plus 25 mL distilled water (final concentration of 75 mM KNO_3), (2) 50 mL of stock plus 50 distilled water (final concentration of 50 mM KNO_3), and (3) 25 mL stock plus 75 mL distilled water (final concentration of 25 mM KNO_3). Store dilutions and stock in closed glass containers at 4°C for up to 6 months.
10. Assay cocktail for nitrate analyses: 100 μL of assay cocktail contains 20 μL TEA buffer (see Subheading 2.4, item 4), 15 μL NAD(P)H solution (see Subheading 2.4, item 5), 10 μL FAD solution (see Subheading 2.4, item 6), 10 μL BSA solution (see Subheading 2.4, item 7), and 45 μL H_2O. Prepare fresh on the day of analyses in a 1-mL microcentrifuge tube. Wrap the tube in tinfoil and keep it on ice. Discard any unused solution at the end of the day.

11. *O*-phtalaldehyde (OPA, Sigma) reagent: use 96 μL of the reagent provided by the supplier and add 4 μL 2-mercapto-ethanol (fume cupboard!). Prepare fresh on the day of analyses in a 1-mL microcentrifuge tube. Wrap the tube in tinfoil and keep it on ice. Discard any unused solution at the end of the day.
12. KH_2PO_4/K_2HPO_4, buffer. Add 2.72 g of KH_2PO_4 to 90 mL of distilled water in a 0.5-L glass beaker. Dissolve on a magnetic stirrer and make up to a final volume of 100 mL. This gives a 200 mM KH_2PO_4 solution. Similarly, proceed in the same way by adding 3.48 g of K_2HPO_4 to 90 mL distilled water and making up the volume to 100 mL; this solution has a final concentration of 200 mM K_2HPO_4. Using a tall 200-mL glass beaker, mix 50 mL each of the KH_2PO_4 and K_2HPO_4 solutions and record the pH of the mixture while adding further K_2HPO_4 solution on a magnetic stirred until the pH is between 7.2 and 7.4. Store this buffer at 4°C for up to 6 months.
13. Amino acid standard (here: glutamic acid; you may also use alanine or aspartic acid). Prepare a 50 mM glutamic acid stock standard by adding 423 mg glutamic acid (monosodium salt; Sigma) to 50 mL of water. Dissolve the glutamic acid on a stirrer. Using this stock make the following dilutions with distilled water: (1) 8 mL of stock plus 2 mL of water (40 mM glutamic acid final concentration), (2) 6 mL of stock plus 4 mL of water (30 mM), (3) 4 mL of stock plus 6 mL of water (20 mM), and (4) 2 mL of stock plus 8 mL of water (10 mM). Store stock and dilutions at 4°C for up to 1 month.

3. Methods

3.1. Preparation of Microcapillaries

1. Pulling of microcapillaries and making constrictions. Glass capillaries are pulled with a horizontal or vertical capillary puller. The load and heat settings have to be adjusted so that the capillary has neither a too narrow nor a too wide tip region. The tip of the capillary can be broken under the stereomicroscope or with a microforge to obtain a tip opening of several micrometers. Pipettes that are used for extraction of cell sap do not have to be modified any further, except that they can be (but do not have to be) silanized (see Note 3). Pipettes used for pipetting small reproducible sample volumes must have a constriction melted into them close to the tip (the proximity to the tip defines the constriction volume for a given tip geometry). A constriction is best made with a microforge, where the capillary is positioned with a *x*-*y*-z micropositioner close to a

platinum wire. This is heated, whilst viewed between ×50 and ×200 magnification (see Note 4).

2. Silanization of microcapillaries. Capillaries used for pipetting and for extracting cell samples from wet surface tissue or tissue surrounded by liquid (e.g., roots) have to be silanized. Make sure that the tip of the microcapillary is broken to the desired tip opening diameter before the capillary is silanized. Place an oven (80–200°C) in a fume cupboard. Place upright capillaries (up to 100) into a capillary holder, which is contained within a glass beaker. Seal the beaker with tin foil and leave it in the oven for 1 h. Next, add about 50 μL of dimethyl-dichlorosilane to the beaker (still in the fume cupboard) and leave the beaker in the oven for a further 30 min. The capillaries can be stored indefinitely.
3. Filling of microcapillaries. For sampling, the capillaries are best back-filled with silicon oil using a hypodermic needle attached to a syringe. The capillary can also be backfilled with liquid paraffin. Alternatively, apply suction to the microcapillary and suck in a sufficient volume of liquid paraffin through the tip opening. The volume should exceed the expected sample volume about 50–100 times.
4. Pipetting with microcapillaries. The microcapillary is mounted on a micromanipulator or micropositioner and held by a capillary holder. Once cell sap has shot into a microcapillary, the sap has to be expelled. Also, samples have to be sucked into and expelled from capillaries during pipetting. The easiest way to achieve this is as follows. Attach silicone tubing to the wide (base) end of the capillary and connect the other end of the tubing to a little T-connector. Attach more silicon tubing to each of the two open ends of the T-connector and attach the end of one of the tubing to a 50-mL disposable syringe and the other tubing to a solenoid-based device (custom-built) that can be switched on and off so that the tube ending is left open or clamped into a "closed" position. If no such device is available, construct some other mechanical device (e.g., strong paper clips) that allows the reversible clamping of tubing; the simplest mechanical device is your teeth! (Beware of Health & Safety in a lab environment). During sampling, the tube ending (e.g., solenoid) is left in the open position. To expel liquid, the ending is in a closed position; the same applies to generate suction for filling the capillary from the tip with sample.

3.2. Preparation of Microscope Slides for Pipetting

1. Once cell sap is extracted, it has to be handled in a microenvironment that prevents evaporative loss of liquid. It also has to be viewed under a (stereo) microscope. To meet both requirements, take some aluminum rings and glue with two-component glue one ring in the middle of a microscope slide. Prepare 5–10 slides.

2. For use, fill the ring with liquid paraffin. After use, remove the liquid paraffin by washing in hand-hot water, then rub (hands) washing-up liquid onto the slide and remove the detergent with further washes with hot water. Dry the slide in an oven (70–80°C).

3.3. Preparation of Cover Slips for Picolitre Osmometry (see Note 5)

Take 10–20 cover slips and rub them each (ca. 1 min) on a glass surface in a slurry of abrasive powder (e.g., Plaster of Paris), which is washed off with water. The cover slips are then dipped shortly in acetone to degrease them and enable later maximum contact between sample and cover slip surface. The dried cover slips are then painted on the non-abraded (shiny) side with a black permanent marker. Once the marker has dried, it is scraped off in lines and crosses with the tip of a scalpel blade such that a black and white (transparent) grid pattern forms.

3.4. Single-Cell Sap Sampling

1. Place a leaf that is still attached to an intact plant in a leaf holder, or in case of roots, mount root in a Petri dish and then add medium. Place the holder or Petri dish onto a little table, mounted on a micropositioner (see Note 6). Next to the tissue, place a microscope slide with a aluminum ring glued to it and that is filled with liquid paraffin.
2. While viewing the illuminated plant tissue under a stereomicroscope, approach the tissue with a microcapillary that is mounted on a micromanipulator and connected to a syringe (T-connector in "Open" position) and filled with silicon oil. Approach the tissue slowly. Successful puncturing of a cell is visible as a fast "shooting in" of sap into the microcapillary. Immediately retrieve the capillary from the cell and move the specimen table so that the slide with the aluminum ring comes into the centre of the viewing field (see Notes 7 and 8). Slowly lower the capillary into the liquid paraffin contained within the aluminum ring, change from "Open" to "Closed" position (T-connector) at the tubing which is attached to the microcapillary, and press on the syringe to expel the single-cell sap onto the surface of the slide beneath the liquid paraffin. The complete emptying of sample becomes visible as silicon oil starts to exit ("streaming") the capillary. Stop applying pressure, change to "Open" position and repeat the procedure to sample another cell sap.
3. Once you have sampled up to ten samples, remove the sampling capillary and take another set of capillaries (as for sampling). Connect a capillary to the micromanipulator and tubing, lower it into the liquid paraffin, and apply suction to fill the capillary from the tip with liquid paraffin (about 50–100 times the sample volume). Then "suck in" a sample, followed by more liquid paraffin (about 5–10 sample volumes), followed

by another sample, more liquid paraffin, a third sample, and a final filling with liquid paraffin (see Note 9). Move the capillary out of the liquid paraffin, put it upright into a closed container, and store the samples for 1–2 weeks in the fridge until analysis (see Note 10). Proceed in the same way with the remaining samples. If you do not want to store samples but analyze them immediately, for example through picolitre osmometry, then place the osmometer stage next to your leaf holder or Petri dish and expel the extracted cell sap directly onto the prepared osmometer stage.

3.5. Picolitre Osmometry

1. Put a small quantity (size of matchstick head) of heat conducting paste onto the stage of the osmometer.
2. Take an abraded and painted cover slip and gently press, with the painted side down, onto the heat conducting paste so that the white paste is evenly squashed. You may also opt to not use an entire cover slip but a quarter of it (see Note 11).
3. Place a droplet of liquid paraffin onto the cover slip so that it covers a large portion of it, yet does not run off at the sides (see Note 12).
4. Switch on the cooling water supply (e.g., running tap water) for the osmometer (see Note 13). Switch on the osmometer (with maximum stage temperature setting) and adjust the airflow so that it is at the minimum flow rate required to remove any condensation droplets from the liquid paraffin. The air flow is generated by a small water tank pump that pumps air through a "washing bottle" filled with dried silica gel (to dry the air) and connected through silicon tubing to the lower and upper inlet on the osmometer stage. The lower inlet guides air directly above the surface of the liquid paraffin, and the upper inlet guides air above a cover slip that is used later to "seal" the osmometer stage during a freezing cycle (see Subheading 3.5, step 7). This prevents condensation from interfering with the viewing of samples (see Note 14).
5. Place standard droplets (up to 1 μL each) onto the microscope slide with the paraffin-filled aluminum ring. You can use a Hamilton syringe (10–25 μL) to pipette the standards.
6. Suck in aliquots of standards into a microcapillary, attached to a micromanipulator, and place the sample beneath the liquid paraffin onto the cover slip. Deposit two droplets for each standard concentration. Thereafter, place up to ten single-cell extracts onto the cover slip (see Note 15).
7. Once all samples and standards have been pipetted, cover the sample stage of the osmometer with a larger, normal cover slip and start a cooling cycle of the osmometer. Adjust the manual temperature regulator of the osmometer to a setting that is

below the anticipated melting temperature (once the osmometer has cooled down the sample stage to below −20°C and reached the lowest possible temperature, it automatically reheats the stage up to the temperature set by the manual regulator).

8. View samples through a stereomicroscope while the stage cools down to below −20°C. Observe any formation of condensation droplets and increase, if necessary, the rate of air flow.
9. When the osmometer reaches its lowest temperature and automatically returned to the temperature set by the manual regulator, start to slowly increase the temperature of the stage while viewing the sample droplets and ice crystals contained in them. Write down the temperature at which the last ice crystal of a sample melts and use this information, together with readings for standards, to calculate the osmolality of standards.
10. You may use the same cover slip again for the analysis of more samples. If it is full, remove it carefully with a pair of curved tweezers, trying not to spill liquid paraffin on the osmometer surface (if it spills, it does not damage the osmometer, it just makes it messy). Then, take a cotton swab ("Q-tip") to remove the heat paste. Take care that only the surface of the sample stage is cleaned and that the surrounding area that contains fine wiring is not touched. The osmometer is now ready for another analysis or can be switched off.

3.6. EDX-Analysis of Picolitre-Sized Droplets

1. Take a microscope slide with a paraffin-filled aluminum ring. Place the slide onto a stage mounted onto a micropositioner that can be viewed with a stereomicroscope. Using a pair of fine, curved tweezers, take an electron microscope double-folding grid (100/200 mesh) coated with, ideally, Pioloform and place it into the center of the well with the shiny surface down.
2. Using a Hamilton syringe (10–25 μL volume), pipette droplets (0.5–1 μL) of internal standard and elemental standards into the same well, but place these droplets at the periphery next to the aluminum ring. Make sure that none of the droplets touch the grid.
3. Connect a microcapillary (connected through silicon tubing to a suction/pressure device such as a syringe) to a micromanipulator. Lower the capillary into the oil well, apply suction to suck in some liquid paraffin from the tip and then suck in some solution of internal standard. While keeping the filled capillary under the liquid paraffin, move the stage so that the grid is now in the center of the field of view. Lower the capillary and empty a droplet on one large square on the 100-mesh half of the grid. About half of the square should be covered by this

"mini-reservoir" droplet (see Note 16). Now, move the stage back so that you have the internal standard stock again in your field of view and empty the standard that remained in the capillary back into this stock (alternatively, you can also expel the excess pipette filling anywhere else on the slide, just make sure that it does not get in contact with the grid).

4. Repeat step 3 for all other standard solutions.
5. Now, it is time to get your single-cell sap samples. If these are stored in microcapillaries, dispel the stored samples, one each, on a large square of the 100-mesh part of grid. If you do not have stored samples, but are harvesting these "fresh," place a leaf holder or Petri Dish on the stage next to the slide holding the grid and start sampling (see Note 17). Once you have punctured a cell and sampled some sap, move the stage quickly so that the grid appears now in the field of view. Lower the capillary quickly into the liquid paraffin. Once this is done, you can be sure that the sample contained in the capillary does not evaporate. Now, you can take your time to slowly lower the capillary and expel the sample onto a large square on the 100-mesh half of grid. Repeat this step with up to ten cell sap samples (see Note 18).
6. You have now all your standard and cell sap samples each on a large square of the grid. Now, change the capillary for a constriction capillary of about 10 pL size. Lower it into the liquid paraffin and start pipetting small droplets (ca. 10 pL) of the internal standard onto the smaller squares, one droplet per small square (see Note 19). Place droplets in rows of five, these serve as technical replicates. Once you have placed down 4–5 rows, use the same capillary to pipette the same capillary filling of a different elemental standard in each row, by adding the elemental standard to the internal standard droplet. Proceed in the same way with your cell sap samples. Remember that you must pipette the same volumes of internal standard and elemental standard (1:1) and of internal standard and cell sap samples (1:1). This means that you must use the same capillary for the pipetting of internal standard droplet and whichever sample is added to it. Therefore, it is advisable not to place too many internal standard droplets down in advance in case the capillary tip breaks off (which can happen all too easily!).
7. While you pipette all your samples, make a note on a piece of paper where each sample is positioned on the grid (see Note 20). Once you have pipetted all your samples, remove your 10-pL constriction capillary and keep it for the next experiment (see Note 21). Now, take a spare microcapillary, it does not have to be silanized, and clear off all the "mini-reservoir" droplets on the 100-mesh squares by sucking these into the capillary. After that, discard the capillary. You now have a grid

where droplets are only present on the 200-mesh half, each droplet consisting of about 10 pL internal standard and 10 pL of either elemental standard or cell sap sample.

8. Take the slide containing the grid and place it in a fume cupboard (no microscope required). Next to the slide, position three 5–10 mL beakers in a row. Fill the first two beakers with hexane and the third one with isopentane. Also position an empty and opened gel capsule in the fume hood in which you can later place the grid.
9. Using a pair of fine, curved tweezers, grab the grid on its 100-mesh side. Remove the grid slowly from the liquid paraffin. While holding the grid in a vertical position, lower it slowly but steadily in the first hexane beaker. Keep holding the grid with your pair of tweezers throughout. You will now see how the liquid paraffin comes off in the beaker and flows to the bottom of it. While keeping the grid submerged in hexane, move the grid slowly up and down to increase the washing effect. After about 10 s, transfer the grid into the second hexane beaker and proceed the same way. After this, transfer the grid into the beaker containing isopentane. Keep it there for about 3–5 s and then rather rapidly remove it. You will see how the grid "dries" and the grid can be transferred into the gel capsule within 10–20 s after removal from the isopentane. Label the capsule and store it in a closed container over dried silica gel. The hexane and isopentane must be disposed of safely.
10. To analyze your grid, mount it onto a carbon stub, typically at an angle of 45°, and view it under a SEM equipped with an EDX analyser. Construct yourself a file for appropriate peaks and keV range for each element (see Note 22). View grids at about 14 keV (see Note 23). Analyze each droplet by fitting an appropriate window, while viewing it at magnifications between ×1,000 and ×3,000. Adjust the total count rate for the rubidium signal to about 3,000–5,000 cps. This means that each droplet will be scanned and the cps counted for each elemental peak until the rubidium signal has reached the desired count rate. Save the values and later transfer these into an Excel spreadsheet. Express all your elemental counts as percentage of the rubidium integral count rate and use this information to construct a calibration curve from your standards and calculate the concentrations of elements in your samples (see Note 24).

3.7. Microfluorometry of pL-Sized Droplets (Nitrate Assay)

1. Start in the same way as for EDX analyses, except that you do not place a grid into the aluminium well, but directly pipette storage droplets of assay cocktail, nitrate reductase, and appropriate standards into the aluminium well, next to the aluminium ring, beneath the liquid paraffin. Place any stored or freshly sampled cell saps onto the glass surface of the slide inside the well.

2. Take a larger constriction capillary of about 5 nL size and pipette rows of 4–5 droplets each of assay cocktail (technical replicates) onto the glass surface of the slide inside the well. If you have four different nitrate standards, plus the 0 mM control, and have five cell sap samples, you will require 10 rows of 4–5 droplets each.
3. Exchange the 5 nL constriction capillary for a 10-pL capillary and pipette into each assay cocktail droplet your respective standard or sample. Make sure that you use the same 10-pL capillary for standards and samples, otherwise readings for standards and cell sap samples cannot be related to each other.
4. Once you have pipetted all samples, lift the slide and gently wipe the glass surface beneath with laboratory tissue paper to remove any loose fibers that could give off-the-scale fluorescence readings. Place the slide onto the stage of an inverted fluorescence microscope and select the appropriate filter combination.
5. Adjust the sample diaphragm so that it covers an entire droplet and keep the size of diaphragm the same for all samples. Switch off any room lighting and darken off the windows (see Note 25). Take fluorescence readings for each droplet and also take background readings of the glass surface without droplets at the start and end of each row of droplets (see Note 26). Make sure that your aluminum ring and oil well is not leaking and that no liquid paraffin touches the objective lenses, these are sensitive to liquid paraffin and may be damaged.
6. After you have taken all readings, return the slide to the micromanipulator stage and add nitrate reductase to each droplet using a 0.5 nL constriction pipette to start the reaction. Transfer the slide to the microscope as before and take readings of all droplets at time intervals of 10 min until the fluorescence does not decrease any further and the reaction (nitrate-dependent conversion of NAD(P)H to NAD(P)) is completed.
7. Construct plots of fluorescence decrease with time. At some point, the fluorescence should have leveled off. Take this reading for your standards to make a calibration curve and use the calibration curve to calculate the concentration of nitrate in your cell sap samples.

3.8. Microfluorometry of pL-Sized Droplets (Total-Amino Acid Assay) (see Note 27)

1. Proceed as for the nitrate assay except that you place down storage droplets of H_2O, OPA-reagent, and buffer into the oil well, followed by your stored or freshly harvested cell sap samples.
2. Using a 0.5 nL constriction pipette, place down rows of droplets of H_2O. Then, using a 10-pL constriction pipette, add amino acid standard or cell sap sample to the water droplets.

3. Read the fluorescence under the fluorescence microscope.
4. On the micromanipulator stage, add 0.5 nL OPA reagent using a 0.5 nL constriction capillary.
5. After 1–3 min, use the same capillary to add 0.5 nL of buffer.
6. Take fluorescence readings and use these readings to construct a calibration curve and calculate the total amino acid concentration in your cell sap samples.

4. Notes

1. Before embarking on any preparations for EDX analyses, check with your local SEM facility that the EDX equipment includes software that is suitable for quantitative analysis of microdroplets. In the past, I have used repeatedly a system from Link, Oxford, and this works fine. These days, some systems are designed mainly for imaging and not for quantitative analysis of X-ray counts for a specified area.
2. When you do microfluorometry of pL-sized droplets, you ideally want an inverted fluorescence microscope equipped with an additional diaphragm to define the size of measuring window around a droplet and a photomultiplier together with the respective software to carry out quantitative analysis of these droplets. In the past, the Leica MPV compact system did exactly that, but unfortunately the MPV system is no longer manufactured—because these days, most researchers are interested in imaging rather than quantitative fluorescence analyses. There are ways around this. You can either use software that is supplied by the microscope manufacturer to measure the fluorescence intensity of droplets; or you can take pictures of droplets and save them in TIFF format and later use a freely downloadable imaging software such as ImageJ, in which you open the pictures as 8-bit images. This effectively changes the fluorescence intensity into a greyscale and provides quantitative information.
3. Silanizing a micropipette has its advantages and its disadvantages. Silanization stops liquid from entering the capillary through capillary forces since it effectively cancels out any adhesion forces and renders the capillary surface very hydrophobic. Therefore, if you lower your sampling capillary into an aqueous medium, for example to puncture and sample the contents of a root cell, you must silanize the capillary. Otherwise, you will get fluid going into the capillary before you sample the cell and the sampled cell contents will be diluted. If you are using a microcapillary that is connected to a cell pressure probe,

you do not have to silanize the microcapillary since you can apply some pressures to stop any unwanted solutions from entering the capillary. A possible disadvantage of silanization is that it makes the inner volume of the capillary, in particular around the tip region, too narrow. This can interfere with getting sufficient cell sample volumes. Also, incoming aqueous solution such as cell sap gets easier mixed up with the liquid paraffin or silicon oil which was used to preload the capillary. It then becomes difficult to later expel a defined cell sap sample that does not have any oil dispersed within it.

4. The way the constriction works is as follows: when a sample is pulled into the capillary through gentle suction (syringe), the liquid enters the capillary easily up to the constriction. This makes it possible to obtain reproducible sample volumes. If the constriction is too narrow, too high suction forces have to be applied but if the constriction is too wide, sap will go passed the constriction. It is impossible to make constriction capillaries of a specified absolute volume, for example 10 pL or 5 nL. Therefore, reference to a 10 or 5 nL pipette in the following text specifies the approximate range of volume. An idea of the approximate constriction volume can be obtained by sucking water into a capillary and discharging the water in liquid paraffin and calculating the volume of the droplet from its diameter.

5. The principle of picolitre osmometery is that small (>5 pL) sample droplets are frozen down and then slowly melted so that the temperature can be recorded where the last ice crystal of the sample melts. The extent of melting point (or freezing point) depression is related to the total concentration of solutes in a sample. The ice crystals that form during a freezing cycle are rather round in shape and transparent, making it difficult to detect them. Supercooling can be a problem that stops ice crystal formation, but is overcome by freezing samples well below –20°C. Formation of ice crystals can be supported by providing crystallization "seeding centers." For these reasons, the glass cover slips on which the samples are placed have to be modified in two ways by rubbing them in a slurry of Plaster of paris (provision of seeding centers) and painting a grid pattern on the back of cover slips (improvement of contrast).

6. Lighting is a key issue in single-cell sampling. A pair of swan necks with a cold-light source are ideal and the swan necks can be arranged so that they shine at the specimen and capillary from almost opposite sides, at an angle of about 30–45° each. Also, the background for your sample and your specimen that you puncture affects the contrast. When you sample a cell, you must be able to see clearly the meniscus between the aqueous

cell solution and the silicon oil/liquid paraffin in the capillary to show that you actually got some sap. Also, you need to see clearly the surface of the plant organ (e.g., leaf and epidermis) to follow the capillary as it approaches the tissue. When you pipette samples, for example in a liquid paraffin oil well (aluminum ring glued onto microscope slide), it helps to have paper on which you paint a black and white grid pattern, placed underneath the slide. This improves the contrast of samples. Many pay slips also have a grid pattern that suits this purpose perfectly.

7. Plant cells can have very short half-times of water exchange, in the range of 0.5–1 s. This means that when sampling, you have to remove the capillary almost instantaneously once you have punctured a cell. With a bit of practice, you will be able to judge a successful puncture as a "shadow" that moves up the capillary rather than as a well-defined meniscus. In addition, when you lower your sample capillary, e.g., directly following sampling into a liquid paraffin oil well, successful sampling becomes immediately apparent due to the different refractory index of the oil and aqueous sample. If you do not retrieve the capillary rapidly enough, water will move from the surrounding tissues into the punctured cell (since the puncture reduced turgor and therefore lowered the cell's water potential) and significantly dilute your cell sap sample (24).
8. A good capillary, whether for sampling or pipetting, is a precious item. The best way to extend the life time of a capillary, and also to have the best possible view of events under the stereomicroscope, is to position the capillary near the center of field (best focus) and to keep it there. Move the specimen or slide with samples from *y*- to *x*- to *z*-direction rather than the capillary. The only exception is when you puncture a plant tissue: first get your capillary positioned and focused in the center of view. Next, move it slightly back (e.g., to the right if you are right-handed and the specimen is on the left side of field of view), move the specimen into the center, then lower it slightly. Now approach it with the capillary and then, in turn, bring up the specimen and adjust the capillary closer to it until the two are in close contact.
9. You can store up to 4–5 cell sap samples in one capillary. You may also decide to only store one sample in each capillary. This depends on individual preferences.
10. You can construct any device that allows you to store capillaries in an upright position without the tip touching anything and in a closed compartment. Onken (yoghurt) used to produce some great yoghurt plastic pots that could be used for this

purpose. Cut a strip of corrugated cardboard (ca. 5 × 12 cm) from a box and fix it with tape to a cubic (3 × 3 × 3 cm) piece of firm foam material and glue this construction on the inside of a (yoghurt) pot plastic lid. You can store the capillaries in the holes of the corrugated paper and close the box with the main body of the (yoghurt) pot.

11. An efficient way to cut a cover slip into four almost evenly sized squares is by putting the cover slip on a piece of dust-free firm rubber material and then press evenly with the single-edge razor blade onto the cover slip to induce a cut. Cut cover slips after you prepared them for picolitre osmometry.

12. You can construct disposable "mini-wells" on a piece of cover slip square by cutting teflon rings of appropriate diameter and glue these with superglue or two-component glue onto the cover slip. It is a little bit of extra work, but it allows you to retain the liquid paraffin on the cover slip and provides a depth of liquid paraffin of defined and reproducible size (and this is good for minimizing evaporation of single-cell saps—see Note 14).

13. Always remember to turn on the cooling water supply! The small gadget that cools down a stage to −20 to −40°C would otherwise overheat.

14. You are cooling down a small stage to −20 to −40°C, so condensation of water from ambient air will always be a problem. You want to flush dry air over the osmometer stage and liquid paraffin containing your sample droplet for as much as required to stop condensation, in particular on the liquid paraffin. Condensation droplets will ultimately sink to the surface of the cover slip and fuse with the much smaller cell sap samples located there. You also want to flush dry air as little as possible, since you do not want to dry out the cell samples, despite being under liquid paraffin. One means to counter the latter problem is to place a ring of water droplets around your samples on the cover slip or to use water-saturated liquid paraffin rather than "normal" liquid paraffin. Take some liquid paraffin, add an equivalent volume of distilled water, stir the mixture for a day, and then let it settle for 1 week. The upper phase (liquid paraffin) should be water-saturated. Finally, the local climate plays a role. Pick a nice, dry day to carry out analyses to minimize condensation. A climatized room with low relative humidity is ideal but mostly not available. Myself, I learned this technique in the laboratory of Prof. Deri Tomos in Bangor, North Wales, UK and later used it for years in the West of Scotland and at the East Coast of Ireland—all geographic areas that are not renowned for their dry, hot climate!

15. When you displace a sample from a capillary that is backfilled with silicon oil into a larger droplet of liquid paraffin, some of the silicon oil comes in contact with the liquid paraffin. Usually, that does not cause any problems. However, I once encountered a persistent problem in a very dry climate in that the silicon oil literally pushed away the liquid paraffin and the samples on the base of the cover slip got exposed to air and dried out almost immediately. I do not know the reason for this "incompatibility" between silicon oil and liquid paraffin; possibly it is of electrostatic nature, but the best way to deal with it is to backfill your capillary not with silicon oil, but with liquid paraffin.
16. The large squares on the 100-mesh half of grid serve as a platform where you place down your "mini-reservoirs" of your standards and also your cell sap samples. The smaller squares on the 200-mesh half of the grid are the actual sites of final droplet deposition for analyses.
17. The specimen and grid-containing slide have to be positioned so that it is possible to move between the two through rapid micropositioner movement, and both within reach of the center of the field of view.
18. The number of cell sap samples depends on your own preference and expertise. If you have two different standard stocks (e.g., A, B) at five different concentrations, with four replicate droplets each, then this amounts already to 40 droplets. Each cell adds another four droplets, so ten cells would result in a total of 80 droplets to be analyzed.
19. The volume specification for constriction capillaries are approximate figures. For example, it is not so important whether your 10-pL constriction pipette really pipettes 10 pL or whether it pipettes 5 or 20 pL. What is important is that it allows you to pipette reproducible volumes, since you compare your sample readings with those of the standards and also use an internal standard during EDX analyses.
20. Draw a large version of a 200-mesh grid on an A4-paper, make many photocopies of it and use it as template during experiments to mark down the position of your droplets.
21. With time, constriction capillaries tend to accumulate some "cell sap residue" and this stops the pipette from having a uniform filling that is easy to pipette. If you have a "precious" 10-pL pipette, you can try to wash/flush it in absolute ethanol; if that does not work, silanize it again. I have used some capillaries for more than 10 years.
22. To increase the Na-sensitivity of your instrument, remove the Be-window. This though makes the SEM more susceptible to any damage from particles. Depending on the setting (17), the

main peak of Ca can partially overlap with the K window (about 10%); this has to be corrected for. Typical settings for the keV windows are as follows: NaKa1 (0.947–1.148 keV), RbLa1 (1.587–1.808), PKa (1.908–2.128), SKa (2.207–2.408), ClKa (2.507–2.727), K (3.207–3.428), Ca (3.467–3.828).

23. The mannitol contained within the internal standard serves as matrix for your samples. You will see during measurements that this matrix starts to "boil"—a sign of the high energy (and efficient vacuum) under which the system operates. If you view the sample at too high magnification or too high accelerating voltage, you may burn a hole into the film on your grid square or the droplet may "pop off." Therefore, use of a 1% Pioloform film is recommended. Before you start any pipetting, take an empty grid with a film coat and test it under the SEM by doing some fake analyses.

24. You can assume that Na, Cl, K, and Ca are present as ions (Na^+, Cl^-, K^+, Ca^{2+}) in cell sap. The same may not apply to S and P.

25. Avoid incident light shining at the microscope stage during fluorescence measurements. The darker, the better! Also, remember that you do not have to carry out the pipetting and fluorescence analyses in the same room.

26. In this assay you measure a decrease in fluorescence. Before you place down many droplets, pipette just one or two droplets of 3–4 different concentrations of NAD(P)H and measure their fluorescence intensity to make sure that your initial reading is well within the linear scale of values and not off scale.

27. Although proline is commonly referred to as an "amino acid," it is chemically an "imino acid" and is not detected by OPA. Thus, if you are, for example, particularly interested in changes in proline content in response to drought or salinity, this is the wrong assay! Unfortunately, I am not aware of any other assay that could be employed through this technique that could detect and measure proline specifically.

Acknowledgments

I would like to thank Prof. Deri Tomos and all his former team members and also Professor Roger A. Leigh for teaching me single-cell sampling and analyzing techniques during my postdoc studies at Bangor University, North Wales, UK. Thanks also to Tamas Visnovitz, UCD, Ireland, for help with ImageJ.

References

1. Munns R, Schachtman DP, Condon AG (1995) The significance of a two-phase growth response to salinity in wheat and barley. Aust J Plant Physiol 22:561–569
2. Fricke W, Akhiyarova G, Wei W et al (2006) The short-term growth response to salt of the developing barley leaf. J Exp Bot 57: 1079–1095
3. Fricke W, Peters WS (2002) The biophysics of leaf growth in salt-stressed barley. A study at the cell level. Plant Physiol 129:374–388
4. Fricke W, Leigh RA, Tomos AD (1996) The intercellular distribution of vacuolar solutes in the epidermis and mesophyll of barley leaves changes in response to NaCl. J Exp Bot 47:1413–1426
5. Cuin TA, Miller AJ, Laurie SA et al (2003) Potassium activities in cell compartments of salt-grown barley leaves. J Exp Bot 54: 657–661
6. Carden DE, Walker DJ, Flowers TJ et al (2003) Single-cell measurements of the contributions of cytosolic Na^+ and K^+ to salt tolerance. Plant Physiol 131:676–683
7. Karley AJ, Leigh RA, Sanders D (2000) Where do all the ions go? The cellular basis of differential ion accumulation in leaf cells. Trends Plant Sci 5:465–470
8. Leigh RA, Storey R (1993) Intercellular compartmentation of ions in barley leaves in relation to potassium nutrition and salinity. J Exp Bot 44:755–762
9. Delane R, Greenway H, Munns R et al (1982) Ion concentration and carbohydrate status of the elongating leaf tissue of *Hordeum vulgare* growing at high external NaCl. I. Relationship between solute concentration and growth. J Exp Bot 33:557–573
10. Dietz KJ, Schramm M, Lang B et al (1992) Characterization of the epidermis from barley primary leaves. 2. The role of the epidermis in ion compartmentation. Planta 187:431–437
11. Karley AJ, Leigh RA, Sanders D (2000) Differential ion accumulation and ion fluxes in the mesophyll and epidermis of barley. Plant Physiol 122:835–844
12. Winter H, Robinson DG, Heldt HW (1993) Subcellular volumes and metabolite concentrations in barley leaves. Planta 191:180–190
13. Outlaw WH, Zhang SQ (2001) Single-cell dissection and microdroplet chemistry. J Exp Bot 52:605–614
14. Brandt S, Kloska S, Altmann T et al (2002) Using array hybridization to monitor gene expression at the single cell level. J Exp Bot 53:2315–2323
15. Roy SJ, Gilliham M, Berger B et al (2008) Investigating glutamate receptor-like gene co-expression in *Arabidopsis thaliana*. Plant Cell Environ 31:861–871
16. Malone M, Leigh RA, Tomos AD (1991) Concentrations of vacuolar inorganic-ions in individual cells of intact wheat leaf epidermis. J Exp Bot 42:305–309
17. Tomos AD, Hinde P, Richardson P et al (1994) Microsampling and measurements of solutes in single cells. In: Harris N, Oparka KJ (eds) Plant cell biology—a practical approach. IRL Press, Oxford
18. Tomos AD, Leigh RA (1999) The pressure probe: a versatile tool in plant cell physiology. Ann Rev Plant Phys 50:447–472
19. Tomos AD, Sharrock RA (2001) Cell sampling and analysis (SiCSA): metabolites measured at single cell resolution. J Exp Bot 52: 623–630
20. Fricke W, Leigh RA, Tomos AD (1994) Concentrations of inorganic and organic solutes in extracts from individual epidermal, mesophyll and bundle-sheath cells of barley leaves. Planta 192:310–316
21. Roy SJ, Cuin TA, Leigh RA (2003) Nanolitre-scale assays to determine the activities of enzymes in individual plant cells. Plant J 34:555–564
22. Bazzanella A, Lochmann H, Tomos AD et al (1998) Determination of inorganic cations and anions in single plant cells by capillary zone electrophoresis. J Chromatogr A 809: 231–239
23. Kehr J (2001) High resolution spatial analysis of plant systems. Curr Opin Plant Biol 4:197–201
24. Malone M, Leigh RA, Tomos AD (1989) Extraction and analysis of sap from individual wheat leaf-cells—the effect of sampling speed on the osmotic-pressure of extracted sap. Plant Cell Environ 12:919–926

Chapter 6

Non-invasive Flux Measurements Using Microsensors: Theory, Limitations, and Systems

Ian Newman, Shao-Liang Chen, D. Marshall Porterfield, and Jian Sun

Abstract

Knowledge of the fluxes of ions and neutral molecules across the outer membrane or boundary of living tissues and cells is an important strand of applied molecular biology. Such fluxes can be measured non-invasively with good resolution in time and space. Two systems (MIFE™ and SIET) have been developed and have become widely used to implement this technique, and they are commercially available. This Chapter is the first comparative description of these two systems. It gives the context, the basic underlying theory, practical limitations inherent in the technique, theoretical developments, guidance on the practicalities of the technique, and the functionality of the two systems. Although the technique is strongly relevant to plant salt tolerance and other plant stresses (drought, temperature, pollutants, waterlogging), it also has rich relevance throughout biomedical studies and the molecular genetics of transport proteins.

Key words: ISE, MIFE, SIET, Salinity, Ion flux, Electrophysiology, Membrane transport, Ion selective microelectrodes, Microsensors, Self-referencing

1. History, Context, Potential Uses, and Previous Reviews

1.1. Origin of the Technique and its Implementation

The technique of measuring a specific ion flux using ion selective microelectrodes that move between two positions near the tissue was proposed by Lucas (1). At that time, it was used in the Lucas laboratory at the University of California Davis (2), then in Kochian's lab at Cornell (3). For both of those labs, the electrodes were moved by hand, the data recorded on a strip chart recorder and the results analyzed using programmable hand calculators. Jaffe and Levy (4) also reported results from a hand-operated system at The Marine Biological Laboratory, Woods Hole MA. The technique was then automated, with computer control of

Sergey Shabala and Tracey Ann Cuin (eds.), *Plant Salt Tolerance: Methods and Protocols*, Methods in Molecular Biology, vol. 913, DOI 10.1007/978-1-61779-986-0_6, © Springer Science+Business Media, LLC 2012

electrode movement and data recording, independently at the University of Tasmania and at the MBL. Papers from these and other labs followed in 1990. The two systems have been developed and refined by these two laboratories and their collaborators. Although a few other laboratories have developed their own system, they are not included in this chapter, which focuses on systems and facilities that are readily available. It also brings together practical information.

1.2. Flux Measurements: Their Context and Research Potential

The measurement of net fluxes is one of a number of techniques to observe the distribution of ions and neutral molecules, as well as their movement and physiological interactions. Flux measurements typically have a time resolution of a few seconds and a space resolution of a few micrometers. Sensitivity threshold levels vary widely, depending on the background concentration of the ion or molecule being measured (see Subheading 3.1).

Shabala (5) has given a detailed evaluation of many other techniques used in the study of membrane transport processes and how they may link to molecular genetics of transport systems. The technique has better time and space resolution than chemical or tracer analysis and is non-destructive, although it does not measure unidirectional fluxes. It cannot provide direct measurement of cell content as can fluorescent probes or impaled ion specific microelectrodes. It can measure specific fluxes, whereas in patch clamp, these must be inferred from currents, and its resolution can approach that of patch clamp for whole-cell mode. The non-destructive nature of ion and biomolecular flux measurements is a particular advantage.

Because of its contrasting capabilities, flux measurements provide a valuable adjunct to the use of other techniques. Kochian et al. (3) is an early example. Specific flux measurements play a key role in studies on transport proteins and the systems that regulate them in vivo. A good example of this in salinity studies is in the report of Chen et al. (6) that used flux measurements with patch clamp, tracers, microelectrode impalement, and an ATPase assay to identify the systems conferring salinity tolerance on barley cultivars. Fluxes can be measured for single cells or protoplasts while they are being patch-clamped in the whole-cell mode (7, 8).

The technique's ability to measure two or more fluxes concurrently (3, 9) can provide valuable stoichiometric information. Within its resolution limits, flux measuring can provide real-time monitoring of the effects of pharmacological agents or other treatments upon transport systems. It can also provide mapping of flux patterns around tissues. This scanning/mapping capability is particularly implemented by the SIET system but not well in the MIFE system that was developed for time-course and multichannel uses.

1.3. Previous Reviews of the Technique and its Implementations

The technique has been described in a number of reviews and research publications. Major reviews are by Smith et al. (10), Newman (11), Kunkel et al. (12), Messerli et al. (13), Porterfield (14), and McLamore and Porterfield (15), written with emphasis on either the SIET/SERIS system or the MIFE system, depending on the system with which the authors were familiar. Each review has a diversity of background information that will be useful to all researchers, whichever system they use. Two early papers (16, 17) also have extensive discussion of system development and of electrode construction, use, and quality evaluation. Sun et al. (18) include a brief overview of the technique's capabilities and of the two systems. "MIFE" is used as a trademark by the University of Tasmania, Hobart Australia. "SIET" functions in a similar way to identify the system made by Applicable Electronics Inc, Forestdale MA USA. Other publications, whether reviews, specific theoretical developments or research articles with useful information, have also been taken into account in preparing this Chapter.

2. The Theory Used by the Two Commercial Systems

"Substances in solution tend to move by diffusion from high concentration to low. For ions, electric forces are also involved, so the movement is from high electrochemical potential to low. If the electrochemical potential gradient can be measured, the net ion movement by diffusion can be calculated from that gradient and the [diffusion coefficient] and concentration of the ion in solution. For H^+, for example, the gradient can be measured using a pH meter, by moving the pH electrode between two positions, with the reference electrode fixed some distance away. For another ion, the principle is the same but the electrode must be sensitive specifically to that ion instead" (11). This kind of measurement of the electrochemical gradient, usually between just two points position 1 (the "near pole") and position 2 (the "far pole"), may be called "self-referencing" because the result is independent of the reference electrode (10, 13). This basic theory underlies all implementations of the flux measuring technique. For neutral molecules, being unaffected by electric fields, this treatment becomes the well-known Fick's Law.

It should be noted that the theory assumes a diffusion-limited solution without significant convective motion or disturbance by probe movement. (This requirement contrasts with the functioning of the Vibrating Probe that was developed earlier (19), and whose rapid vibration mixes the solution in its vicinity, thereby collapsing all local concentration gradients and replacing them by an electrically equivalent voltage gradient that the probe measures to allow net bioelectric currents to be calculated.)

Net fluxes are measured by the technique, not unidirectional influx or efflux. Radioisotope tracer studies are required to measure unidirectional fluxes (influx vs. efflux), therefore some in the field have suggested that the terms "influx" and "efflux" should be reserved for unidirectional fluxes, not to describe net fluxes that are inwardly or outwardly directed. For these net fluxes, the words "uptake" and "release" may be more appropriate.

The two main systems use opposite sign conventions for reporting fluxes. MIFE system users, following the initial publication (2) when uptake of nutrients was the focus, adopt the convention that the inward direction of the net flux is positive. SIET users, following Kunkel et al. (12) and Xu et al. (20), have generally adopted the reverse convention. Experimental reports should make explicit which convention is being used.

Because the ion selective electrodes measure the electric voltage *plus* the electric (Nernst) equivalent of the chemical potential, use of electrochemical theory provides a straightforward description. The use of Fick's Law, applied to ion concentration alone as the theoretical starting point, requires the (unnecessary) assumption that there is negligible electric field in the region of movement of the electrodes. The ISE does not measure concentration alone. The difference in solution voltage between the near and far positions is automatically part of the ISE measurement. The voltage difference does not need extra, separate evaluation, as Messerli et al. (13) have attempted. Whether the electric component is small or large, it is an explicit part of the electrochemical treatment, the following being a modified summary from Newman (11). The final Eq. 4 below is applicable to both systems.

2.1. Ion Flux Theory

The output of the flux measuring instrumentation is a voltage change ΔV as the electrode is moved the distance Δx between the near position and the far position, x being distance of the near position from the tissue surface. The average voltage V_{av} of the electrode at these positions is also given. The theory provides the method of calculating the net ion flux from these results. It may be incorporated into the system's analytical software or implemented in a user-generated spreadsheet.

The electrochemical potential μ (Joule/mole) at a point in the bathing medium near the tissue is given by the equation ((13), Eq. 1, and references therein),

$$\mu = \mu_0 + RT \ln \gamma c + zFV_b \qquad (1)$$

Here, μ_0 is a reference value of the electrochemical potential, the valence is z, the activity coefficient is γ, T the temperature (K), and V_b the voltage at that point in the bathing solution. The gas constant R is 8.3 J mol^{-1} K^{-1} and the Faraday number F is 96,500 C mol^{-1}.

The well-known Nernst relationship for electrochemical equilibrium between two points is derived from this equation.

The inward ion flux J (mol m^{-2} s^{-1}) depends on the concentration c, (mol L^{-1}), the diffusion coefficient D (m^2 s^{-1}) at the measurement temperature and the electrochemical potential gradient $\Delta\mu/\Delta x$. It is directed opposite to the electrochemical gradient. From Newman ((11), Eq. 2, replacing u with D/RT), the flux equation is

$$J = cD\,/\,(RT)(\Delta\mu\,/\,\Delta x). \tag{2}$$

The equation has a negative sign if the positive flux is considered to be away from the surface.

For an uncharged analyte molecule, e.g. O_2, $z=0$ and $\gamma=1$ in Eq. 1, the flux equation reduces to Fick's Law.

In the ISE calibration, the electrode voltage V (mV) is measured at three or more concentrations of the ion. The resulting graph of V against $\log_{10}(c)$ is the calibration graph whose slope is referred to as the "Nernst slope" (mV per decade concentration ratio) and whose intercept is V_0,

$$V = V_0 + (\textit{Nernst slope})\log_{10}(c). \tag{3}$$

This equation, as for Messerli et al. ((13), Eq. 5) or Kunkel et al. ((12), Eq. 4), assumes that the *Nernst slope* is the same at the near and far positions. The reason that this assumption is required is because the original system developed at the MBL does not directly measure the near-position and far-position electrode output, instead it measures only ΔV and the average background output. The commercial SEIT system does not suffer from this limitation. See McLamore and Porterfield (15) for detailed discussion of this issue.

The equations that describe the calibration graph also allow the average concentration c to be found. From the flux Eq. 2, and using the *Nernst slope*, Newman ((11), Eq. 7) showed that the flux is given by:

$$J = cD(\ln(10)/\textit{Nernst slope})\Delta V/\Delta x.(\text{mol m}^{-2}\text{s}^{-1}) \tag{4}$$

If the flux measurements are done at a temperature T_f that is substantially different from the temperature of the calibration solutions T_c, the flux in Eq. 4 must be multiplied by T_c/T_f.

When more convenient units are used in practice (e.g. Δx in μm or J in pmol cm^{-2} s^{-1}), care must be taken to use the correct powers of 10 in calculations.

During experimental data acquisition, as the electrode is moved the distance Δx between the near and far positions, the system records a suitable average ΔV over each cycle. The averaging strategy, designed to ignore settling time at each position after movement, is provided by each system. See Newman (11) and another

Chapter herein for MIFE and see Kunkel et al. (12) for SIET. The system also records the settled voltages at the two positions from which the mean V_{av} is calculated. Together with the calibration graph *Nernst slope* and the concentration c calculated from V_{av} using the calibration graph, these values allow the calculation of the net flux J using the Eq. 4. The MIFE and SIET systems both provide this information. This calculation of c assumes that the electric voltage between the bath reference electrode and the ISE may be neglected.

The above description is valid for a planar diffusion profile i.e. for a flat surface or for measurements close to a curved surface. If the measurements are done on some tissue with a curved surface (e.g. a plant root, a protoplast or oocyte) using an x and a Δx that are not small compared with the radius of the curved surface, the Eq. 4 must be modified. Geometrical analysis, with integration of the electrochemical potential gradient over the finite distance x to $x+\Delta x$ shows that Eq. 4 may be used with its Δx being replaced by the expressions in Eq. 5 below for spherical and cylindrical geometry respectively.

$$\begin{aligned} \Delta x &= r^2(1/(r+x) - 1/(r+x+\Delta x)) \text{ (sphere)}; \\ \Delta x &= r\ln((r+x+\Delta x)/(r+x)) \text{ (cylinder)}. \end{aligned} \tag{5}$$

When one of these expressions replacing Δx is included in Eq. 4 for tissue of radius r, the flux value calculated is the flux at the tissue surface, while x is the distance from the surface to the near position of movement. Henriksen et al. (17) give the derivation in detail for a cylinder. They also discuss the effect of water uptake by the tissue and other complicating factors.

In all of this theory, it has been assumed that the direction of electrode movement is normal to the underlying surface. Electrodes can be moved in other directions in order to measure fluxes in those directions. In all cases, the resulting calculated flux is the component of flux in the direction of movement, because the flux may also have some lateral component.

2.2. Neutral Molecules Flux Theory

The measurement of neutral molecule fluxes is included here briefly for completeness, although it figures little in salinity studies. It is useful in other stress studies, notably waterlogging. The probes may use a variety of sensing mechanisms, depending on the analyte, but they are not based on electrochemical equilibrium across the probe-sensing tip. Analytes like oxygen (21–23), peroxide (14), auxin (24), and nitric oxide (25) have been measured using redox based amperometry, where that analyte is either reduced or oxidized by a polarized microelectrode. Peroxide amperometry is also the foundation for oxidase-based biosensors for analytes like glucose and glutamate (26–28).

For amperometry, the redox current is recorded by the system electronics. This output is related to a concentration differential based on the slope of the calibration curve. For redox probes, this is linear with the analyte concentration in the solution (14). Thus, the calibration provides a graph of current vs. concentration c, rather than against $\log(c)$ as for ions. While the final derivation of the math for converting sensor output to net flux is different from that for ions, both approaches conform to the fundamental laws of diffusion. Amperometric electrodes do not have the voltage noise problem of ISEs and calculating flux is fairly straightforward given the linearity of the electrode calibration. Using the Fick equation to calculate flux, the difference in current ΔI, measured over the electrode excursion, Δx, is related to concentration differential based on the linear slope calibration of the sensor S. For details see McLamore and Porterfield (15).

$$J = DS\Delta I/\Delta x \tag{6}$$

See also Pang et al. (29) for an equivalent approach, and see Porterfield (14) for a basic review of amperometric microsensors for measuring flux and a detailed description of the theory. Both the commercial systems, MIFE and SIET, implement this capability and can readily be used for neutral molecule fluxes. Messerli et al. (13) provide informative discussion on amperometric selectivity and positional artifacts.

Chatni and Porterfield (30) have made a significant advance by using an optical method of flux measurement using a microscale "optrode" as a microsensor. This offers the advantage of increasing sensitivity by diminishing electrical noise. While optrodic sensing can be coupled to the sensitivity of a reporter dye (31) via fluorescence intensity or lifetime, lifetime based measurements offer significant advantages. These include immunity from drift associated with fluorescence bleaching, and when frequency domain fluorescence lifetime methods are employed, the sensor can be operated in nominal lighting conditions (32). For an overview of the technology and examples of applications for physiology studies see Chatni and Porterfield (30).

New approaches based upon optrodic sensing approaches have been developed and used for plant root research (15, 33). These sensors report analyte concentration, and flux by monitoring changes in the fluorescence of a reporter dye (32) that is immobilized on the tip of a microscale optical fiber. This new class of microscale flux measurement sensors is still in the infancy stage, but they offer great promise because they are immune from electromagnetic interference and do not require a reference electrode. They also offer the promise of multi-analyte detection using a single probe.

3. Theoretical Developments and Practical Limitations of the Technique

Some limitations and problems are inherent in the technique itself so are common to all systems that implement it. This section brings together information about these. The technique has been extensively validated in general terms (11–14, 17). Kunkel et al. (12) also describe a useful test source of analyte flux. It remains for individual experimenters to make sufficient validation of their own systems under their own conditions.

The SIET and MIFE systems in use can be tested for correct functioning. Known V and ΔV inputs (from a mV source, measured by a voltmeter) can be traced through the system to its display and output. Validity of flux computations can be verified for the system analytical software or system-supplied spreadsheets, using the published theory (e.g. in Subheading 2). It is crucial to check that the *sign* of the fluxes is correct, e.g. cation fluxes are away from the tissue if the near position has the higher value of V.

3.1. Ion-Selective Electrode Resistance and Sensitivity Limit due to Noise

The ion-selective LIX in its tip gives the ISE a high resistance—usually in the range 1–10 GΩ. Thermal movement of electric charges in this resistance produces electric "noise" that sets an absolute limit to the sensitivity threshold of the ISE.

Standard electrical theory shows that the thermal electronic RMS noise ΔV in the resistor R is given by

$$\Delta V = \sqrt{(kT\beta R)}. \tag{7}$$

Here k is Boltzmann's constant, T the temperature, and β the system bandwidth. This ΔV is a measure of about the smallest real signal that can be detected above the noise. If this value of ΔV is used in the flux Eq. 4, the threshold flux sensitivity for an ion can be calculated for the experimental conditions.

Using a β of 10 Hz (typical for MIFE and SIET electronics) and a resistance of 1 GΩ at room temperature, Ryan et al. (9) calculated ΔV to be 13 μV. Using this value in their equivalent of the flux Eq. 4 and using a Δx of 40 μm, they tabulated the flux sensitivity thresholds for several ions. For example, for H^+ at pH = 6, the minimum flux that can be detected is 0.11 nmol m^{-2} s^{-1}. For K^+ at concentration of 1 mM this flux sensitivity is 24 nmol m^{-2} s^{-1}. It is easy to scale these numbers for different values of β, R, Δx, and concentration or for different ions (via their mobilities or diffusion coefficients). It should also be noted that, when signals are averaged over a few seconds (as is done by both MIFE and SIET), the averaging makes the effective bandwidth $\beta < 1$ Hz, thereby decreasing the ΔV by a factor of 3 or 4. For salinity research, Na^+ concentration is a particular problem: at 100 mM the flux sensitivity is 1,600 nmol m^{-2} s^{-1} compared with only 16 nmol m^{-2} s^{-1} at 1 mM.

Thus, to obtain the best flux measurements, try to keep R low, Δx large, and concentrations low. Of course, in practice these preferences must be balanced against other requirements for the particular experiments.

There are other sources of noise that limit sensitivity in ISE systems: (a) Noise in the amplifying electronics. Both MIFE and SIET systems have been designed so that this noise is $< 1\ \mu V$. This is not normally a problem. (b) Electrochemical instabilities in the LIX and other junctions of the measuring circuit including the Ag/AgCl half cells of the bath reference electrode and the ISE. These can produce substantial noise. (c) Pickup of radiated electrical mains or high frequency signals. If such signals are large, even though outside the bandwidth, they can be rectified by the electronics and appear as small, unwanted voltage offsets. Suitable earthing of equipment can minimize this noise. One key principle is that the grounding wires from each component should go to one central point, at the analog/digital converter or at the computer chassis. Trial may show which of these is better. The wires join at that point only, so that there are no loops of wire that can act as aerials to pick up radiated mains or high frequency signals.

3.2. Time Resolution and Response Time; Space Resolution

Time resolution of the technique is determined by the period of the square-wave movement of the electrodes. This is at the experimenter's choice and is typically in the range 3–10 s. As an electrode is moved from the near position to the far position and back in each measurement cycle, it is observed that the voltage takes some time to settle to its steady value. All past reviewers including Messerli et al. (13) and Kunkel et al. (12) agree that there are several reasons for this, in addition to the electrode movement time. They discuss the reasons in detail, so they are not discussed here. It is also recognized that the ΔV to be used for flux calculations should be the difference between the "settled" values at the two positions of movement. Both systems allow the experimenter to choose a suitable settling time, a time period that is ignored when the average electrode voltage at the position is calculated.

In practice, settling time is greater when Δx is larger and when electrodes are of poor quality or have large resistance. Settling time ranges from about 0.2 s to several seconds. This sets the practical minimum time resolution of the technique to about 1 s.

Spatial resolution has also been well discussed previously (13, 17) and it is shown how the highest spatial resolution can be obtained if it is needed. When the electrode tip is placed very close to the tissue, the spatial resolution depends also on the nature of the electrode and its size and shape (12, 13). For many measurements, particularly in salinity studies, the spatial resolution on the tissue surface is determined mostly by the distance of the near position from the surface. It is reasonable to suggest that the observed fluxes arise mostly from the region whose diameter is equal to the closest distance of the electrode from the surface. The choice of this distance is part of the experimental design.

3.3. Sensor Selectivity and Interferents

These issues are extensively described by Henriksen et al. (16) and Messerli et al. (13), including the poisoning effect of some buffers. Electrode construction and calibration have been well described in previous reviews (10, 13) as well as another chapter herein.

Sensor Selectivity. For ions, a LIX may show a good Nernst slope when the ion of interest is alone. However, that particular LIX may also respond to another ion of the same sign that is also present in the measuring solution. Notoriously, the commercial LIX for Na^+ are responsive to Ca^{2+} and K^+. Specifications have been provided (34) for making a new Na^+ LIX that shows good discrimination against other ions. In the absence of such a LIX, the experimenter may have four options. The first two of them require the use of the selectivity ratio of the LIX for Na^+ to the other ion. These ratios can be obtained from the Fluka Web site http:\\www.sigmaaldrich.com.

The options are: (a) Use the Nicolsky-Eisenman analysis (13) to evaluate the correction to apply to the recorded flux data. This tedious, somewhat unreliable process is rarely attempted. (b) Measure the flux of the interfering ion concurrently with the ion of interest. For Na^+ in a solution containing also K^+, use the K^+ flux and the Na^+/K^+ selectivity ratio of the Na^+ LIX to estimate the K^+ flux contribution to the measured Na^+ flux. (c) If it is physiologically realistic, keep the measuring solution free from the other ion. (d) Sometimes a useful, consistent comparison can be made between responses to different treatments, even though the fluxes are not correct in their absolute values. Similar strategies and approaches can be employed to deal with similar problems with other types of sensors.

Interfering substances. Some pharmacological agents chosen to affect membrane transport, or other substances in solution, may also affect the ion carriers of the LIX. If there are alternatives, choose a LIX that is least affected. All substances in the proposed measuring solution, other than simple inorganic ions, should be tested for an effect on the calibration Nernst slopes. The general principle is that calibrations should be done by changing the ion's concentration in a background of the proposed measuring solution, including any pharmacological agents.

3.4. Solution Buffering

Buffered solutions are used in research to define and stabilize H^+ concentrations. Consequently, buffers also affect H^+ fluxes. Some of the protons that cross the tissue surface will not be measurable as free protons due to the action of the buffer molecules. Demarest and Morgan (35) discussed and analyzed this. It was analyzed with greater theoretical coverage by Arif et al. (36). Refer to that paper for details. The practical outcome from Arif et al. ((36), Eq. 13) is that the total membrane flux ${}^{m}J_{H}$ of H^+ can be found from the H^+

flux J_H measured by the ISEs. The measured J_H must be multiplied by a factor that depends on the solution pH, the buffer concentration C_B (mol L^{-1}), its diffusion coefficient $D_{B,}$ and pK (or dissociation constant K_B).

$$^{m}J_H = J_H(1 + r_B) \text{ where}$$
$$r_B = (D_H/D_B)C_B 10^{pK}(10^{pH}/(10^{pH} + 10^{pK}))^2. \quad (8)$$

This quantity can be incorporated easily into a spreadsheet to obtain the H^+ fluxes at the tissue boundary. The calculation is built into the MIFE analytical software, to include water at pH > 6.5, and for a single buffer whose C_B, $D_{B,}$ and pK are provided by the experimenter. Arif et al. (36), for a mixture of buffers, note that the flux ratios are additive. Hence, the r_B in Eq. 8 must be replaced by the sum of the equivalent r_B values for each buffer. Messerli et al. (13) also have noted this, and they have given an alternative writing of Eq. 8 in terms of the buffer dissociation constant K_B (where $K_B = 10^{-pK}$) instead of using the pK. For buffers with more than one pK value, Arif et al. ((36), Eq. 36) provide the expression for r_B, with examples of phosphate and acetate. They also note that the situation is more complicated if the buffer is itself being transported (e.g. phosphate, or carbonate in respiration).

Porterfield et al. (37) have also developed an approach based on modeling the buffer disassociation constant activities in the linear range. This model does offer an approach for approximating the total flux in buffered media for analyzing the data offline. This model does not consider the potential role of buffer diffusion, and instead considers this to be negligible in comparison to the biologically mediated proton flux. This issue has yet to be resolved. Either analysis is difficult to test, mainly because of the near impossibility of measuring the actual membrane flux of protons into the bathing medium. The disagreement lends force to the suggestion that, in practice, buffers with pK at least 0.5 unit above the chosen solution pH should be used and their concentration should be kept low.

3.5. Cell Wall (Donnan System) Ion Exchange

The plant cell wall, with its many fixed anions, forms a Donnan ion-exchange system. This wall system can store or release cations and it acts in the flux pathway between the plasma membrane and the external solution, thereby causing a time-dependent difference between membrane flux and externally measured flux.

It is possible to model this effect if the wall thickness and the concentration of fixed anions are known. Ryan et al. (38) have done this for the alga *Chara*. In addition, they incorporated a particular binding mechanism "condensation" that can affect fluxes of H^+ and di or tri-valent ion fluxes in the pH range 4–6. They modeled Ca^{2+} flux from a solution of constant 0.05 mM Ca^{2+} concentration when external pH or K^+ were changed. At the change, a large Ca^{2+} flux took place that decayed with a time constant of about 10 min. The modeled results were confirmed by flux measurements on *Chara* cells and cell walls.

These wall exchange effects are hard to quantify. Arif and Newman (39) did so successfully for the effect of fusicoccin or auxin on fluxes of Ca^{2+} and H^+ from oat coleoptiles. Shabala and Newman (40) also applied the model to salinity-induced transient Ca^{2+} fluxes. Models that are better than the 20-year-old model of Ryan et al. (38) may now be possible. Even if exchanges are not modeled, ion flux researchers should consider them, at least qualitatively, when making pH changes in the range 4–6 or large changes in other cation concentrations. This applies particularly to Na^+ changes in salinity studies.

4. MIFE and SIET Systems' Details

Each of these commercial systems is an integrated combination of hardware and software. Their use is an art as well as a science. As with patch clamping, it is necessary to gain experience in an existing laboratory. Each system may be set up with a slightly varied configuration to suit a particular experimental requirement.

Each system implements the general theory in Subheading 2.1 and 2.2 above. Both systems essentially measure the electrical output from an electrode or sensor that is moved repetitively between two positions in solution: position 1 (the "near pole") and position 2 (the "far pole"). This output may be the voltage output from an ion-selective electrode for ion fluxes, or the current output (or its voltage equivalent) from a polarographic or other electrode for fluxes of neutral molecules. Each kind of sensor requires the appropriate theory and calibration procedures to obtain the fluxes from the electrical measurements. Each kind of sensor requires the provision of the appropriate analytical software or spreadsheet to accommodate them and their characteristics. Each system requires that the microscope has a long working-distance objective to accommodate the sensor probes. The following gives the specific features, including advantages and limitations, of each system.

4.1. MIFE System Features

Following the report by Ryan et al. (9), which was the first using the automated system developed in Tasmania, the system was improved. It was referred to as "MIFE" from 1995, the name being derived from "MFET" ((17), and Henriksen personal communication). The hardware control and data acquisition software is referred to as CHART, as it was designed to emulate a strip chart recorder, with all raw data stored on the hard disk. The main display of the graphic user interface appears like a chart of the electrode voltages. It has drop-down menus, keyboard-encoded control buttons, and dialog windows to provide quick interaction with hardware controlled by the program. This interaction includes electrode position adjustments and changes to the electrometer offsets to keep measurements within the 50 mV data window. Another chapter herein gives protocols for using the MIFE system, including calibrating electrodes. Electrode resistance can also be measured.

To measure fluxes, the electrodes are driven in a square-wave cycle of amplitude and frequency specified by the user. The data averaging to calculate fluxes is done according to Newman ((11), Fig. 2). CHART then produces a file of experimental parameters and of average V and ΔV values. This file and the calibration file of V_0 and *Nernst slope* are subsequently imported into the analytical program MIFEFLUX to produce files of concentrations and fluxes. MIFEFLUX also implements the buffer corrections for H^+ fluxes (see Subheading 3.4). All files are tab-delimited text files, allowing importation into a spreadsheet.

The CHART software incorporates a "day-book" feature that allows recording of experimental configuration and conditions, as well as "chart annotations" as the experiment progresses. The CHART control software allows for three or four positions of measurement, as suggested by Messerli et al. (13). This allows fluxes to be calculated at different distances from the tissue. The software development began in 1987, so it is DOS based, with a data acquisition card (DAS-08) in an ISA-bus slot in a computer running Windows 98 or ME. Direct access to the computer hardware that the software and card require are precluded by Windows 2000, XP, or later. A modern system, which will retain the same user functionality, is planned.

Because its electronics has eight channels, MIFE can link with microelectrode impalement (for which it is also suitable), patch clamp (8), or other techniques, and record their data concurrently with the flux data for subsequent comparative display and analysis. It can likewise accommodate as many electrodes as can be physically mounted on a multiple electrode mount for concurrent flux measurements. The MIFE system has several options for the microscope-manipulator configuration and for the controlled micromanipulator. These are described in another chapter in this Volume.

The MIFE system also has the ability to measure fluxes with high time resolution—perhaps 0.2 s—using a pair of "stationary" electrodes. The two ISEs with the same LIX are placed with their tips at the near and far positions of the Δx of interest. They are first used for normal flux measurements for a time. Then, when a fast transient is to be observed, electrode movement is halted for the duration of the transient, before being continued. While the movement is halted, the ΔV for use in flux calculations is the difference between the absolute V values of the pair of ISEs. Thus the stationary phase is bounded by standard flux measurements that can normalize the initial voltage difference and quantify any relative drift of the pair of electrodes during the transient.

MIFE is designed to be simple to use, to measure several fluxes in one chosen direction. The experimenter has a clear display of the incoming and recorded data. There is easy control over planning and conducting the experiment and the subsequent analysis of the data. MIFE systems have been used in more than nine laboratories from which have come over 100 publications, reporting on plant, microbiological, and animal studies. The Web site http://www.mife.com provides information for users or potential purchasers of the system, as well as possible collaborative opportunities with existing laboratories.

4.2. SIET System Features

The SIET (scanning ion-electrode technique) system was developed by Shipley and Feijó (41) and it is manufactured by Applicable Electronics http://www.applicableelectronics.com. This system was originally developed with the operation of ion-selective sensors in mind, but is also suitable for amperometric sensors, scanning electrochemical microscopy (amperometric or impedance). The software system for the SEIT is referred to as the Automate Scanning Electrode Technique (ASET) system and better reflects the derivation of the theoretical approach to measure biophysical flux of other molecules besides ions. Recently the software and motion control subsystems were modified for the operation of optrodic sensors by adapting a frequency domain fluorometer to operate with the system.

The system features and software systems are especially well suited for mapping and scanning large features, and also include capabilities for multi-probe measurements. The system is widely used in both biological and material science applications (especially corrosion). The ASET system is equipped with two head mounting assemblies for the micromanipulator for simultaneous two electrode measurements. Microsensors are stepped from one position to another in a highly programmable 3-D sampling routine using an orthogonal microstepper motor system with controllers. ASET has established specific sampling rules for 1-, 2-, or 3-D measurements using one ion-selective microelectrode or sequential or interleaved measurements of multiple ion-selective microelectrodes (42). SIET also includes basic amperometry for redox based polarographic microelectrodes for direct detection of non-ionic analytes, as well as enzyme based biosensors. The system is capable of simultaneous measurement of net fluxes of both ionic (e.g. H^+) and non-ionic molecules (such as oxygen and nitric oxide).

As previously mentioned the WINDOWS based software for manipulator control and data acquisition is abbreviated as ASET. The ASET software, which was developed by E. Karplus (Science Wares, E. Falmouth, MA, USA http://www.sciencewares.com), significantly advances the programmability of the user defined sampling rules for both stationary measurements and three-dimensional flux measurements. ASET is capable of not only operating a single probe, but also multiple probes in a range of micrometers to measure differential concentrations. Scanning probe operations can be programmed over a large dynamic range in the centimeter scale. Three-dimensional data obtained by the ASET software are converted into 3-D plotted maps using a web based online system MageFlux developed by Yue Xu (http://www.youngerusa.com/mageflux or http://xuyue.net/mageflux). The calculated fluxes can be plotted as stereo pairs of 3-D flux vectors or total flux.

The SIET is equipped with a video monitor for a quick and easy visualization of the samples under study. The images can also be captured and stored by the software for further analyses. In addition, video microscopy feedback can be used to operate a "point-and-click" software routine to dynamically control the probe position relative to the sample, or to program large-scale sampling routine maps. Video feedback also allows for user input to set probe movement angles around a curved surface in order to ensure that the sampling positions are normal to the surface tangent.

4.3. NMT

As previously mentioned, the operation of the system and mastery of these technical approaches require substantial skill and experience. In order to decrease the learning curve and increase the productivity of this promising technology, joint efforts have been made by YoungerUSA and Sciencewares to (1) simplify setup and instrumentation; (2) standardize sampling rules for specific sensors, and improve user feedback relative to common artifacts. The technology in China is marketed as the Non-Invasive Microtechnique approach that utilizes a new software system, named imFlux, which addresses the aforementioned issues. Because of the attention to user simplicity, and system reliability, the use of these approaches has been greatly expanded within the Chinese academic community. This system works seamlessly with the online software MageFlux to ease the burden of data calculation as well as to automate the integration of improved calculating methods developed by different research groups led by Newman (11), Kunkel et al. (43), and Porterfield et al. (37) etc. Facilities and useful resources are offered to researchers by Younger USA (http://www.youngerusa.com and http://www.xuyue.net).

Acknowledgements

We thank Mr. Wenjun Wang for providing equipment information of SIET and NMT:SIET system. We thank Mr. Yue (Jeff) Xu and Professor Sergey Shabala for information, advice and critical reading of the manuscript.

References

1. Lucas WJ, Kochian LV (1986) Ion transport processes in corn roots: an approach utilizing microelectrode techniques. In: Gensler WG (ed) Advanced agricultural instrumentation: design and use 402-425. Martinus Nijhoff, Dordrecht
2. Newman IA, Kochian LV, Grusak MA et al (1987) Fluxes of H^+ and K^+ in corn roots—characterization and stoichiometries using ion selective microelectrodes. Plant Physiol 84:1177–1184
3. Kochian LV, Shaff JE, Lucas WJ (1989) High affinity K^+ uptake in maize roots. A lack of coupling with H^+ efflux. Plant Physiol 91:1202–1211
4. Jaffe LF, Levy S (1987) Calcium gradients measured with a vibrating calcium-selective electrode. Proc IEEE/EMBS Conf 9:779–781
5. Shabala SN (2006) Non-invasive microelectrode ion flux measurements in plant stress physiology. In: Volkov AG (ed) Plant electrophysiology—theory and methods. Springer, Berlin
6. Chen Z, Pottosin II, Cuin TA et al (2007) Root plasma membrane transporters controlling K^+/Na^+ homeostasis in salt-stressed barley. Plant Physiol 145:1714–1725
7. Shirihai D, Smith P, Hammar K et al (1998) Microglia generate external proton and potassium gradients utilizing a member of the H/K ATPase family. Glia 23:339–348
8. Tyerman SD, Beilby M, Whittington J et al (2001) Oscillations in proton transport revealed from simultaneous measurements of net current and net proton fluxes from isolated root protoplasts: MIFE meets patch-clamp. Aust J Plant Physiol 28:591–604
9. Ryan PR, Newman IA, Shields B (1990) Ion fluxes in corn roots measured by microelectrodes with ion-specific liquid membranes. J Membr Sci 53:59–69
10. Smith PJS, Hammar K, Porterfield DM et al (1999) Self-referencing, non-invasive, ion selective electrode for single cell detection of trans-plasma membrane calcium flux. Microsc Res Tech 46:398–417
11. Newman IA (2001) Ion transport in roots: measurement of fluxes using ion-selective microelectrodes to characterize transporter function. Plant Cell Environ 24:1–14
12. Kunkel JG, Cordeiro S, Xu Y et al (2006) Use of non-invasive ion-selective microelectrode techniques for the study of plant development. In: Volkov AG (ed) Plant electrophysiology—theory and methods. Springer, Berlin
13. Messerli MA, Robinson KR, Smith PJS (2006) Electrochemical sensor applications to the study of molecular physiology and analyte flux in plants. In: Volkov AG (ed) Plant electrophysiology—theory and methods. Springer, Berlin
14. Porterfield DM (2007) Measuring metabolism and biophysical flux in the tissue, cellular and sub-cellular domains: recent developments in self-referencing amperometry for physiological sensing. Biosens Bioelectron 22:1186–1196
15. McLamore ES, Porterfield DM (2011) Non-invasive tools for measuring metabolism and biophysical analyte transport: self-referencing physiological sensing. Chem Soc Rev 40(11): 5308–5320
16. Henriksen GH, Bloom AJ, Spanswick RM (1990) Measurement of net fluxes of ammonium and nitrate at the surface of barley roots using ion selective microelectrodes. Plant Physiol 93:271–280
17. Henriksen GH, Raman DR, Walker LP et al (1992) Measurement of net fluxes of ammonium and nitrate at the surface of barley roots using ion-selective microelectrodes. II. Patterns of uptake along the root axis and evaluation of the microelectrode flux estimation technique. Plant Physiol 99:734–747
18. Sun J, Chen S, Dai S et al (2009) Ion flux profiles and plant ion homeostasis control under salt stress. Plant Signal Behav 4:261–264
19. Jaffe LF, Nuccitelli R (1974) An ultrasensitive vibrating probe for measuring steady state extracellular currents. J Cell Biol 63:614–628
20. Xu Y, Sun T, Yin L (2006) Application of non-invasive microsensing system to simultaneously measure both H^+ and O_2 fluxes around the pollen tube. J Integr Plant Biol 48:823–831
21. Porterfield DM, Trimarchi JR, Keefe DL et al (1998) Metabolism and calcium homeostasis during development of the mouse embryo to the blastocyst stage in M2 culture medium. Biol Bull 195:208–209
22. Land SC, Porterfield DM, Sanger RH et al (1999) The self-referencing oxygen-selective microelectrode: detection of transmembrane oxygen flux from single cells. J Exp Biol 202:211–218
23. Porterfield DM, Smith PJS (2000) Characterization of trans-cellular oxygen and proton fluxes from *Spirogyra grevilleana* using self-referencing microelectrodes. Protoplasma 212:80–88

24. McLamore ES, Diggs A, Marzal PC et al (2010) Non-invasive quantification of endogenous root auxin transport using an integrated flux microsensor technique. Plant J 63:1004–1016
25. Porterfield DM, Laskin JD, Jung S-K et al (2001) Proteins and lipids define the diffusional field of nitric oxide. Measurement of nitric oxide fluxes from macrophages using a self-referencing electrode. Am J Physiol 281: L904–L912
26. McLamore ES, Shi J, Jaroch D et al (2011) A self referencing platinum nanoparticle decorated enzyme-based microbiosensor for real time measurement of physiological glucose transport. Biosens Bioelectron 26:2237–2245
27. Shi J, McLamore ES, Jaroch D et al (2011) Oscillatory glucose flux in INS 1 pancreatic β cells: a self-referencing microbiosensor study. Anal Biochem 411:185–193
28. McLamore ES, Mohanty S, Shi J et al (2010) A self-referencing glutamate biosensor for measuring real time neuronal glutamate flux. J Neurosci Methods 189:14–22
29. Pang JY, Newman I, Mendham N et al (2006) Microelectrode ion and O_2 fluxes measurements reveal differential sensitivity of barley root tissues to hypoxia. Plant Cell Environ 29:1107–1121
30. Chatni MR, Porterfield DM (2009) Self-referencing optrode technology for non-invasive real-time measurement of biophysical flux and physiological sensing. Analyst 134:2224–2232
31. Chatni MR, Maier DE, Porterfield DM (2009) Optimization of oxygen sensitive optical dye membrane polymers for fluorescent lifetime based physiological biosensing. Sens Actuators B 141:471–477
32. Chatni MR, Li G, Porterfield DM (2009) Frequency domain fluorescence lifetime optrode system design and instrumentation without a concurrent reference LED. Appl Opt 48:5528–5536
33. McLamore ES, Jaroch D, Chatni R et al (2010) Self-referencing optrodes for measuring spatially resolved, real-time metabolic oxygen flux in plant systems. Planta 211:384–389
34. Jayakannan M, Babourina O, Rengel Z (2011) Improved measurements of Na^+ fluxes in plants using calixarene-based microelectrodes. J Plant Physiol 168:1045–1051
35. Demarest JR, Morgan JLM (1995) Effect of pH buffers on proton secretion from gastric xyntic cells measured with vibrating ion-selective microelectrodes. Biol Bull 189:219–220
36. Arif I, Newman IA, Keenlyside N (1995) Proton flux measurements from tissues in buffered solution. Plant Cell Environ 18:1319–1324
37. Porterfield DM, McLamore ES, Banks MK (2009) Microsensor technology for measuring H^+ flux in buffered media. Sens Actuators B 136:383–387
38. Ryan PR, Newman IA, Arif I (1992) Rapid calcium exchange for protons and potassium in cell walls of *Chara*. Plant Cell Environ 15:675–683
39. Arif I, Newman IA (1993) Proton efflux from oat coleoptile cells and exchange with wall calcium after IAA or fusicoccin treatment. Planta 189:377–383
40. Shabala S, Newman I (2000) Salinity effects on the activity of plasma membrane H^+ and Ca^{2+} transporters in bean leaf mesophyll: masking role of the cell wall. Ann Bot 85:681–686
41. Shipley AM, Feijó JA (1999) The use of the vibrating probe technique to study steady extracellular currents during pollen germination and tube growth. In: Cresti M, Cai G, Moscatelli S (eds) Fertilization in higher plants: molecular and cytological aspects. Springer, Berlin Heidelberg New York, pp 235–252
42. Faszewski EE, Kunkel JG (2001) Covariance of ion flux measurements allows new interpretation of *Xenopus laevis* oocyte physiology. J Exp Zool 290:652–661
43. Kunkel JG, Lin L-Y, Xu Y et al (2001) The strategic use of good buffers to measure proton gradients around growing pollen tubes. In: Geitmann A, Cresti M, Heath IB (eds) Cell biology of plant and fungal tip growth. Ios Press, Amsterdam, pp 81–94

Chapter 7

Quantifying Kinetics of Net Ion Fluxes from Plant Tissues by Non-invasive Microelectrode Measuring MIFE Technique

Sergey Shabala, Tracey Ann Cuin, Lana Shabala, and Ian Newman

Abstract

Non-invasive microelectrode ion flux measuring (the MIFE system) allows concurrent quantification of net fluxes of several ions with high spatial (several μm) and temporal (*ca* 5 s) resolution. Over the last 10 years, the MIFE system has been widely used to study various aspects of salt stress signaling and adaptation in plants. This chapter summarizes some major findings in the area such as using MIFE for deciphering the specific and non-specific components of salinity stress, resolving the role of the plasma membrane H^+-pump in salinity responses, proving K^+ homeostasis as a key feature of salinity tolerance, and discovering the mechanisms behind the ameliorative effects of Ca^{2+} and other mitigating factors (such as polyamines or compatible solutes). The full protocols for microelectrode fabrication, calibration, and use are then given, and two basic routines for measuring net K^+ and Na^+ fluxes from salinity stressed roots are described in the context of plant screening for salt stress tolerance.

Key words: Electrophysiology, Salinity stress, Sodium, Potassium, Membrane transport, Ion-Selective microelectrodes

1. Introduction

1.1. Background Information

Ion transport across plant membranes is a major factor governing plant performance and crop yield. At the very least, ion transport systems provide the means by which plants acquire nutrients from the soil and transport them internally. Even more importantly, membrane transport systems also often act as *receptors,* so are crucial in mediating plant–environment interactions. Plant adaptive responses to salinity are a classic example of such interactions. As well as playing a critical role in the uptake and efflux of Na^+ and other ions under saline conditions, membrane transporters also have a crucial role in salinity perception and in the resulting

Sergey Shabala and Tracey Ann Cuin (eds.), *Plant Salt Tolerance: Methods and Protocols*, Methods in Molecular Biology, vol. 913, DOI 10.1007/978-1-61779-986-0_7, © Springer Science+Business Media, LLC 2012

signaling pathways. Indeed, changes in membrane potential and modulation of ion fluxes are amongst the earliest cellular events in response to salinity (1–5). The SOS1 putative Na^+/H^+ exchanger is suggested to act as a sodium sensor, triggering a downstream signal transduction pathway (6). Also, the plasma membrane H^+-ATPase pump may operate as an osmosensor (7), so adjusting cell turgor by modulating the activity of a range of K^+ and Cl^- transport systems under hyperosmotic stress conditions imposed by salinity (8, 9). Therefore, it is obvious that ion transport systems are central targets for genetic manipulation in the quest for salt tolerant crop varieties.

Numerous methods exist to investigate the kinetics of ion transport in plants (reviewed in (10)). Membrane potential measurements (11), patch-clamp (12), ion and pH-sensitive dyes alongside imaging techniques (13, 14), ion-selective intracellular microelectrodes (15), nuclear magnetic resonance (NMR) spectroscopy (16), and single-cell sampling and analysis (SISCA) (17, 18) have all provided information on ion distribution and movement. However, there are drawbacks to many of these techniques. For example, many are technically demanding (intracellular microelectrodes, patch-clamp, SISCA), are highly destructive (patch-clamp), do not allow the evaluation of transport kinetics (SISCA), or have poor resolution and limitations to the various ions that can be measured (radioactive tracers, NMR).

1.2. Basic MIFE Principles and Features

The introduction of the MIFE system and others for non-invasive ion flux measurements (see the previous chapter for the historical overview, theoretical background, and review of MIFE and SIET systems) has revolutionized the entire concept of studying ion flux kinetics in plant systems. The technique measures *net* ion fluxes across a membrane in intact "in situ" plants. This allows the assessment of ion fluxes under relatively physiologically realistic conditions and permits real-time recordings of fluxes after an imposed treatment. This particular feature has been crucial in examining the immediate effects of salinity on ion transport in plants. Additionally, the technique has a high spatial resolution ($< 2\ \mu m$), so ion fluxes across the cell surface can be mapped (19). Its high temporal resolution enables insights into very early and fast events associated with NaCl imposition, and due to the technique being non-invasive, measurements can be carried for an extended period of time. Finally, the technique permits simultaneously recordings of fluxes of several ions. This is vital in achieving stoichiometry ratios between different ions; necessary for a full understanding of the underlying ionic mechanisms of cell responses to an imposed stress. The MIFE system therefore represents a powerful technique that will help in "pyramiding" the physiological traits behind salinity tolerance. This will contribute towards the goal of developing salinity tolerant crops.

The theory of MIFE measurements is given in detail in the previous chapter as well as in previously published papers (20, 21). In brief, if an ion is taken up by root cells, its concentration in the proximity of the root surface will be lower than that further away. Vice versa, if the ion is extruded across epidermal root cells, there will be a pronounced electrochemical potential gradient directed away from the root surface. The principle of the method is to measure this electrochemical potential gradient by slow, square-wave movement of ion-selective microelectrode probes between two positions, close to (position 1), and distant from (position 2) the sample surface (Fig. 1). At each position, the electrode voltage, representing the electrochemical potential, is recorded. From that potential difference, the ion flux is calculated using also the approximate concentration at those points obtained from the calibrated Nernst slope of the electrode. It is assumed that convection and water uptake are negligible and unstirred layer conditions are met.

1.3. Deciphering the Specific and Non-specific Components of Salinity Stress

The two principal adverse effects of salinity in non-tolerant plants are osmotic stress and specific ion (Na^+ or Cl^-) toxicity (22). Using the MIFE system, we have shown that the ionic basis of plant adaptive responses to each of these components of salinity is strikingly different (23, 24). While NaCl promotes a net K^+ efflux, isotonic mannitol treatment induces a gradual increase in the net K^+ uptake leading to the turgor recovery (8). A pharmacological approach

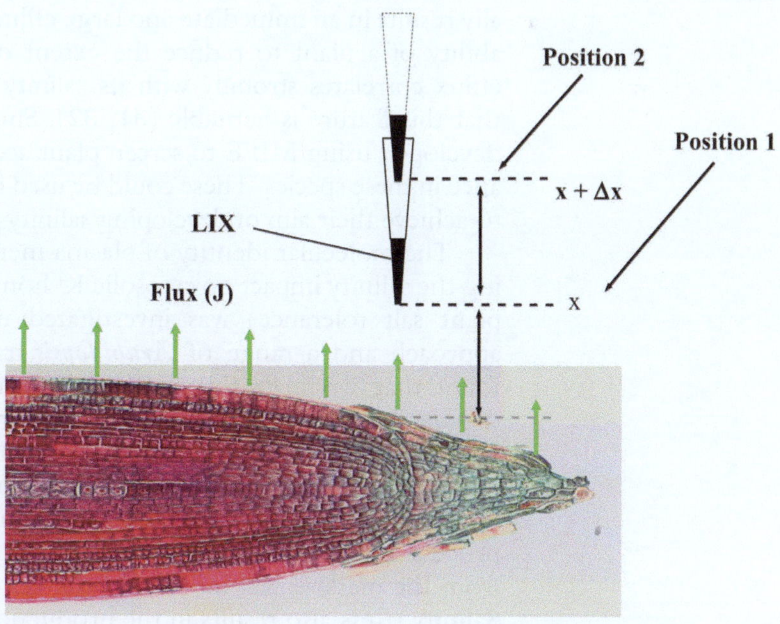

Fig. 1. The principle of the MIFE ion flux measurements (modified from (54)).

has shown that NaCl-induced K^+ efflux is mediated essentially by depolarization-activated K^+ outward-rectifying channels (KOR), while inward-rectifying channels (KIR) are responsible for K^+ uptake in response to mannitol treatment. Thus, two oppositely directed signals appear to initiate K^+ fluxes from salinity stressed plant cells: (1) K^+ efflux resulting from NaCl-induced plasma membrane depolarization, and (2) K^+ uptake resulting from an as yet unknown "osmosensing mechanism". Under mild salinities, the latter component would dominate, whilst at higher NaCl levels, the result would be a net loss of K^+.

1.4. Resolving the Role of the Plasma Membrane H^+-Pump in Salinity Responses

A salinity-induced increase in H^+-pump activity has been reported in many halophytic species (25, 26). The stimulation of H^+-ATPases by salinity stress would provide a driving force for a plasma membrane H^+/Na^+ exchanger to export Na^+ from the cytoplasm and out of the cell. This would increase salinity tolerance. Nonetheless, the role of the H^+-ATPase in salinity tolerance has been questioned (27) and we resolved this using the MIFE. In both leaf mesophyll tissue (23) and in roots of glycophyte species (28), NaCl application did result in an increased net H^+ efflux. Furthermore, higher H^+-pump activity correlated significantly with a cultivar's tolerance, as shown in experiments on barley (29). This work establishes a role for the H^+-pump in salinity tolerance.

1.5. Proving That K^+ Homeostasis is a key Feature of Salinity Tolerance

Application of the MIFE system to a wide range of barley and wheat genotypes contrasting in their salinity tolerance has revealed a critical role for cytosolic K^+ homeostasis in plant salinity tolerance (reviewed in ref. (9)). When salinity is imposed on a plant, it generally results in an immediate and large efflux of K^+ from cells and the ability of a plant to reduce the extent of its salinity-induced K^+ efflux correlates strongly with its salinity tolerance (24, 30), and that this feature is heritable (31, 32). Simple protocols have been developed using MIFE to screen plant accessions for salinity tolerance in these species. These could be used by plant breeders in order to achieve their aim of developing salinity tolerant crop varieties.

The molecular identity of plasma membrane channels mediating the salinity impact on cytosolic K^+ homeostasis (and, ultimately, plant salt tolerance) was investigated using a pharmacological approach and a range of *Arabidopsis* transport mutants. It was found that two major K^+-permeable channels, namely a K^+-selective depolarization-activated outward-rectifying channel (GORK) and a non-selective weakly rectifying cation (NSCC) channel, mediate NaCl-induced K^+ efflux under saline conditions. In barley and *Arabidopsis*, GORK channels play a dominant role in the above process and are activated by membrane depolarization resulting from the massive Na^+ influx into the cell under saline conditions. Salinity stress also results in the production of significant amounts of reactive oxygen species (ROS) that activate NSCC channels (33) and result in a further K^+ leak from the cytosol (34, 35). The resultant depletion of the cytosolic K^+ pool was found to result in activation of caspase-like proteases and trigger programmed cell death in plant roots under saline conditions (36, 37).

1.6. Discovering the Mechanisms Behind the Ameliorative Effects of Ca^{2+} and Other Mitigating Factors

The application of Ca^{2+} significantly ameliorates salinity stress in many species (38–40). It has traditionally been accepted that the dominating mechanism behind this was the Ca^{2+} restriction of Na^+ uptake via NSCC, the likely main pathway for Na^+ uptake into the cell (41, 42). But is NSCC blockage by Ca^{2+} the only mechanism involved? Using MIFE, we have shown that supplemental Ca^{2+} efficiently reduces or even prevents NaCl-induced K^+ efflux through GORK channels (11, 23, 43, 44). We have also found a similar role for other mitigating factors in reducing NaCl-induced K^+ efflux to provide salinity tolerance. These are factors for which previously a role was largely unknown or only conjectured. They include compatible solutes such as glycine betaine, proline, and certain amino acids (45–47), and polyamines (48, 49). We found that this reduction in NaCl-induced K^+ efflux resulted from one or more of the following mechanisms: (1) directly targeting the K^+ efflux pathways, (2) blocking NSCC so reducing the amount of Na^+ influx that would lead to (3) less salinity-induced membrane depolarization, a depolarization that would activate GORK channels, and (4) increasing the activity of the plasma membrane H^+-pump, which would repolarize the membrane and reduce GORK activation. The final result is the maintenance of a high K^+/Na^+ ratio in the cytosol, so improved salinity tolerance.

1.7. Quantifying the Kinetics of Na^+ Transport in Salinity Stressed Plants

Microelectrode studies of Na^+ kinetics were significantly handicapped by poor selectivity of the Na^+ sensor (50); the Na^+ LIX, in addition to Na^+, is also sensitive to K^+ and Ca^{2+}, with an almost ideal Nernst response for each (24). Thus, any apparent Na^+ flux recorded after the imposition of NaCl is confounded by the massive efflux of K^+ (see above) and Ca^{2+} (the result of Donnan exchange in the cell wall, (51, 52)) that occur in response to salinity treatment, and in turn, any measured net Na^+ uptake will be substantially underestimated. Furthermore, high levels of Na^+ in the bathing medium results in a large amount of noise and the resultant signal-to-noise ratio seriously impedes the detection of potentially small but critical differences in net flux in different plant lines. This has resulted in some limitations to the technique regarding its application to salinity tolerance; Na^+ transport is obviously an important factor.

To overcome the above limitations, a new method defined as a "recovery protocol" was developed. This method involves recording the net efflux of Na^+ in a Na^+-free medium, immediately after the removal of NaCl. Plants are treated with NaCl for a certain period of time, then NaCl is quickly removed and Na^+ flux responses recorded. With little Na^+ in the bathing medium and without the confounding effects of either K^+ or Ca^{2+} fluxes, any measured net Na^+ flux is likely to represent the actual Na^+ movement across the plasma membrane. We have recently used this method to show that a more salinity tolerant wheat variety has a far superior ability to export Na^+ out of the plant root than more susceptible varieties (53). Pharmacology revealed that this flux is active; a SOS1-like Na^+/H^+ antiporter appears to be responsible for the export (53). This MIFE protocol provides a relatively reliable and straightforward method for rapidly assessing a plant's ability to actively export Na^+ and it contributes towards knowledge regarding the transporters responsible; critical factors that contribute significantly to our knowledge of salinity tolerance in a vitally important crop species.

2. Materials

2.1. MIFE™ System Components (see Fig. 2)

1. MIFE main amplifier/controller and preamplifier.
2. Multi-manipulator providing three-axis positioning, with hydraulic option *or* PatchMan option (see Note 1).
3. PC with ISA-bus, running Windows 98, for system control and data acquisition.
4. CIO-DAS08 card for analogue to digital conversion.
5. MIFE CHART and MIFEFLUX software.
6. Handbook DOC files, explaining the hardware and software.

2.2. Auxiliary Equipment Required for Flux Studies

1. Microscope with long-working distance objectives (×100 or ×200). This can be an inverted microscope (Fig. 2) or a compound microscope lying on its back.
2. Anti-vibration table.

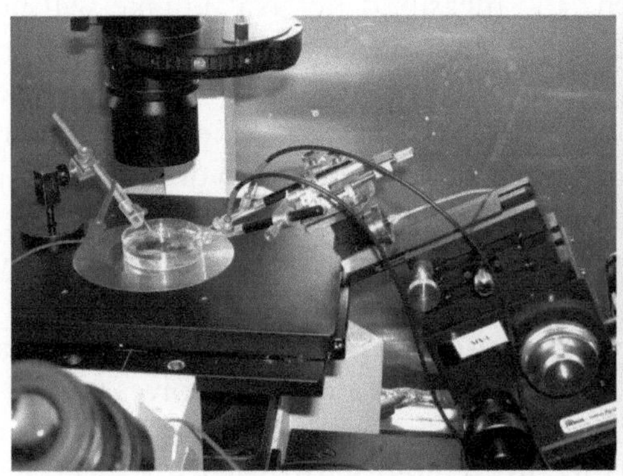

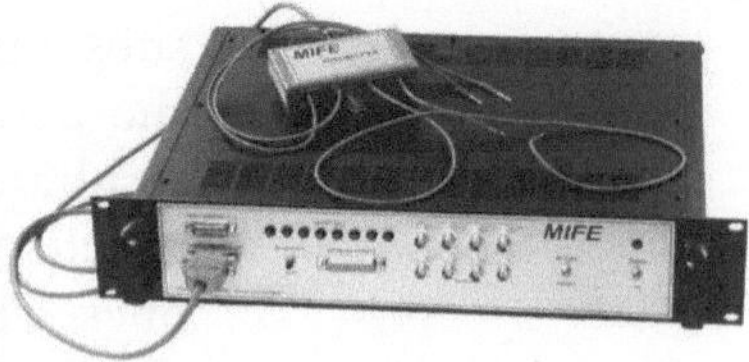

Fig. 2. MIFE hardware and electronics.

3. Faraday cage.
4. Vertical electrode puller (see Note 2).
5. Borosilicate glass capillaries, without filling fiber, OD = 1.5 mm (GC150-10; Harvard Apparatus Ltd, Kent, UK).
6. Specific ion-selective cocktails (LIX; see Note 3).
7. Silanizing agent (tributylchlorosilane; Fluka Chemicals 90796).
8. Electrode holders (e.g., E45W-F15PH; Warner Instruments, USA).
9. Non-metallic syringe needle for filling micropipettes (MF34G-5, 0.1 mm ID; WPI, Sarasota, FL, USA).
10. Electrode filling station consisting of two simple micromanipulators and a stereomicroscope.
11. Small oven, to 250°C, with gloves and metal electrode racks with metal covers.
12. Fume cabinet.
13. Measuring chambers to immobilize plant roots.

2.3. Solutions and Plant Material

1. Basic measuring solution (see Note 4).
2. Equipment to grow plants in aerated hydroponic solutions under sterile conditions.
3. Temperature-controlled room.

3. Methods

3.1. Pulling Electrode Blanks

1 Insert non-filamentous borosilicate glass capillaries into a vertical pipette puller.

2 Pull the blanks to <3 μm diameter tips.

3 Store pulled electrodes in a stainless steel- or aluminum-covered rack in a vertical position.

3.2. Baking and Silanizing the Blanks

1. Place electrode blanks uncovered in a rack with tips upright and base down.
2. Oven-dry the blanks at 220°C overnight. They remain in the 220°C oven for the next three steps.
3. Ten to fifteen minutes before silanization, place a steel cover over the electrode blanks.
4. Add two drops (approximately 50 μL) of tributylchlorosilane on the rack under the cover using a yellow-tip pipettor.
5. Ten minutes later, remove the lid and bake electrode blanks for a further 30 min.

6. Switch the oven off and let the blank electrodes cool down.
7. Take electrodes from the cold oven and keep them in a closed container.

3.3. Filling the Electrodes with an Appropriate Liquid Ion Exchanger (LIX)

1. Prepare a LIX-containing tube by dipping a broken-tip glass micro-capillary (tip diameter approximately 50 μm) into the LIX, taking up a column of cocktail of approximately 1 mm.
2. Mount the pulled microelectrode blank horizontally on a three-dimensional micromanipulator. Likewise also mount the LIX tube on another micromanipulator. Observe both under a dissecting microscope.
3. Break back the blank tip to achieve external tip diameter of 2–3 μm by placing it against a flat glass surface.
4. Position the blank co-axially with the LIX-containing tube at 100–200 μm distance.
5. Fill the blank with appropriate backfilling solutions (see Note 5) using a syringe with a non-metallic needle.
6. Front-fill the blank tip with LIX by putting it briefly into contact with the LIX-containing tube to achieve the column length between 100 and 150 μm.
7. Store electrodes having their tips immersed in the proposed measuring solution until use (up to 8–10 h).

3.4. Calibrating Electrodes

1. Mount electrodes in the holder.
2. Fabricate a reference electrode by inserting a chlorided silver wire (galvanized at 1.5 V in 0.25 N HCl for 15 min) into a broken glass micro-capillary (*ca* 50 μm tip diameter) filled with 1 M KCl in 2% agar and seal it with parafilm. Alternatively, use a commercial reference electrode.
3. Prepare an appropriate set of at least three standards covering the expected range of the ion in question, using about equal factor change between their concentrations.
4. Calibrate electrodes using the MIFE CHART calibration routine below (see Note 6).

3.5. Calibration Routine

1. Put electrodes into your first standard solution, close the Faraday cage, and start data acquisition.
2. Press the function key F7. In the window that opens, correct the temperature, type the names of all the ions to be used, and press <ENTER>.
3. When the data for the ion of the first solution is stable and in range, press F7 again and enter the concentration for that ion.

After you press <ENTER> the data recorded will be used in the calibration calculation.

4. After about 20 s, a third press of F7 terminates that set, giving you the option of keeping it or rejecting it. Details are given under F7 in the Handbook file THECHART.DOC.
5. Repeat the process of three presses of F7 for the other concentrations and the other ions.
6. When all needed calibration data have been recorded, type ALT-H to stop data acquisition.
7. From the <E>lectrometer menu, choose the <A>verage option and the <C>alibration average. CHART will do the least squares fit to the data for each ion, create the AVC file, and display in a window the parameters of the electrodes. The slope and intercept are also saved in the DAY file in the main MIFE directory.

3.6. Conducting Measurements

1. Mount roots in the measuring chamber filled with an appropriate solution.
2. Under the microscope, position microelectrode tips in one plane, separated laterally by 1–2 μm, 20–30 μm away from the root surface.
3. From the DOS prompt or from Windows Explorer, select your working directory.
4. Start the CHART program. This is often done by a batch file (see Note 7), which may also select the working directory.
5. Set up parameters (including buffer pK and concentration) for the run using the function key F9. These can be confirmed or modified at the end of the run.
6. Set or check the electrometer offsets for each channel, using the <S>et Offsets option in the <E>lectrometer menu.
7. When electrodes are in solution, you may measure their resistance with the <R> option of the <E>lectrometer menu. Resistance values are recorded in the DAY file.
8. Start data acquisition using ALT-S or the <S> option in the <M> ain Menu. A small window opens to let you set a filename. It is easiest to accept the default that encodes date and time, by typing S. Then a window opens to let you set the starting time, duration of measurements, and other values. Accept all the defaults by typing G.
9. A one-line window opens in which you should type your name and other identifying information about the experiment you have planned. Press <ENTER> to close it and begin the data acquisition. This is recorded in the DAY file.

10. Navigate the Chart Display screen, using the "buttons" at the bottom and right of the screen, to view any or all of the data as desired. Many more instructions and options are given in the MIFE Handbook.
11. Stop data acquisition by pressing ALT-H or the <H> option in the <M>ain Menu.

3.7. MIFE Protocols for Salinity Studies: Screening Plants for K^+ Retention Ability

1. Sterilize seeds and grow in a sterilized aerated hydroponic solution until they are the required size (see Note 8).
2. Prepare and calibrate K^+ electrode using 60031 K^+ LIX.
3. One hour prior to measurement, immobilize a root in a measuring chamber.
4. Add NaCl stock to the measuring chamber to achieve the final Na^+ concentration of 100 mM.
5. One hour after NaCl exposure, transfer the measuring chamber onto microscope stage.
6. Position K^+ selective microelectrode 20 μm away from the root surface (see Note 9).
7. Record net K^+ flux for 5–10 min as described in Subheading 3.6.
8. Analyze data as described below in Subheading 3.10.
9. You can set up a "production line", repeating steps 3–7, to measure 1 root every 10 min.

3.8. MIFE "Recovery" Protocol for Quantifying plants' Na^+ Efflux Ability

1. Sterilize seeds and grow in a sterilized aerated hydroponic solution until they are the required size.
2. Expose roots to NaCl for a certain time period (see Note 10).
3. One hour prior to measurement, transfer a root to a measuring chamber containing the bathing medium with a concentration of NaCl equal to that used during the exposure.
4. Prior to measurement, pour off this bathing solution and rapidly rinse the root three times in 10 mM $CaCl_2$ to remove surface NaCl.
5. Transfer the root to a clean chamber containing the bathing medium, minus NaCl (see Note 11).
6. Make ion flux measurements of Na^+ (and other ions of interest) for the required time, as described above.

3.9. Data Analysis: Produce Files of Average Voltage V and ΔV

1. If you are not analyzing the run that you have just made, load and display the required data file using the CHART software (use <P>lot option under <F>iling menu).
2. To produce an AVM file, use the <M>anipulator cycle average routine in the <A>veraging option in the <E>lectrometer menu.

3. Specify a "Valid Time" at the end of each stage of manipulator movement for which the actual "Stage Time" is provided. Pressing <ENTER> moves the highlight to the next stage. Pressing UPARROW moves up to the previous one to allow you to correct an error.
4. Press <ENTER> to accept the order of the curve ("Kind of <F>it") to fit the data (typing F to cycle through the orders). Linear is probably what you need. As the averaging is done, the chirp and the Line Count indicate progress. An AVM file (for "AVerage Manipulator") is created with the same name as the DAT and LOG files of the data. The AVM file, with its DAT file and selected AVC file(s), is read by the MIFEFLUX program to calculate the fluxes. You can also import it into a spreadsheet or view it in a text editor.
5. Quit the CHART program by pressing ALT-Q.

3.10. Data Analysis: Produce the Flux File

1. Run the MIFEFLUX program under DOS in your current directory that contains the AVC and AVM files for the flux calculations. The batch file MFLUX.BAT, which is in the C:\MIFE directory, will do this just by typing MFLUX in your current directory.
2. The program first asks you for the AVC file(s) you want, then for the AVM file of your averaged flux data (see Note 12), and finally for the kind of analysis (planar, cylindrical, or spherical). The most often used defaults are indicated by [] and are obtained just by pressing <ENTER>.
3. After calculation is done, you can calculate more fluxes using the same calibration files.
4. Once calculations are completed, the resulting FLX file (ASCII text format, with same name as the DAT, LOG, AVM files) can be opened by any spreadsheet and evaluated in a conventional way.

4. Notes

1. Each of these configurations has been used in our lab. *Hydraulic option* uses Narishige components: SM-17 3-electrode holder, mounted on hydraulic manipulator MHW-4 (one-axis) or MHW-3 (three-axis), mounted on MX-2 (three-axis) mechanical micromanipulator. A custom-made MIFE stepper motor control unit drives the hydraulic manipulator to measure fluxes. Manipulator details are available at http://www.narishige.co.jp/english/. *PatchMan option* uses the Eppendorf PatchMan NP2. This gives three-axis positioning for a single electrode

with an inverted microscope. An alternative is: the NP2 supports a vertical sample chamber in place of the stage of a microscope lying on its back; the Narishige MMT-5 micromanipulator (holding three electrodes in the vertical plane) is mounted on the microscope body. PatchMan details are available at http://www.eppendorf.com.

2. Any standard puller may be used. Because the electrode tip diameter is controlled at a later stage, the pulling process is much less demanding than for the patch-clamp and can be performed in a single step.
3. The list of commercial ionophore cocktails (LIX) for the ion-selective electrodes can be found at http://www.sigmaaldrich.com/. The cocktails listed in Table 1 have been routinely used in our laboratory.
4. The specific ionic composition of measuring solutions depends on the specific purpose of the experiment. A few general principles, however, have to be observed: (1) solution ionic composition should be as simple as possible to avoid potential confounding effects of interfering ions or LIX poisons; (2) concentrations of ions whose fluxes are to be measured should be kept as low as practically (physiologically) possible to maximize signal-to-noise ratio of the measured signal; (3) for H^+ fluxes, avoid alkaline pH, keep buffer concentrations as low as possible and choose a pK at least 1 unit above the pH.
5. The backfilling solutions routinely used in our laboratory are shown in Table 2. Three criteria for choosing these are given in Chapter 6.

Table 1
Commercially available ion-selective LIX routinely used in our laboratory (all from Fluka, Buchs, Switzerland)

Ion	Catalog number
K^+	60,031
Na^+	71,138
Ca^{2+}	21,048
Mg^{2+}	63,048
H^+	95,297
NH_4^+	09,882
Cl^-	24,902

Table 2
Backfilling solutions

Ion	Backfilling solution (mM)
H^+	15 NaCl + 40 KH_2PO_4
K^+	200 KCl
Na^+	500 NaCl
Ca^{2+}	500 $CaCl_2$
Mg^{2+}	500 $MgCl_2$
Cl^-	500 NaCl
NH_4^+	500 NH_4Cl

6. It is not prudent to use calibrations showing less than 50 mV per decade slope for monovalent ions and 25 mV per decade for divalent cations, or having correlation coefficients less than 0.999 (for the three-point calibration). If you are unhappy with the calibrations, you can repeat the calibration routine for as many ions as you wish. No need to quit CHART. Eventually, in running MIFEFLUX, you will have the option of selecting just those calibrations you want from the AVC file(s) you have made, either before or after the experimental run.
7. Batch files allow convenient selection of working directory and running CHART or MIFEFLUX there. Batch files are placed in the main C:\MIFE directory, with a shortcut on the Windows quick task bar. The file MFLUX.BAT (to run MIFEFLUX) and SAMPLE.BAT (to change to the SAMPLE directory and to run CHART there) are supplied with the software. To make a batch file for work in another directory, SAMPLE.BAT may be used as a template; use a text editor.
8. It is important that seedlings are grown under sterile conditions to prevent infection that would otherwise confound the result. We found that autoclaving all solutions and growth containers, alongside standard sterilization of seeds, sufficed.
9. Due to methodological reasons, it is recommended that NaCl-induced K^+ fluxes are measured in the mature root zone. For cereal plants such as wheat and barley, measuring at about 10 mm from root tip seems to be the most appropriate.
10. We used 150 mM added to the hydroponic solution for pretreatment with NaCl (53). This "salt shock" should ensure that a large amount of Na^+ is taken into the plant (55). Pretreatment for 24 h was sufficient to induce SOS1-like activity in wheat.

11. It is important to remove K^+ from the bathing medium if there is the danger of K^+ fluxes after Na^+ removal. Due to the poor selectivity of the Na^+-LIX, fluxes of K^+ would confound recordings of Na^+ fluxes (see above). In *Arabidopsis*, we found an influx of K^+ into the roots after the removal of NaCl, but a similar influx did not occur in wheat (53).
12. The AVP averaged file is used with a paired electrode configuration to follow transients with 0.2 s time resolution (see the MIFE HANDBOOK). It may be ignored for salinity studies.

Acknowledgements

This work was supported by the ARC and GRDC grants to Sergey Shabala.

References

1. Blumwald E, Aharon GS, Lam BCH (1998) Early signal transduction pathways in plant-pathogen interactions. Trends Plant Sci 3: 342–346
2. Sanders D, Brownlee C, Harper JF (1999) Communicating with calcium. Plant Cell 11:691–706
3. Zimmermann S, Ehrhardt T, Plesch G et al (1999) Ion channels in plant signaling. Cell Mol Life Sci 55:183–203
4. Spalding EP (2000) Ion channels and the transduction of light signals. Plant Cell Environ 23:665–674
5. Knight H, Knight MR (2001) Abiotic stress signalling pathways: specificity and cross-talk. Trends Plant Sci 6:262–267
6. Zhu JK (2003) Regulation of ion homeostasis under salt stress. Curr Opin Plant Biol 6: 441–445
7. Palmgren MG (1991) Regulation of plasma membrane H^+-ATPase activity. Physiol Plant 83:314–323
8. Shabala SN, Lew RR (2002) Turgor regulation in osmotically stressed Arabidopsis epidermal root cells. Direct support for the role of inorganic ion uptake as revealed by concurrent flux and cell turgor measurements. Plant Physiol 129:290–299
9. Shabala S, Cuin TA (2008) Potassium transport and plant salt tolerance. Physiol Plant 133:651–669
10. Shabala S (2006) Non-invasive microelectrode ion flux measurements in plant stress physiology. In: Volkov AG (ed) Plant electrophysiology—theory and methods. Springer-Verlag, Berlin Heidelberg, pp 35–71
11. Shabala S, Shabala L, Van Volkenburgh E (2003) Effect of calcium on root development and root ion fluxes in salinised barley seedlings. Funct Plant Biol 30:507–514
12. Tester M (1997) Techniques for studying ion channels: an introduction. J Exp Bot 48:353–359
13. Knight MR, Campbell AK, Smith SM et al (1991) Transgenic plant aequorin reports the effects of touch and cold-shock elicitors on cytoplasmic calcium. Nature 352:524–526
14. Roos W (2000) Ion mapping in plant cells—methods and applications in signal transduction research. Planta 210:347–370
15. Miller AJ (1996) Ion-selective microelectrodes for measurement of intracellular ion concentrations. Methods Cell Biol 49A: 275–291
16. Ratcliffe RG (1997) In vivo NMR studies of the metabolic response of plant tissues to anoxia. Ann Bot 79:39–48
17. Tomos AD, Hinde P, Richardson P et al (1994) Microsampling and measurements of solutes in single cells. In: Harris N, Oparka KJ (eds) Plant cell biology. A practical approach. IRL Press, Oxford, pp 297–314
18. Tomos AD, Leigh RA (1999) The pressure probe: a versatile tool in plant cell physiology. Annu Rev Plant Physiol Plant Mol Biol 50: 447–472
19. Shabala S, Newman I, Whittington J et al (1998) Protoplast ion fluxes: their measurement

and variation with time, position and osmoticum. Planta 204:146–152

20. Newman IA (2001) Ion transport in roots: measurement of fluxes using ion-selective microelectrodes to characterize transporter function. Plant Cell Environ 24:1–14
21. Shabala L et al (2006) Non-invasive microelectrode ion flux measurements to study adaptive responses of microorganisms to the environment. FEMS Microbiol Rev 30:472–486
22. Greenway H, Munns R (1980) Mechanisms of salt tolerance in nonhalophytes. Annu Rev Plant Physiol 31:149–190
23. Shabala S (2000) Ionic and osmotic components of salt stress specifically modulate net ion fluxes from bean leaf mesophyll. Plant Cell Environ 23:825–837
24. Chen Z, Newman I, Zhou M et al (2005) Screening plants for salt tolerance by measuring K^+ flux: a case study for barley. Plant Cell Environ 28:1230–1246
25. Ayala F, Oleary JW, Schumaker KS (1996) Increased vacuolar and plasma membrane H^+-ATPase activities in *Salicornia bigelovii* Torr in response to NaCl. J Exp Bot 47:25–32
26. Vera-Estrella R, Barkla BJ, Bohnert HJ et al (1999) Salt stress in *Mesembryanthemum crystallinum* L cell suspensions activates adaptive mechanisms similar to those observed in the whole plant. Planta 207:426–435
27. Serrano R, Mulet JM, Rios G et al (1999) A glimpse of the mechanisms of ion homeostasis during salt stress. J Exp Bot 50:1023–1036
28. Shabala L, Cuin TA, Newman IA et al (2005) Salinity-induced ion flux patterns from the excised roots of *Arabidopsis sos* mutants. Planta 222:1041–1050
29. Chen Z, Pottosin II, Cuin TA et al (2007) Root plasma membrane transporters controlling K^+/Na^+ homeostasis in salt-stressed barley. Plant Physiol 145:1714–1725
30. Chen Z, Zhou MX, Newman IA et al (2007) Potassium and sodium relations in salinised barley tissues as a basis of differential salt tolerance. Funct Plant Biol 34:150–162
31. Chen Z, Shabala S, Mendham N et al (2008) Combining ability of salinity tolerance on the basis of NaCl-induced K^+ flux from roots of barley. Crop Sci 48:1382–1388
32. Cuin TA, Zhou MZ, Parsons D et al (2012) Genetic behaviour of physiological traits conferring cytosolic K/Na homeostasis in wheat. Plant Biol 14(3):438–446
33. Demidchik V, Shabala SN, Coutts KB et al (2003) Free oxygen radicals regulate plasma membrane Ca^{2+} and K^+- permeable channels in plant root cells. J Cell Sci 116:81–88
34. Shabala S, Cuin TA, Prismall L et al (2007) Expression of animal CED-9 anti-apoptotic gene in tobacco modifies plasma membrane ion fluxes in response to salinity and oxidative stress. Planta 227:189–197
35. Cuin TA, Shabala S (2007) Compatible solutes reduce ROS-induced potassium efflux in Arabidopsis roots. Plant Cell Environ 30: 875–885
36. Shabala S (2009) Salinity and programmed cell death: unravelling mechanisms for ion specific signalling. J Exp Bot 60:709–711
37. Demidchik V, Cuin TA, Svistunenko D et al (2010) *Arabidopsis* root K^+-efflux conductance activated by hydroxyl radicals: single-channel properties, genetic basis and involvement in stress-induced cell death. J Cell Sci 123:1468–1479
38. Cramer GR, Lynch J, Lauchli A et al (1987) Influx of Na^+, K^+ and Ca^{2+} into roots of salt-stressed cotton seedlings. Effects of supplemental Ca^{2+}. Plant Physiol 83:510–516
39. Reid RJ, Smith FA (2000) The limits of sodium/calcium interactions in plant growth. Aust J Plant Physiol 27:709–715
40. Rengel Z (1992) Role of calcium in aluminium toxicity. New Phytol 121:499–513
41. Tyerman SD, Skerrett M, Garrill A et al (1997) Pathways for the permeation of Na^+ and Cl^- into protoplasts derived from the cortex of wheat roots. J Exp Bot 48:459–480
42. Demidchik V, Tester M (2002) Sodium fluxes through nonselective cation channels in the plasma membrane of protoplasts from *Arabidopsis* roots. Plant Physiol 128:379–387
43. Shabala S, Shabala L, Van Volkenburgh E et al (2005) Effect of divalent cations on ion fluxes and leaf photochemistry in salinized barley leaves. J Exp Bot 56:1369–1378
44. Shabala S, Demidchik V, Shabala L et al (2006) Extracellular Ca^{2+} ameliorates NaCl-induced K^+ loss from *Arabidopsis* root and leaf cells by controlling plasma membrane K^+-permeable channels. Plant Physiol 141: 1653–1665
45. Cuin TA, Shabala S (2005) Exogenously supplied compatible solutes rapidly ameliorate NaCl-induced potassium efflux from barley roots. Plant Cell Physiol 46:1924–1933
46. Cuin TA, Shabala S (2007) Amino acids regulate salinity-induced potassium efflux in barley root epidermis. Planta 225:753–761
47. Chen Z, Cuin TA, Zhou M et al (2007) Compatible solute accumulation and stress-mitigating effects in barley genotypes contrasting in their salt tolerance. J Exp Bot 58: 4245–4255

48. Shabala S, Cuin TA, Pottosin I (2007) Polyamines prevent NaCl-induced K^+ efflux from pea mesophyll by blocking non-selective cation channels. FEBS Lett 581:1993–1999
49. Pandolfi C, Pottosin I, Cuin T et al (2010) Specificity of polyamine effects on NaCl-induced ion flux kinetics and salt stress amelioration in plants. Plant Cell Physiol 51:422–434
50. Carden DE, Walker DJ, Flowers TJ et al (2003) Single-cell measurements of the contributions of cytosolic Na^+ and K^+ to salt tolerance. Plant Physiol 131:676–683
51. Ryan PR, Newman IA, Arif I (1992) Rapid calcium exchange for protons and potassium in cell walls of *Chara*. Plant Cell Environ 15: 675–683
52. Shabala S, Newman I (2000) Salinity effects on the activity of plasma membrane H^+ and Ca^{2+} transporters in bean leaf mesophyll: masking role of the cell wall. Ann Bot 85:681–686
53. Cuin TA, Bose J, Stefano G et al (2011) Assessing the role of root plasma membrane and tonoplast Na^+/H^+ exchangers in salinity tolerance in wheat: *in planta* quantification methods. Plant Cell Environ 34:947–961
54. Wherret T (2006) Aluminium toxicity, tolerance and amelioration in wheat. PhD Thesis. University of Tasmania.
55. James RA, Munns R, Von Caemmerer S, Trejo C, Miller C, Condon AG (2006) Photosynthetic capacity is related to the cellular and subcellular partitioning of Na^+, K^+ and Cl^- in salt-affected barley and durum wheat. Plant Cell Environ 29: 2185–2197

Part II

Imaging Techniques

Chapter 8

Quantitative Cryo-Analytical Scanning Electron Microscopy (CEDX): An Important Technique Useful for Cell-Specific Localization of Salt*

Margaret McCully and Martin Canny

Abstract

Advances in the techniques required for the X-ray microanalysis of cryo-fixed, naturally hydrated plant tissues in the cryo-scanning electron microscope have reached the stage that accurate, cell-specific localization and quantification of the nutrient and toxic elements can be achieved. Advances are described in the successive processes of cryo-fixation, cryo-planing to produce flat surfaces, monitored minimal etching to reveal cell outlines, coating with aluminum, spectrum collection, and quantification by comparison with comparable frozen standard solutions of the elements.

Key words: Cell-specific quantitative elemental analysis, Cryo-scanning electron microscopy, Energy dispersive X-ray microanalysis of frozen specimens (CEDX), Preparation of inherently hydrated frozen specimens for CEDX, Na quantitation

1. Introduction

The elusive goal of botanical microscopy is to visualize plant structures, from organs to cells, in their real functioning mode in the intact plant. Improvement in methods of specimen preparation (use of fresh tissues, chemical fixations, and embedments) that better preserve cytoplasm, organelles, and macromolecules, and which reduce tissue distortion for use with either simple or sophisticated instrumentation, has greatly enhanced the ability of microscopists to obtain physical and molecular insight into how plants function.

*A version of the information in this chapter has been available on the website PrometheusWiki

Sergey Shabala and Tracey Ann Cuin (eds.), *Plant Salt Tolerance: Methods and Protocols*, Methods in Molecular Biology, vol. 913,
DOI 10.1007/978-1-61779-986-0_8, © Springer Science+Business Media, LLC 2012

Important as these advances in specimen preparation are, they do not allow plant biologists to address the basic fact that functioning plants are comprised mainly of water and soluble nutrients. Most of these vital components are lost during all conventional preparative treatment, leaving the roughly 10–15% of plant component composed of macromolecules that are immobilized by chemical fixation, and not extracted during preparation by dehydration and/or solvent and resin monomers (1). To date, the only way to immobilize water and soluble components in situ, and to preserve the location of inherent gas and water-filled spaces and tissue dimensions as they were in the living intact plant, is by cryo-fixation. The simplest (and most informative way) to observe these frozen specimens is in a scanning electron microscope equipped with a cryo-stage (CSEM) that maintains the inherently hydrated specimen at liquid nitrogen (LN_2) temperature during observation. Scanning electron microscopes reveal either surface features of intact whole mounts or anatomical details of internal surfaces produced by fracturing or cryo-planing the frozen specimens.

Because organs, cells, and tissues can be viewed undistorted, unextracted, in their inherent state of hydration and in situ, the CSEM has produced a wide spectrum of newly revealed plant properties (1). Importantly, it has also opened the way for the previously unobtainable cell-specific localization and quantitation of endogenous elements in situ at the concentrations present at the moment of cryo-fixation. This analysis (CEDX) is made possible by equipping a CSEM with an energy dispersive X-ray analyser.

It was the need to locate and quantify accumulations of sodium in salt-stressed plants that inspired Läuchli (2) to combine for the first time the techniques of cryo-SEM and X-ray microanalysis to investigate sodium and chloride distribution in maize roots. Läuchli and coworkers and Schwander (3) were already familiar with the use of X-ray microanalysis technology whereby X-rays of element-specific energies are emitted when samples are bombarded in the electron beam in a standard SEM. The X-rays are collected and characterized in a spectrometer, and the energy spectra used to identify elements present in the sample. All the early work using this technology to locate sodium and chloride was, by necessity, done with dried samples, and results were unreliable due to solute displacement and/or loss. The newly available cryo-attachments for the SEM made possible the analysis of hydrated frozen samples and eliminated this problem. Läuchli has incorporated the many technological advances in CSEM instruments, specimen preparation, and analysers that have developed since the pioneer study of Yeo et al. (2) and has recently published a similar study of Na location in wheat roots (see Fig. 1) using the most reliable technologies currently available (4). Another effective user of the evolving technology for accurate Na and Cl localization and quantitation at the cell level has been Storey and

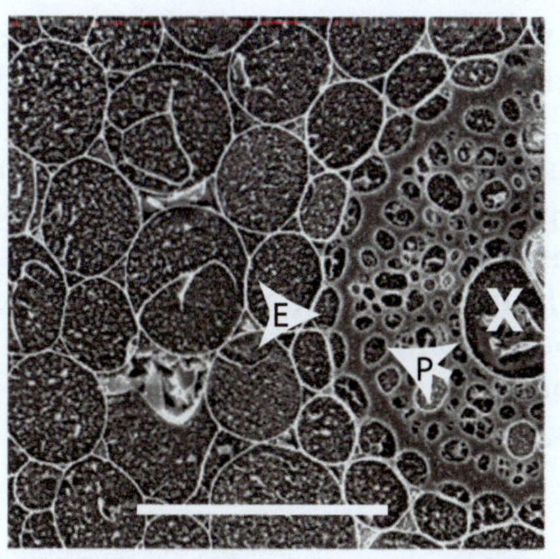

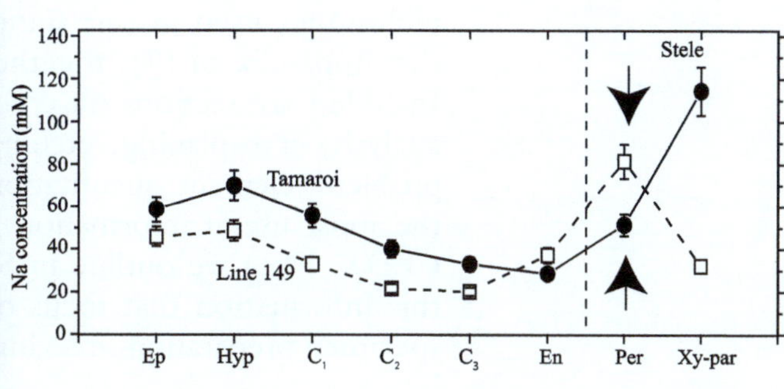

Fig. 1. A CSEM image of a typical transverse, cryo-planed face of a seminal root of a durum wheat seedling prepared as described in Subheading 3. Seedlings of two genotypes, Tamaroi (high rate of Na transport from root to shoot) and Line149 (low rate of transport) were exposed to 50 mM NaCl for 10 days. The cell-specific concentrations of Na were determined by CEDX for epidermis (Ep), hypodermis (Hyp), cortex (C), endodermis (En), pericycle (Per), and xylem parenchyma (XP). *Arrows* draw attention to the surprising concentrations in pericycle cells. Modified from (4). Bar = 100 μm.

coworkers (5). They have analyzed hydrated, cryo-fixed, cryo-planed, and appropriately coated specimens of salinized grapevine and used frozen standards, essential for accurate quantification.

Many researchers have used CSEM in different forms to localize Na and Cl since Läuchli and coworkers in 1977 (2). However, they have generally not achieved the accurate quantification now possible with the methods described in this chapter.

Combining CSEM with quantitative energy dispersive X-ray microanalysis of the cryo-fixed specimens (CEDX) allows cell-specific localization of elements heavier than atomic number four (depending on concentration and the type of detector available). With suitable preparations and standards, the elements can be quantified, thus greatly extending the possibility of linking specific structure with function. The most fundamental general observation that has been revealed by this technique is marked heterogeneity of element type (both essential and toxic, including Na and Cl) and their concentrations, not only in specific tissues, but even in adjacent cells (apparently structurally identical) within the same tissue. These observations, together with those obtained with other techniques (6), raise basic questions regarding mechanisms directing this heterogeneous patterning and its significance in many important plant functions (7). These include tolerance of salinity and the role of silicon in plant function, especially its reported role in promoting salinity tolerance (8, 9).

Advances in the understanding of plant function achieved so far and possibilities for further fruitful use of CSEM and CEDX are reviewed in (1, 7).

Detailed information on background and rationales for the techniques used in our standard method (below) are given in the Appendix of (7), together with many additional references. Included are sections on cryo-fixation, specimen preparation for analysis, cryo-planing, etching, specimen coating, the sampling problem, element quantitation using frozen standards, achieving the most useful information from specimens, and limitations of CEDX. Here we outline in Subheading 4 some highlights from this information that focus on the four most critical aspects of specimen preparation, freezing, cryo-etching, and coating.

2. Materials

2.1. Instrumentation

1. A scanning electron microscope fitted with an energy dispersive X-ray microanalyser, a cold stage and cryo-transfer unit. The cryo-preparation chamber of the microscope must be capable of achieving a vacuum of $<10^{-6}$ Torr for aluminum coating of specimens. The commercial evaporator in the cryo-preparation chamber must be fitted with a spiral of resistance wire to hold a rolled piece of high purity Al foil of predetermined mass.
2. Cryo-microtome.

2.2. Accessory Materials

1. Liquid nitrogen (LN_2), suitable containers for it, cryo-storage for frozen specimens, and if frozen specimens are to be shipped, an approved "dry shipper."
2. Cryo-pliers (Fig. 2).
3. Suitable specimen stubs and holders (Fig. 3) and (10), compatible with the cryo-microtome, transfer unit, and cold stage.
4. Low-temperature cryo-glue for affixing specimens to stubs.
5. Glass and diamond knives.

2.3. Frozen Standards

1. Analytical grade salts of Na, Cl, K, P, S, Ca, Mg.
2. 5% colloidal graphite.
3. Ethane, cooled with LN_2.
4. Make up a series of aqueous solutions at 0, 12.5, 25, 50, 100, 300, and 500 mM, each solution containing equimolar concentrations of Na, Cl, K, Ca, Mg, P, and S, and 5% colloidal graphite.

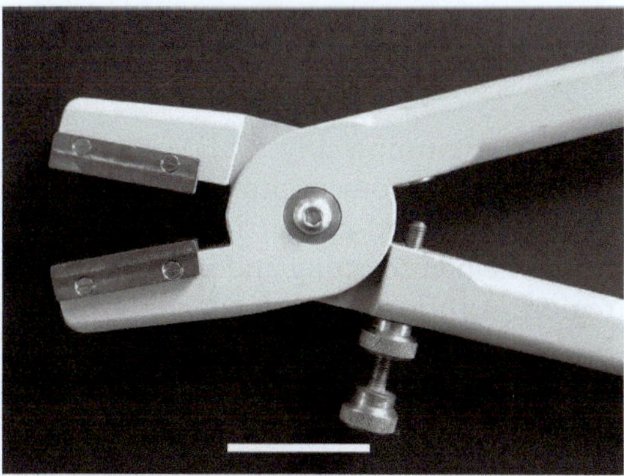

Fig. 2. Purpose-made cryo-pliers with thick, polished copper jaws. The spacing between the jaws is adjustable with the screw system to accommodate specimens of different thickness. Bar = 4 c m.

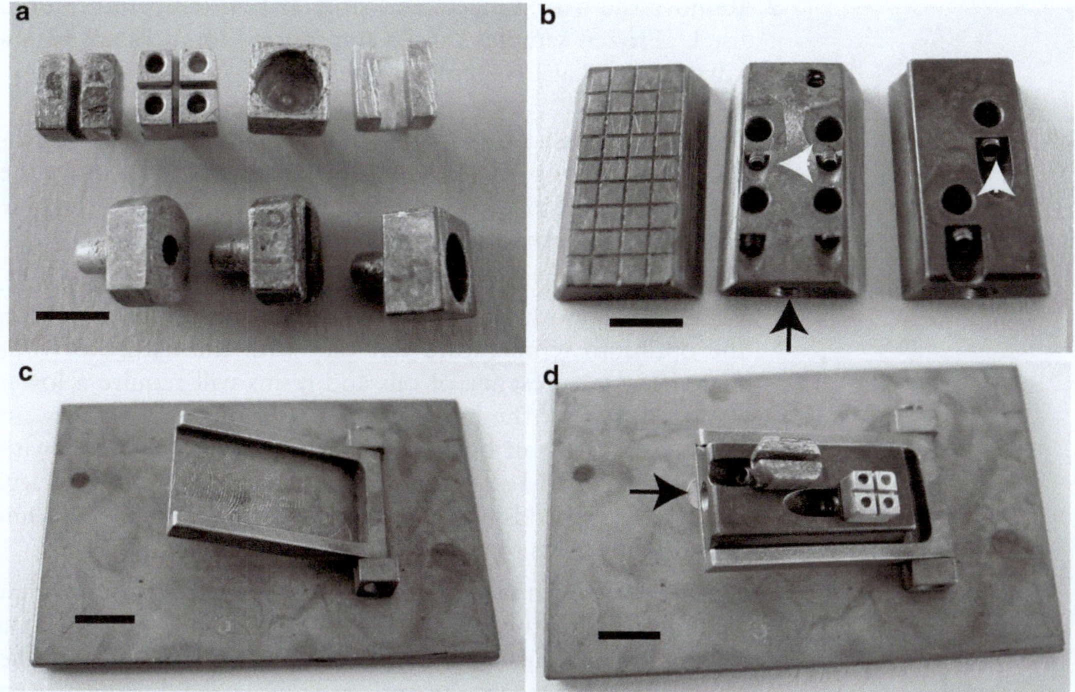

Fig. 3. Purpose-made hardware for handling frozen specimens. (**a**) Aluminum stubs that can accommodate a variety of specimen sizes for either transverse or longitudinal planing. Stubs fit into the cryo-microtome chuck, and into the holes in the adapters in (**b**). Bar = 6 mm. (**b**) Adapters that accommodate cryo-specimens and slot into the microscope cryo-stage. The one on the *left* is for whole mounts, the others have holes for two or four stubs, each secured by a grub screw (*arrowheads*). The small hole at the *top* end of the *middle* adapter can hold a standard. Each adapter accommodates the rod of the cryo-transfer device in a threaded hole in one end (*arrows*). Stubs with the frozen, planed specimens are inserted into the adapters under LN_2. Bar = 6 mm. (**c**, **d**) The mount (**c**) for the adapters (**b**) is slanted upwards so that the cryo-transfer rod clears the edge of the LN_2 bath, for threading into the adapter (**b**, **d**). The specimens can then be transferred to the CSEM. Bars (**c**, **d**) are 6 and 10 mm, respectively.

3. Methods

3.1. Preliminary Study

Before beginning any CSEM or CEDX analysis of unfamiliar material, carry out preliminary optical microscopy of the region of interest. Hand-cut sections and even thin sections of embedded tissues help in selecting and interpreting SEM images of the desired region. A thorough understanding of the tissue architecture saves both time and money, particularly when doing analysis.

3.2. Selection of Material

Ensure that the samples are representative of the conditions for which you want the observations and/or analyses (e.g., from well-watered, or desiccating plants in the field, or seedlings in a Petri dish).

3.3. Cryo-Fixation

1. Freeze the desired regions with LN_2, or cryo-pliers (Fig. 2) cooled with LN_2, disturbing the functioning plant as little as possible.
2. Remove the frozen pieces and store at LN_2 temperature until used. (Frozen samples can be transported long distances in a "dry shipper" at LN_2 temperature) (10).

3.4. Cryo-Planing

1. Trim frozen tissue to ~2 mm lengths under LN_2. Place cryo-glue on an appropriate Al stub (Fig. 3) and quickly remove the sample from the LN_2 using cooled forceps, orient it in the glue and re-freeze immediately.
2. Transfer the stub and specimen in LN_2 to the cryo-microtome. Plane to a smooth face in the desired plane at ~ −90°C. Note, the optimum temperature can vary with the nature of your material, e.g., presence of oils and resins will require a lower temperature. For the initial "rough" planing, the sections cut (and discarded) can be relatively thick (ca. 0.5–0.2 μm) with cutting speeds of 1–1.5 mm/s. For final polishing of the face, much slower speeds are used (e.g., 0.3 mm/s) and only thin (60–80 nm) slices removed.
3. During the slow final polishing, gently remove any debris that accumulates on the specimen face or the knife edge with a fine sable-hair brush. If debris is accumulating in spaces in the tissue at the block face, an anti-static gun used during cutting may help.
4. Use glass knives to trim to the desired plane. For many specimens, glass-planed surfaces are satisfactory. For hard specimens, use a diamond knife for the final polishing.
5. Planed specimens can be stored in LN_2 or used immediately.

3.5. Transfer to the Microscope

1. Remove the stub with its planed sample from the microtome and quickly affix it in place on a stage adapter (Fig. 3) under LN_2. Slide the adapter into the sloped mount (Fig. 3) under LN_2.
2. Attach the rod of the cryo-transfer apparatus to the adapter and quickly move it to the CSEM preparation chamber and then to the cold stage. During transfers, avoid exposure of specimens to air as much as possible.

3.6. Etching

1. While watching the specimen at 1 kV, raise the temperature of the cold stage to −90°C. Leave at this temperature until any accumulated frost has sublimed, but cell outlines are *barely* detectable. During this process the decontaminator should be held at least 25°C colder than the specimen.
2. Immediately turn off the stage heater and cool the stage rapidly to −160°C by increasing the flow of LN_2.
3. Return the specimen to the preparation chamber for coating.

3.7. Coating

1. Increase the vacuum in the preparation chamber to $<10^{-6}$ Torr.
2. Coat the specimen with Al by evaporating the foil previously placed in the resistance coil, while moving the specimen about in the Al shower to get a shiny mirror coating.
3. Return the coated specimen to the microscope cold stage and set up the microscope for analysis.

3.8. Microscope Settings Recommended for CEDX Analysis

1. Choose an appropriate accelerating voltage. It needs to be high enough to excite X-ray lines for the elements you are interested in observing. We use 15 kV for Na and Cl.
2. Use the working distance as dictated by the geometry of your EDX detector. This information should be provided by the manufacturer. For our detector, for example, this is 15.3 mm.
3. It is important to use an instrument with good beam stability. Maintain a fixed beam current; if possible, measure this using a Faraday cup and a nano-ammeter. Check regularly during your session to ensure the chosen value is maintained. Any change in this value will translate into changes in measured X-ray intensities that are independent of any element concentration you are trying to measure. We use a probe current of 300 pA.

3.9. Collecting Spectra

1. Make sure the sample is not charging when collecting X-ray data, otherwise the measured intensities can be severely in error. You may need to recoat the sample.
2. Count for a fixed live-time (e.g., 100 s).
3. Collect spectra using raster dimensions that cover the tissue compartment that you want to analyze (e.g., vacuole, cytoplasm, wall, nucleus, plastid, vessel lumen) without overlap of

adjoining compartments. This will eliminate error due to the heterogeneity produced by ice crystals and sequestered solutes by giving a mean reading for the whole compartment (11).

3.10. Quantitation

1. The use of frozen standards is critical for accurate quantification. Quickly set individual drops of thoroughly mixed solution (use a vortex mixer) over small holes in stubs (the drops should be convex) and immediately freeze in liquid ethane.
2. Cryo-plane, etch, and coat the frozen drops of standards identically to the specimens. Unplaned drops can be stored indefinitely at LN_2 temperature and prepared for recalibration when required.
3. Collect ten spectra per drop of frozen standard of each concentration, using the same instrument parameters used for the tissue samples. Peak intensities (I) are determined by either using the net integrated counts for each of the peaks of interest. Alternatively, it may also be possible to derive a parameter known as the *k*-ratio for each of the elements of interest depending on your software. The *k*-ratio, defined as $k = I_{\text{unknown}} / I_{\text{standard}}$ is simply a scaled version of the above-mentioned net integrated counts. The advantage of using *k*-ratios over net integrated counts is that peak fitting, as well as background removal routines, are sometimes more sophisticated when using the *k*-ratio approach (depending on your software). This leads to more accurate estimates of (relative) peak intensities.
4. Construct calibration curves (e.g., Fig. 4) for each element by plotting the mean and SD of the *k*-ratios vs. element concentrations.

3.11. Special Procedure for Na Quantification

1. Of the elements listed above for calibration, sodium, which has been from the beginning a major focus of plant CEDX analysis, is the least rewarding. The energy of the emitted X-rays is the lowest, so its detection is the least efficient. The threshold concentration at which it may be distinguished from the background noise is higher than for any of the others. From Fig. 4 for the Ca signal as a function of concentration, the mean and error for the signal at 12.5 mM is clearly distinct from zero and provides an estimate of the lower limit of detection. For sodium, the lowest detectable threshold is around 40 mM.
2. A second limitation is the nearness of the sodium $K\alpha$ peak to that of oxygen, which, in hydrated specimens, is very large. Moreover, the secondary $K\beta$ peak of oxygen coincides with the sodium $K\alpha$ peak. A correction must be made by subtracting an estimated oxygen $K\beta$ peak from the sodium $K\alpha$ peak.

Despite these limitations, cell-specific quantification of sodium in hydrated plant cells remains an active field and is directing attention to specific cell types involved in salt tolerance. For example, CEDX

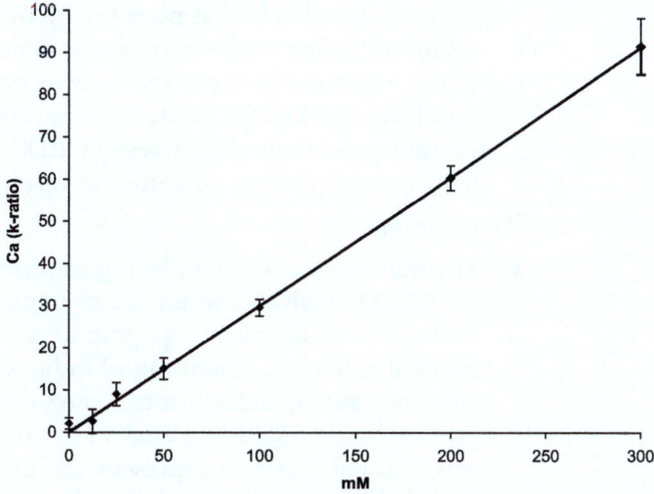

Fig. 4. A typical calibration curve for quantification relating *k*-ratio to [Ca] in a cryo-planed and Al-coated standard.

analysis has identified two cell types, previously unsuspected to accumulate sodium, pericycle in durum wheat (4) and grape vine roots (5), and cells in the inner cortex that develop phi thickenings in salt-stressed *Brassica oleracea* (9).

4. Notes

1. Cryo-fixation.
   (a) Cryo-microscopists have emphasized the importance of very rapid freezing that produces no ice crystals in specimens (12). While this vitrification is desirable for retention of maximum fine cellular detail, it can be realized only with minute tissue pieces since, at best, it is attained only within a few μm from the specimen surface. Cutting out such small pieces for freezing is clearly not compatible with obtaining specimens from intact, functioning plants. If plant parts must be severed before freezing, it is important to consider that liquids will have been displaced and wound responses may have occurred. Most intact plant parts are too large to be frozen quickly. The ice crystals formed will commonly be a few micrometers wide, with layers of the solutes sequestered between them (eutectics). These patterns can be useful since they indicate the presence of solutes and they are usually absent in senescing or dead cells.
   (b) Since the ice crystals are confined within cell compartments, their presence is not a limitation for quantitative CEDX, so long as the X-rays are collected from an area that covers most of the vacuole or other compartment

being analyzed. This has been shown by analyses of known solutions loaded into empty vessels in the interior of roots. These experiments have also shown no significant variation from the known solute concentration of the infusing solutions and their within-vessel CEDX analyses following freezing with a range of common cryogens (11).

2. Cryo-planing.
   (a) The importance of cryo-planing to produce a smooth face for CEDX analysis cannot be overemphasized (13, 14). Surfaces with irregular topography result in scattering of incident electrons, obscuring of induced X-rays by surface hills and valleys, and spurious X-rays that distort the emission spectrum. Goldstein et al. (12) summarized the problem: "Rough surfaces represent the greatest challenge to quantitative XR microanalysis." Fracture faces are rarely suitable for quantitative analysis.
   (b) Interior, smoothly planed surfaces in any desired plane are also excellent for tissue structural investigations. They also have the great advantage that they can be returned to the cryo-microtome repeatedly and replaned, making serial observations and analyses possible.
3. Specimen etching.
   (a) Etching at –90°C should not last longer than about 3 min. Over-etched specimens cannot be used for accurate quantitation. The observer is aiming for the minimum change compatible with certain knowledge of the anatomy of the specimen. The etching protocol has been shown to not affect elemental quantification significantly, but higher temperatures do (Table 1, (14)).
   (b) Etching should not be regulated by a set time because some parts of a complex plant tissue etch faster than others. Visual monitoring of etching of specimens on the cold stage is therefore critical. The judgement of etching progress is personal and subjective and should be done by a practiced operator, constantly watching the loss of ice, varying the area observed, and the magnification.
4. Specimen coating.
   (a) Metal (or carbon) coating of specimens is essential to prevent charge buildup that obscures regions of the specimen and distorts analysis (12). The usual coatings for structural studies (gold, platinum, etc.) are unsuitable for specimens to be analyzed because, (i) elements may be differentially sublimed by the high heat required for coating with these metals; and (ii) these coatings differentially absorb emitted X-rays of different energy Na and K are especially volatile. Only three elements are suitable for X-ray analysis, Al, Cr,

and C, and of these, Al has the least distortion of the X-ray spectrum (15), Tables 1 and 2 of (16), Table 1 of (12). The relatively low temperature of the Al coating minimizes element loss.

(b) Coating with Al is also an art that needs practice, maintaining the right current through the heating coil for the right time, while the specimen is moved about in the Al shower. This usually needs two operators. For a discussion of the effect of the thickness of Al on measured concentrations, see (7, 15).

5. Quantitation.

(a) The solutions of mixed elements and graphite are used because the signal received from one element is influenced by the presence of other elements that may be present in the plant together with the Na and Cl. The colloidal graphite provides a carbon matrix to mimic the carbon content of plant tissue, and also aids dispersion of the elements. An additional benefit of the multielement solutions is that one series of measurements provides calibration curves for all these elements.

(b) Once prepared, unplaned frozen standard drops can be stored indefinitely in a cryo-store, and planed and processed later as required. Repeat calibrations are done about twice annually.

(c) In our experience, the calibration curves for each element are linear only up to 300 mM. The 500 mM points usually fall below the projected line, and we do not claim the same accuracy for concentrations above 300 mM.

(d) To ensure accurate quantification we recommend maintaining a cobalt, copper, or nickel standard in the chamber of the microscope. These standards all emit low ("L" lines) and high energy ("K" lines): X-rays that are useful for checking for drift in the offset and gain settings of the spectrometer. Your software will probably provide a calibration routine for this purpose. Check the settings at the end of each day.

References

1. McCully ME, Canny MJ, Huang CX (2009) Cryo-scanning electron microscopy (CSEM) in the advancement of functional plant biology: morphological and anatomical applications. Funct Plant Biol 36:97–124
2. Yeo AR, Läuchli A, Kramer D et al (1977) Ion measurements by X-ray microanalysis in unfixed, frozen, hydrated plant cells of species differing in salt tolerance. Planta 134:35–38
3. Läuchli A, Schwander H (1966) X-ray microanalyser study on the localization of minerals in native plant tissue sections. Experimentia 31:503–505
4. Läuchli A, James RA, Huang CX et al (2008) Cell-specific localization of Na^+ in roots of durum wheat and possible control points for salt exclusion. Plant Cell Environ 31: 1565–1574

5. Storey R, Schachtman DP, Thomas MR (2003) Root structure and cellular chloride, sodium and potassium distribution in salinized grapevines. Plant Cell Environ 26:789–800
6. Conn S, Gilliham M (2010) Comparative physiology of elemental distribution in plants. Ann Bot 105:1081–1102
7. McCully ME, Canny MJ, Huang CX et al (2010) Cryo-scanning electron microscopy (CSEM) in the advancement of functional plant biology: energy dispersive X-ray microanalysis (CEDX) applications. Funct Plant Biol 37:1011–1040
8. Hashemi A, Abdolzadeh A, Sadeghipour HR (2010) Beneficial effects of silicon nutrition in alleviating salinity stress in hydroponically grown canola, *Brassica napus* L., plants. Sci Plant Nutr 56:244–253
9. Fernandez-Garcia N, Lopez-Perez L, Hernandez M et al (2009) Role of phi cells and the endodermis under salt stress in *Brassica oleraceae*. New Phytol 181:347–360
10. Erbe EF, Rango A, Foster J et al (2003) Collecting, shipping and imaging snow crystals and ice grains with low-temperature scanning electron microscopy. Microsc Res Tech 62:19–32
11. McCully ME, Shane MW, Baker AN et al (2000) The reliability of cryoSEM for observation and quantification of xylem embolisms and quantitative analysis of xylem sap *in situ*. J Microsc 198:24–33
12. Goldstein JI, Newbury DE, Joy DC et al (2003) Scanning electron microscopy and X-ray microanalysis, 3rd edn. Springer, Berlin
13. Hess FD, Falk RH, Bayer DE (1975) The influence of topography on X-ray microanalysis element mapping. Am J Bot 62: 246–253
14. Huang CX, Canny MJ, Oates K et al (1994) Planing frozen hydrated plant specimens for SEM observation and EDX microanalysis. Microsc Res Tech 28:67–74
15. Reid AP, Oates K, Potts WTW (1993) The effect of aluminium coating on elemental signals in X-ray microanalysis. Microsc Res Tech 24:168–172
16. Marshall AT. X-ray microanalysis of frozen hydrated biological bulk specimens. Mikrochim Acta. 1998; suppl 15:273–282

Chapter 9

Fluorescence Lifetime Imaging (FLIM) Measurements in Salinity Research

Olga Babourina and Zed Rengel

Abstract

Plant response to salt stress involves changes in intracellular ion concentrations that can be estimated by ion-selective fluorescent dyes. Conventional confocal/fluorescent imaging does not always accurately reflect the process, since it is dependent not only on the ion concentration but on the dye concentration itself. Fluorescent lifetime imaging (FLIM) can resolve this problem: images derived under this technique are based on the fluorescent lifetime distribution of the bound/unbound state of the fluorophore, and not on the fluorescent intensity. On the example of 2′-7′-bis(carboxyethyl)-5(6)-carboxyfluorescein (BCECF) and Ca Green, we demonstrate the potential of FLIM technique in salinity studies.

Key words: Fluorescence lifetime imaging, Ion selective fluorophore, Ca^{2+}, H^+

1. Introduction

1.1. Background Information

Under salt stress, transport of all ions across the plasma membrane of plant cells is affected either directly or indirectly. The most important ions in NaCl-induced salt stress are Na^+, K^+, and Cl^-. However, transport of Ca^{2+}, H^+, and many other ions are also affected because salt stress in plants is always linked to changes in major regulators of ion transport such as the plasma membrane potential (E_m), ATP-generation capacity, H^+ gradient, and other parameters (1–4).

Changes in ion transport in plants under salt stress have been successfully documented by many techniques, such as isotopes (5), electrophysiology (non-invasive ion flux measurement, patch clamp, etc.) (3, 6) and microscopy (7, 8). Microscopy studies included conventional fluorescent and confocal imaging that allow

Sergey Shabala and Tracey Ann Cuin (eds.), *Plant Salt Tolerance: Methods and Protocols*, Methods in Molecular Biology, vol. 913, DOI 10.1007/978-1-61779-986-0_9, © Springer Science+Business Media, LLC 2012

changes in intracellular concentrations of Ca^{2+}, H^+, Na^+, and K^+ to be estimated (7, 9–11).

The majority of studies dealing with intracellular ion concentrations are based on the estimation of changes in fluorescent intensity of ion-binding fluorescent dyes. However, measuring fluorescent intensity using conventional ion imaging in plants is linked with several problems. The *first problem* is based on the dependence of fluorescent intensity being not only on the ion concentration in the cytoplasm but also on the dye concentration itself (12). Unfortunately, there are numerous processes regulating dye concentration in the cytosol of plant cells. If dye microinjection (suitable for single cells or a single cell derivative such as root hairs or guard cells) cannot be used, a dye is loaded under low temperature and low pH, so-called "acid loading" (13). As the first step, a fluorescent dye can penetrate the plasma membrane only as an uncharged molecule, the acetoxymethyl (AM) ester, which is not fluorescent for many dyes. Once the dye is in the cytosol, nonspecific esterases break the ester bond and the dye becomes fluorescent. Hence, during this first step, at least two processes are regulating the dye concentration: the rate of dye transport across the plasma membrane and the rate of an esterase activity. Both are highly dependent on temperature and substrate concentration. The second step involves excretion of the dye from the cytosol to the vacuole and the apoplast. These two steps are multiplied in a tissue with multiple cell layers, such as plant roots.

The dye concentration is stable in the cell cytosol for a relatively short time (generally <6 h) as it has been shown for a mouse brain tissue (14). This problem of unstable dye concentration in the cytosol can be resolved by the ratiometric dyes that have dual excitation or emission wavelengths. However, many ratiometric dyes (e.g., indo or fura range for Ca^{2+} imaging) are excited only by the wavelength in the UV range. This is linked with the *second problem* in ion imaging of plant cells, i.e., autofluorescence. The source of autofluorescence in plant tissue is variable and is not completely understood. Moreover, autofluorescence can vary in different batches of plants. When dye loading or de-esterification in the cytosol is poor, it is unclear whether measured fluorescence emanates from the dye or from cell autofluorescence. Knowing the source of measured fluorescence is vital for measurements of ion concentrations in the cytosol because of the need to determine the initial (generally low) concentration of an ion (e.g., Ca^{2+}) before tissue/cell exposure to a signal/stress. If cell autofluorescence contributes to the weak initial fluorescence from the dye, the subsequent measurements would be compromised.

There are techniques available to overcome the above-mentioned problems associated with attempts to quantify the fluorescent intensity of a dye. One such approach is fluorescent lifetime imaging (FLIM) of a dye. FLIM has been known for more than two decades (15). It is based on a specific time-of-decay profile of an excited fluorescent molecule rather than on fluorescence intensity of these molecules, which is used in conventional microscopy. This chapter provides a brief overview of the theory and practice of applying FLIM in salt stress research, with an emphasis on ion-binding fluorophores.

1.2. FLIM Theory

After dye excitation is ceased, a dye fluorophore starts to return to a non-excited energetic level with a certain decay rate. This is called the fluorescence lifetime (FL). As shown in Fig. 1, after a pulse of light, fluorophore molecules reach a new excited energy level. Within ~10^{-8} s interval, the molecules return to the ground state.

In the formula below, N indicates the number of molecules in the excited state at a certain time (T_n) after the pulse. The number dN of molecules decaying back to the ground state in the time interval dT is proportional to the number N available in the excited state at that time.

$$\mathrm{d}N/\mathrm{d}T = -kN(T)$$

$N(T) = N_0 e^{-kT}$, where k is the rate constant for a particular fluorophore (expressed in s^{-1}), and N_0 is the quantity of excited molecules at $T=0$. Therefore, $t=1/k$ is the lifetime of the excited state (τ, expressed in s); this is the time during which the population of the molecules in the excited state decreases in multiples of e (e~2.72). In practice, the fluorescence lifetime is defined as the time in which the fluorescence intensity decays to $1/e$ of the intensity immediately following excitation (16).

Fluorescence lifetime (FL) can be estimated by (1) the time-domain method, (2) the frequency-domain method, and (3) the pump-probe technique. The time-domain method is based on measurements of fluorescence after a pulse excitation at varying times of gate operation within a picosecond interval, e.g., at T_2 and T_4 in Fig. 1. The frequency-domain method is based on calculation of the phase shift of fluorescence and the reduction in its amplitude using a detector with a gain modulator. Under the pump-probe technique, after exciting the sample with one (strong) pulse, the relaxation is later probed with another (weak) pulse at a variable delay (15).

Time-correlated single photon counting (TCSPC) can be considered as a variation of a time-domain gating approach. This method is based on an assumption that an intensity distribution of each photon generated by a single excitation over time (FL) equals probability distribution of the first photon generated by the single excitation over time. The Time-to-Amplitude Converter (TAC) starts accumulating electric charges in the internal condenser from the Start Time ($T=0$) at which the excitation light source emits light. TAC stops accumulating electric charge in the condenser at

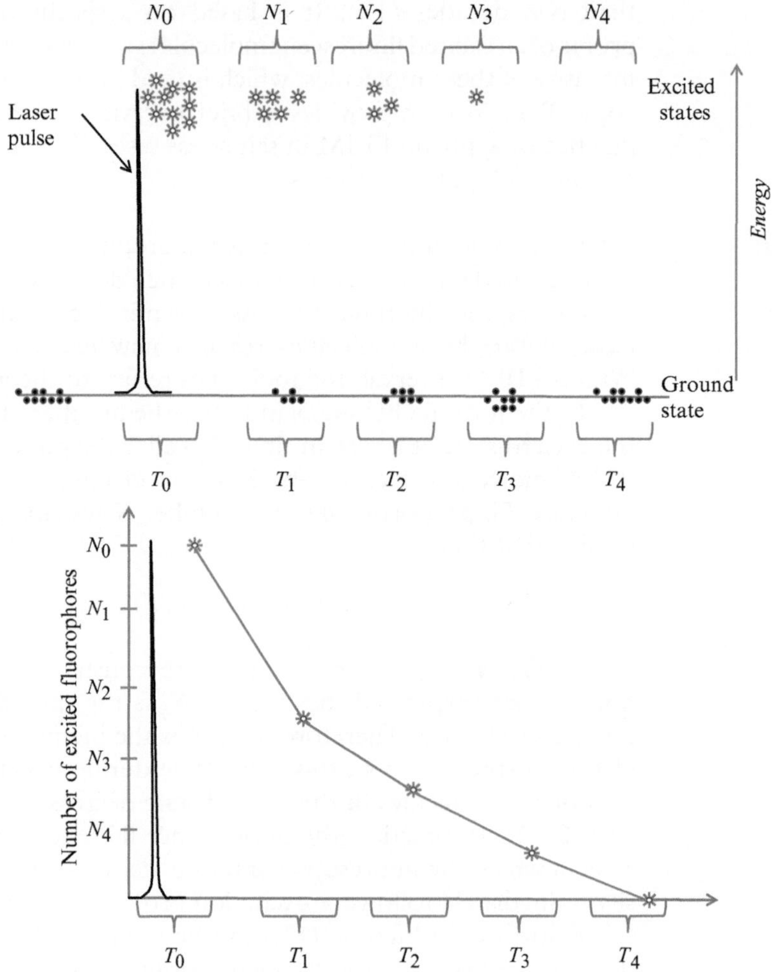

Fig. 1. The diagrams demonstrate that the number of fluorophore molecules in the excited state declines within certain time intervals following a laser pulse. The rate of this decline is specific for a molecule and is called fluorescence lifetime (FL).

the Stop Time, the time at which the first fluorescent photon is identified by the detector, and outputs the charge voltage at this moment. In short, TAC records the time required to detect the first fluorescent photon as voltage. The recordings are repeated for additional pulses and, after enough recorded events, a histogram is calculated to show the number of events across all these recorded time points. This histogram can be fitted to an exponential function that contains the lifetime decay function of interest, and the lifetime parameter can be calculated (17). This approach has become popular in recent years because of its commercial availability.

Fluorescence lifetime images are shown in pseudocolor, according to their lifetimes. In a FLIM image, pixel values vary according to the fluorescence lifetime, not fluorescence intensity, with the differences in the latter potentially due to differential fluorophore concentration. Figure 2 demonstrates the FLIM

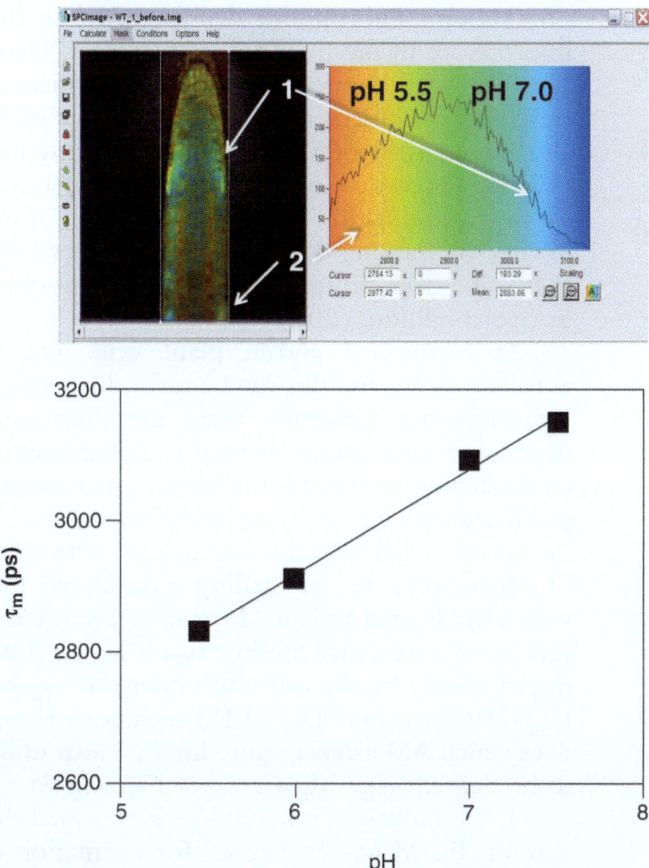

Fig. 2. A 5-day-old Arabidopsis seedling (Col-0) loaded with BCECF. The meristem zone and distal elongation zone of the root are relatively alkaline (1); and the phloem vessels in the proximal elongation zone are relatively acidic (2).

image of the root tip of a 5-day-old Arabidopsis seedling loaded with 2′-7′-bis(carboxyethyl)-5(6)-carboxyfluorescein (BCECF) dye. The fluorescence lifetime (FL) distribution is shown in the right pane of the image. It is calculated for the area bordered by the white lines in the left image. Arrows indicate the relatively alkaline zone (1) and the relatively acidic root zone (2).

1.3. Advantages and Limitations of FLIM Measurements

Advantages. Fluorescent lifetime (FL) is dependent on intra- and inter-molecular interactions, but independent of dye concentration, fluorescence intensity, and photobleaching, the factors that are important for all biological studies where an ion-specific dye is loaded into the tissue. Fluorophore/dye binding to a ligand affects FL of this particular fluorophore. In most cases for ion-sensitive dyes, FL of unbound/bound dye shifts from a shorter to a longer lifetime (18). Using FLIM, the non-ratiometric dyes according to their excitation/emission spectra become "ratiometric" in terms of their lifetimes. With an increased ion concentration, the quantity of unbound fluorophore decreases, while the quantity of the fluorophore in the bound state increases. Hence, ion concentration can be calibrated for quantitative measurements based on different parameters, including the ratio of intensities of unbound and bound state of a fluorophore. Although earlier publications (19) suggested that in vitro calibrations of a dye would be suitable for in vivo estimation of ion concentration, it was later shown that in vivo calibration is highly desirable. Several publications indicate that some dyes demonstrate different calibration for in vitro and in vivo conditions (20).

As mentioned above, plant cells and tissues are highly autofluorescent and this can be easily distinguished with the FLIM measurements. Generally, tissue autofluorescence increases with decreasing excitation wavelengths. Sometimes under conventional confocal microscopy, it is unclear whether measured fluorescence is produced by a dye or by cell autofluorescence. The FLIM estimation produces absolutely clear results, where fluorescence lifetime of a tissue prior to dye loading is distinctive from the same tissue with a dye loaded and an AM ester cleaved. We have demonstrated that autofluorescence lifetime has at least a 2-nm shift and is produced mostly by the cell wall under our experimental conditions (λ_{Ex}=940 nm) (4). The FLIM technique is especially helpful for dyes which AM esters require long periods of time (several hours) to be cleaved (e.g., Mg-Green or FluoZin-3).

To be unbiased, it should be mentioned that in animal tissue studies, FLIM has been used for estimation of autofluorescent flavoproteins, i.e., in some studies autofluorescence can be helpful (21). For example, the FL of free NAD(P)H is distinguishable from protein-bound NADH (22). This feature has been used as a measure of cell oxidative stress, as free NAD(P)H increases in these circumstances (23). Measurements of NAD(P)H-related fluorescence lifetime of adipocytes and fibroblasts demonstrated that glucose can be measured non-invasively in cells by monitoring changes in the lifetimes of cell autofluorescence (24).

In studies with conventional confocal microscopy, ion concentration in different cell compartments can be estimated by simultaneous loading of several dyes that have different excitation or emission wavelengths. The usual combination for protein fluorescent labeling is Fluorescein Isothiocyanate/Tetramethyl Rhodamine Isothiocyanate (FITC/TRITC) or Oregon Green/Texas Red, with dyes emitting green and red fluorescent light, correspondingly. In ion concentration studies, the concentrations of different ions (Ca^{2+} and H^+) were simultaneously estimated by Ca Green and seminaphtharhodafluor (SNARF) (25). Similarly, dyes with different fluorescent lifetimes can be used for simultaneous estimation of different ligands. The FLIM measurements are especially helpful if dyes cannot be separated according to their emission spectra (19, 26).

Other advantages of FLIM include the ability of this technique to characterize local environment changes (e.g., viscosity) and in Forster resonance energy transfer (FRET) studies. Under FRET, the energy is non-radiatively transferred from a fluorescent donor molecule to an acceptor when they are located in a close proximity (1–10 nm) (27). This energy transfer can be monitored by conventional confocal microscopy, but often there is a spectral overlap between donor and acceptor fluorophore causing FRET signal contamination. This contamination can be avoided using FLIM (28).

Limitations. For each FL image, multiple images have to be collected to produce enough information for calculations of FL for every pixel of the image, especially for calculations of the double and triple exponential decays (17). This information collection requires time: at least 2–3 min when the signal is strong and 5–10 min (or more) when the signal is weak. This time frame is not suitable for studies pursuing quick changes (under 1 min) in ion concentrations within a cell. However, in comparison with animal tissues, plant tissues are slightly slower in signal response (29), and the time frame of 3–5 min is reasonable.

Another problem with the FLIM measurements is photobleaching because a specimen is constantly exposed to laser illumination during information collection. Preferably, there should be time allocated for tissue recovery from illumination.

A specific limitation related to measurements of ion concentrations in live tissue/cells is that a specimen should not move during FLIM image capture. In relation to salt stress, a specimen should not change its size/volume, but this always occurs during the first 5–10 min after plant tissue exposure to high concentrations of NaCl (shrinkage/recovery). Consequently, FLIM images can only be taken only 5–10 min after plant tissue exposure to salt stress.

2. Materials

1. Loading solution: 0.2 mM $CaCl_2$ and 50 mM mannitol, pH 4.2, made up using MilliQ water (~18.2 MΩ cm). The bulk solution can be poured into 10 mL plastic tubes and frozen until required.
2. Dimethyl sulfoxide (DMSO) and poly-L-lysine (Sigma, Castle Hill, Australia).
3. Recovery solution may vary, depending on the experiment. In the majority of salt stress experiments, we used Basic Salt Solution (BSM) containing 0.2 mM $CaCl_2$ and 1 mM KCl.
4. Dyes. We present the data on Calcium Green and BCECF as case studies (both produced by Molecular Probes, Eugene,

OR, USA), but any fluorescent dye is suitable as long as it is cell permeable. In our studies we use acetoxymethyl esters (AM) of the dyes. Usually, these esters are sold in 50 μg freeze-dried aliquots in special packaging of 10 or 20. For in vitro calibrations, water-dissolved salts of the same dyes can be purchased from the same supplier. Note that these salts are not cell permeable.

5. A protonophore, carbonyl cyanide *m*-chlorophenylhydrazone (CCCP), and the calcium ionophore A23187 (both produced by Sigma, Castle Hill, Australia) are dissolved in DMSO and diluted to 100 μM with a required Ca^{2+} buffer (Calcium calibration buffer Kit #1, Molecular Probes, Eugene, OR, USA). These solutions have to be filtered through 22-μm Millipore filters.

3. Methods

3.1. Dye Loading

1. Dissolve fluorescent dyes BCECF-AM or Calcium Green™ AM in DMSO and dilute by loading solution to the final concentration of 20 μM (an aliquot of 50 μg dye was dissolved in 10 μL DMSO and transferred into a 2-mL Eppendorf tube). The final concentration of DMSO in loading solution is 0.5%v/v.
2. Place plant seedlings (e.g., Arabidopsis or wheat) into 2-mL Eppendorf tubes with 200–500 μL of loading solution.
3. Place these tubes on ice for 2 h in dark.
4. Remove plants from the tube after 2 h and allow them to float onto a recovery solution (usually BSM) for 1 h (for some other dyes, e.g., FluoZin-3 or Magnesium Green, this recovery period could be significantly longer, up to 12 h).

3.2. Plant Mounting

In our experiments we use an inverted microscope, so these mounting suggestions are related to this microscope configuration. Seedlings should be always held in solutions; otherwise, they quickly dry up.

1. Attach seedlings to a cover glass that was polarized with a drop of poly-L-lysine solution for about 30 min before rinsing with de-ionized water. To localize solution in a certain area, an Aqua-Hold Barrier Pap Pen can be used to draw a line so creating a barrier for solution. The cover glass is placed in a specially designed holder allowing easy change of cover glasses.
2. Alternatively, place seedlings between two cover glasses in a drop of the BSM solution (see Note 1).

3.3. Setting Up Confocal Microscope

1. Before starting the FLIM measurements, use conventional confocal microscopy for focusing and finding regions of interests (ROIs). Also, it is advisable to obtain a good quality conventional confocal image before starting FLIM experiments (see Note 2).
2. Set up excitation wavelength at 940 nm (preliminary experiments demonstrated that this wavelength minimized contribution from autofluorescence). Fluorescence emission is recorded by a photo multiplier between 500 and 692 nm for both dyes (BCECF and Ca Green™).
3. With our microscope setup, we can take FLIM images only with the multiphoton laser. For many dyes, the spectrum of excitation by wavelengths between 700 and 1,000 nm (multiphoton laser range) is not known. Hence, it is highly advisable to determine the dye spectrum before starting any experiments and choose the longest wavelength (to prevent cell and tissue damage and reduce autofluorescence) that would still allow the fluorescent signal from the dye.

3.4. Dye Calibration

For in vitro calibration, dissolve the BCECF potassium salt in different buffers. The median lifetime is used for creating a calibration curve, assuming a single exponential decay for BCECF (30). For in vivo calibrations of BCECF, CCCP is used as a protonophore to eliminate a gradient between external and internal H^+ concentrations. We found a shift (~0.45 ns) between in vitro and in vivo calibrations for BCECF, similar to other studies (20, 31). However, in vitro calibration of a new dye is highly desirable, to estimate a slope. The calibration curve of BCECF is shown in Fig. 2 (the lower pane); the upper pane demonstrates a quick estimation of pH values.

1. For FLIM calibration of $[Ca^{2+}]_{cyt}$, prepare stock solution of Ca-Green potassium salt in 0 mM Ca^{2+} buffer (Calcium Calibration Kit #2, Molecular Probes, Eugene, OR, USA).
2. Dilute stock solution to 2 μM in different Ca^{2+} buffers (free Ca^{2+} concentration ranged from 0 to 1 mM), immediately after preparation. The intensity of the first component of Ca-Green lifetime is used for calibration, assuming a double exponential decay for this dye (33). (see Note 3).
3. For in vivo calibrations, dissolve the calcium ionophore A23187 in different Ca^{2+} buffers (see Note 4).

3.5. FLIM Measurements

1. Load seedlings or any other plant specimen (mesophyll tissue, suspension cells, etc.) with a dye as described earlier (Subheading 3.1).
2. Mount a specimen to a cover slip as described in Subheading 3.2.

3. Focus and select region of interest (ROI) by conventional confocal microscopy. Once ROI is selected, take a high quality image (e.g., 1,256 × 1,256, 3 averages).
4. Take the first FLIM image before any solution change. Usually, several initial images are taken to ascertain that the concentration is stable and the minimal time of exposure is established. It is highly desirable to have a specimen exposed for 2–3 min for FLIM measurements, and then allocate at least 2 min for the recovery of the specimen in the dark, before taking the next FLIM image.
5. Put microscope settings back to the conventional microscopy before changing solutions.
6. Change solution by a pipette, adding new solution from one side of the cover slip and sucking the old solution with another pipette from the other side (alternatively, it could be done with a tissue, e.g., Kimtech wipes). The volume of the new solution should be at least three times larger than the initial volume of the old solution (see Notes 5 and 6).

4. Notes

1. Usually we use longer cover glasses (20 × 40 mm) as a base, and a shorter one (20 × 20 mm) at the top. The stable position is achieved by a weight (e.g., a metal washer) placed on the edges of the top cover glass, making the central part of the slide (with the seedling or tissue) visible and stable during solution changes. For bigger objects (wheat seedlings), the upper cover slip can be replaced with the washed nylon mesh.
2. For conventional microscopy and FLIM measurements, we use an inverted confocal microscope (Leica TCS SP2 AOBS, Leica Microsystems GmbH, Wetzlar, Germany). Light pulses are generated at a frequency 80 MHz by the Mai Tai Laser (Spectra Physics, Mountain View, CA, USA). Fluorescence is recorded by photo multipliers, and FLIM analysis is performed using electronics (SPC-730; Becker & Hickl, Berlin, Germany) and software (SPC7.22; Becker & Hickl) for time-correlated single-photon counting (34). Lifetime images are analysed using SPCImage 2.6 (Becker & Hickl).
3. Figure 3 demonstrates changes in median fluorescent lifetime (τ_m) with increased Ca^{2+} concentration. At higher Ca^{2+}, there is an increase in intensity of the slower component. The fast component is usually designated as the first component, and all parameters are labeled "1": τ_1 = FL of the first component and α_1 = intensity of the first component. Correspondingly, the

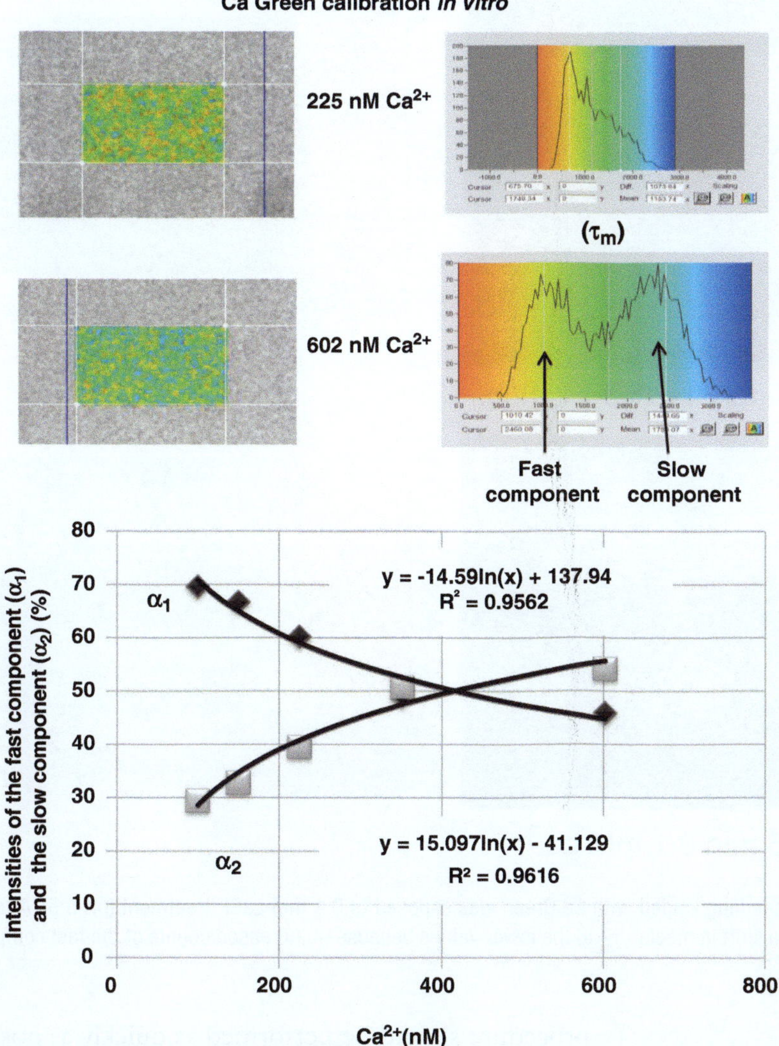

Fig. 3. In vitro calibration of Ca Green. The median fluorescence lifetime (FL) of the dye dissolved in Ca^{2+} buffers increases with increasing Ca^{2+} concentration because the counts of the slow component are increasing (the *upper pane*). The *lower pane* demonstrates distribution of intensities of fast and slow components: their sum should equal 100%.

parameters of the slow component under double exponential decay are labeled "2." Therefore, with an increase in Ca^{2+} concentration, we observed α_1 decrease and α_2 increase of Ca-Green, and these can be calibrated (Fig. 3). As mentioned earlier, we can calibrate the ratio α_2/α_1 as well; similar to conventional ratiometric dyes (32).

4. We found that in vitro and in vivo calibrations of Ca-Green are very close.

5. This procedure should be performed very delicately, keeping the specimen in the same place, and this is checked later using conventional confocal microscopy. On the other hand, this

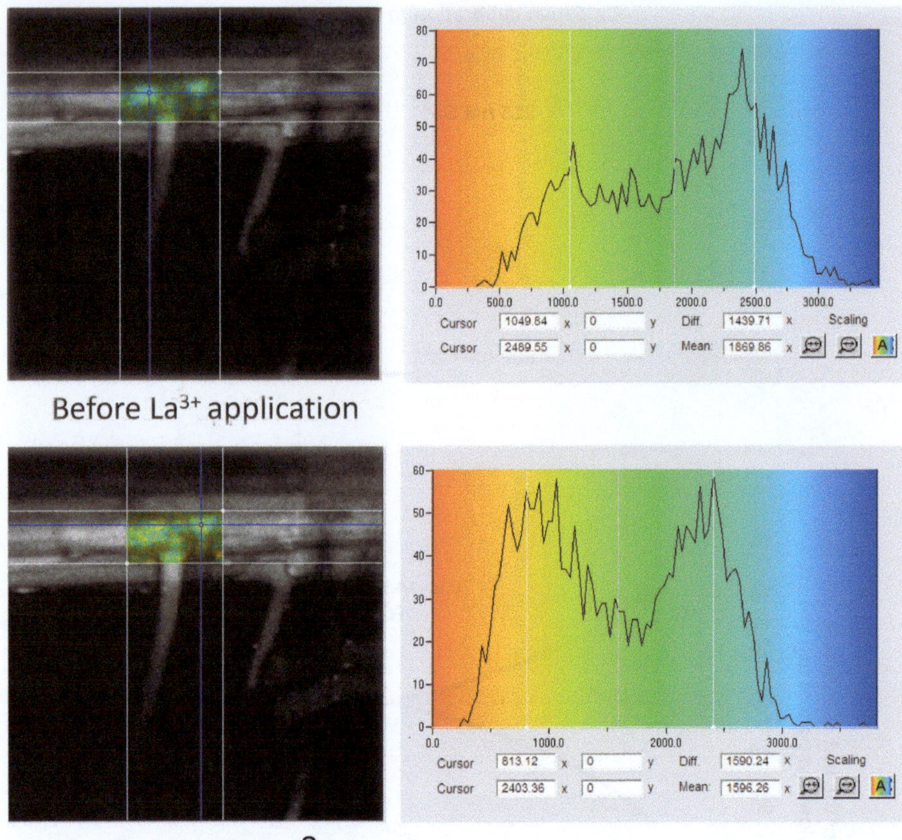

Fig. 4. A wheat seedling loaded with Ca Green was exposed to 0.1 mM $LaCl_3$ treatment (pH 5.5). After 5 min of exposure, there was a shift in median τ_m to the lower values because of increased counts of the fast component (the *lower right pane*).

procedure should be performed as quickly as possible, to start collecting information on the induced concentration changes occurring in the tissue.

6. Figure 4 demonstrates an example of the Ca^{2+} concentration change in the basal part of the root hair of a 5-day-old wheat seedling. After 5 min exposure to 0.1 mM $LaCl_3$, Ca^{2+} concentration declined, with an increase in the fast component shown in the right pane.

References

1. Katsuhara M, Yazaki Y, Sakano K et al (1997) Intracellular pH and proton-transport in barley root cells under salt stress: *in vivo* P-31-NMR study. Plant Cell Physiol 38: 155–160
2. Rengel Z (1992) The role of calcium in salt toxicity. Plant Cell Environ 15:625–632
3. Shabala S, Babourina O, Newman I (2000) Ion-specific mechanisms of osmoregulation in bean mesophyll cells. J Exp Bot 51:1243–1253
4. Guo K-M, Babourina O, Rengel Z (2009) Na^+/H^+ antiporter activity of the SOS1 gene: lifetime imaging analysis and electrophysiologi-

cal studies on Arabidopsis seedlings. Physiol Plant 137:155–165

5. Davenport R, James RA, Zakrisson-Plogander A et al (2005) Control of sodium transport in durum wheat. Plant Physiol 137:807–818
6. Tyerman SD, Skerrett IM (1999) Root ion channels and salinity. Sci Hortic 78:175–235
7. Halperin SJ, Lynch JP (2003) Effects of salinity on cytosolic Na^+ and K^+ in root hairs of Arabidopsis thaliana: in vivo measurements using the fluorescent dyes SBFI and PBFI. J Exp Bot 54:2035–2043
8. Kader MA, Lindberg S (2005) Uptake of sodium in protoplasts of salt-sensitive and salt-tolerant cultivars of rice, *Oryza sativa* L. determined by the fluorescent dye SBFI. J Exp Bot 56:3149–3158
9. Gao DJ, Knight MR, Trewavas AJ et al (2004) Self-reporting Arabidopsis expressing pH and Ca^{2+} indicators unveil ion dynamics in the cytoplasm and in the apoplast under abiotic stress. Plant Physiol 134:898–908
10. Knight H, Trewavas AJ, Knight MR (1997) Calcium signalling in *Arabidopsis thaliana* responding to drought and salinity. Plant J 12:1067–1078
11. Park M, Lee H, Lee JS et al (2009) *In planta* measurements of Na^+ using fluorescent dye CoroNa Green. J Plant Biol 52:298–302
12. Tadrous PJ (2000) Methods for imaging the structure and function of living tissues and cells: 2. Fluorescence lifetime imaging. J Pathol 191:229–234
13. Zhang WH, Rengel Z, Kuo J (1998) Determination of intracellular Ca^{2+} in cells of intact wheat roots: loading of acetoxymethyl ester of Fluo-3 under low temperature. Plant J 15:147–151
14. Stosiek C, Garaschuk O, Holthoff K et al (2003) In vivo two-photon calcium imaging of neuronal networks. Proc Natl Acad Sci USA 100:7319–7324
15. O'Connor DV, Phillips D (1984) Time-correlated single photon counting. Academic, London
16. Lakowicz JR (1999) Principles of fluorescence spectroscopy. Kluwer Academic/Plenum Publishers, New York, p 725
17. Becker W, Bergmann A, Hink MA et al (2004) Fluorescence lifetime imaging by time-correlated single-photon counting. Microsc Res Tech 63:58–66
18. Terpetschnig E, Jameson DM (2011) Fluorescence lifetime (FLT). Technical note: http://www.iss.com/resources/pdf/technotes/K2CH_FLT.pdf
19. French T, So PTC, Dong CY et al (1998) Fluorescence lifetime imaging techniques for microscopy. Methods Cell Biol 56:277–304
20. Nakabayashi T, Wang H-P, Tsujimoto K et al (2007) A correlation between pH and fluorescencelifetimeof2′,7′-Bis(2-carboxyethyl)-5(6)-carboxyfluorescein (BCECF) in vivo and in vitro. Chem Lett 36:206–207
21. Konig K, Riemann I (2003) High-resolution multiphoton tomography of human skin with subcellular spatial resolution and picosecond time resolution. J Biomed Opt 8:432–439
22. Lakowicz JR, Szmacinski H, Nowaczyk K et al (1992) Fluorescence lifetime imaging of free and protein-bound NADH. Proc Natl Acad Sci USA 89:1271–1275
23. Konig K, So PTC, Mantulin WW et al (1996) Two-photon excited lifetime imaging of autofluorescence in cells during UVA and NIR photostress. J Microsc 183:197–204
24. Evans ND, Gnudi L, Rolinski OJ et al (2005) Glucose-dependent changes in NAD(P)H-related fluorescence lifetime of adipocytes and fibroblasts in vitro: potential for non-invasive glucose sensing in diabetes mellitus. J Photochem Photobiol B Biol 80:122–129
25. Martinez-Zaguilan R, Parnami G, Lynch RM (1996) Selection of fluorescent ion indicators for simultaneous measurements of pH and Ca^{2+}. Cell Calcium 19:337–349
26. Pepperkok R, Squire A, Geley S et al (1999) Simultaneous detection of multiple green fluorescent proteins in live cells by fluorescence lifetime imaging microscopy. Curr Biol 9:269–272
27. Jares-Erijman EA, Jovin TM (2003) FRET imaging. Nat Biotechnol 21:1387–1395
28. Wallrabe H, Periasamy A (2005) Imaging protein molecules using FRET and FLIM microscopy. Curr Opin Biotechnol 16:19–27
29. Babourina O, Newman I, Shabala S (2002) Blue light-induced kinetics of H^+ and Ca^{2+} fluxes in etiolated wild-type and phototropin-mutant Arabidopsis seedlings. Proc Natl Acad Sci USA 99:2433–2438
30. Niesner R, Peker B, Schlusche P et al (2005) 3D-Resolved investigation of the pH gradient in artificial skin constructs by means of fluorescence lifetime imaging. Pharm Res 22:1079–1087
31. Wang HP, Nakabayashi T, Tsujimoto K et al (2007) Fluorescence lifetime image of a single halobacterium. Chem Phys Lett 442:441–444
32. Shabala S, Bækgaard L, Shabala L et al (2011) Plasma membrane transporters mediating virus-induced acquired resistance to oxidative stress. Plant Cell Environ 34:406–417
33. Sanders R, Gerritsen H, Draaijer A, Houpt P, Levine Y (1994) Fluorescence lifetime imaging of free calcium in single cells. Bioimaging 2: 131–138
34. O'Connor DV, Desmond P (1984) Time-correlated single photon counting. Academic Press, London, pp 288

Chapter 10

Cytosolic Ca^{2+} Determinations in Studying Plant Responses to Salinity and Oxidative Stress

Anuphon Laohavisit, Renato Colaço, and Julia Davies

Abstract

Salinity and oxidative stress can transiently elevate cytosolic free Ca^{2+} ($[Ca^{2+}]_{cyt}$) of plant cells. The $[Ca^{2+}]_{cyt}$ increase may be part of a signaling cascade or cell death, depending on cell type, the magnitude and the duration of stress exposure. Several approaches for determining $[Ca^{2+}]_{cyt}$ responses are available to plant biologists, but some require highly specialized equipment. Here we describe protocols for using aequorin as a standard $[Ca^{2+}]_{cyt}$ reporter, with output detected with a plate-reader luminometer.

Key words: Aequorin, Calcium, Epidermis, Protoplast, Root, Stress

1. Introduction

Salinity and oxidative stress can transiently elevate cytosolic free Ca^{2+} ($[Ca^{2+}]_{cyt}$) of plant cells (1–6). The $[Ca^{2+}]_{cyt}$ increase may be part of a signaling cascade or cell death, depending on cell type, the magnitude and the duration of stress exposure (3, 6, 7). Salinity (usually taken to mean NaCl or sodicity) stress is thought to elevate $[Ca^{2+}]_{cyt}$ in roots to trigger adaptive pathways, in particular, the activation of the plasma membrane Na^+-H^+ antiporter encoded by SOS1 (salt overly sensitive1) (7, 8). However, for some cells, such as root hairs, salinity stress *decreases* $[Ca^{2+}]$ (9), which may be indicative of high Ca^{2+} influx under control conditions then block of the Ca^{2+} channels. Reactive oxygen species are now acknowledged as growth regulators and important components of signal transduction, including that of salinity stress (8, 10). Testing their effects is not straightforward as they can interconvert, be extremely short-lived (hydroxyl radicals), membrane-permeant (H_2O_2), and

Sergey Shabala and Tracey Ann Cuin (eds.), *Plant Salt Tolerance: Methods and Protocols*, Methods in Molecular Biology, vol. 913, DOI 10.1007/978-1-61779-986-0_10, © Springer Science+Business Media, LLC 2012

react with nitric oxide (10, 11). You can certainly try generating a controlled dose but be circumspect about what species is having the effect.

There are several options for quantifying the effects of salinity and oxidative stress on $[Ca^{2+}]_{cyt}$. Impalement with Ca^{2+}-selective microelectrodes could be applied to accessible cells, but the majority of studies use some sort of light-emitting Ca^{2+} indicator. The use of fluorescent Ca^{2+} dyes and the FRET-based reporter cameleon in plant cells is comprehensively addressed by Gilroy and colleagues in ref. (12). Loading dyes into the cytosol can be problematic and may require microinjection or microbombardment. Use of cameleon is an elegant and robust alternative but, in common with dyes, also requires prohibitively expensive imaging equipment to which all labs may not have access. Here, we describe the use of aequorin to determine changes in $[Ca^{2+}]_{cyt}$. It too has limitations, but the equipment may be more within reach.

Aequorin luminometry is a non-invasive technique. It was first applied to plants to measure changes in $[Ca^{2+}]_{cyt}$ in response to touch, cold shock and elicitors (13). Since then, it has been used to measure changes in $[Ca^{2+}]_{cyt}$ in a variety of short- and long-term stress responses (e.g., (1–6)). The apoaequorin protein from *Aequorea victoria* is expressed under the control of a 35S constitutive promoter. The cytosolic protein becomes the bioluminescent aequorin upon binding coelentrazine. Binding of up to three Ca^{2+} ions at EF hands initiates aequorin oxygenase activity that converts coelentrazine to (liberated and excited) coelenteramide. Blue light is emitted at 469 nm on the return to ground state. The light emitted is used to estimate $[Ca^{2+}]_{cyt}$ (13–15). Expression of apoaequorin has been targeted to individual cell types (1) and subcellular locations including the stroma and nucleus (15, 16). Its use is not limited to *Arabidopsis*; lines available also include *Nicotiana* and *Medicago* (15).

Aequorin luminometry has been used to explore the $[Ca^{2+}]_{cyt}$ response to both salinity and oxidative stress (e.g., (1–6)). The main problem with aequorin is that insufficient light is emitted to resolve single cell responses. However, specific organs can be used (leaves or roots) and real-time changes in whole seedling $[Ca^{2+}]_{cyt}$ in response to salinity or oxidative stress have been captured with a photon-counting camera (3, 5). With this approach, it has been possible to quantify regional responses of roots and shoots. Use of whole organs means that output is from a range of cell types in different developmental states; detail is lost (17). Cell-specific apoaequorin expression in *Arabidopsis* allows the response of root epidermal, endodermal and pericycle cells to be quantified in situ, but this is still a population level response (1). Here, we describe protocols for examining effects of salinity and oxidative stress on single roots. Additionally, we describe a basic protocol for using root epidermal protoplasts as an important cell type in detecting environmental change (2, 4).

2. Materials

Use molecular grade reagents; high purity D-sorbitol is advised to minimize metal contamination. We use D-sorbitol rather than mannitol as an osmoticum because the latter can quench the radicals that might be produced during the stress response.

2.1. Plant Material

The approach described here uses *Arabidopsis thaliana* Col-0, transformed to express apoaequorin constitutively, under the control of a 35S promoter. Such transgenic plants should be freely available from published authors. Plants described here were transformed by an *Agrobacterium tumefaciens* strain GV301 harboring a modified vector pJahwohl8-RNAi gateway compatible vector (pJAA) containing a 35S::AEQ construct and the *PAT* (dl-*phosphinothricin acetyl transferase*) gene, which confers resistance to the herbicide phosphinothricin (PPT) (ref. (17)). Transformants were selected by growing the plants under PPT throughout. A T3 generation is needed, but we find less reliable results beyond that generation.

2.2. Specialist Equipment

1. Plate reader luminometer such as the FLUOstar OPTIMA (BMG labtech; (http://www.bmglabtech.com), see Note 1).
2. White, flat-bottomed 96-well plates (Greiner Bio-One http://www.greinerbioone.com/; catalogue number 655075).
3. Osmometer.

2.3. Luminometry Solutions

1. Coelentrazine solution: Dissolve 1 mg of native coelentrazine (http://www.biosynthe.com/; catalogue number C-7001) in 1,250 μL HPLC grade methanol. Aliquot 50 μL into 24 lightproof microfuge tubes and vacuum concentrate for 15–20 min on medium heat. Store powder at −80°C.
2. Assay buffer: 2 mM Tris-base, 10 mM $CaCl_2$, 0.1 mM KCl, adjust to pH 5.8 with MES. Filter sterilize with a 0.22 μm filter. For use with root protoplasts, include 4.5% (w/v) D-sorbitol to give a final osmolarity of 270 mmol/kg (see Note 2).
3. Test solutions. It is to the assay buffer in step 2 that NaCl or oxidative stress agents will be added (see Note 3).
4. Discharge solution: 30% (v/v) ethanol (HPLC grade), 3 M $CaCl_2$.

2.4. Protoplasting Solutions

1. Buffer: 10 mM $CaCl_2$, 10 mM KCl, 2 mM $MgCl_2$, 0.1% (w/v) bovine serum albumin (BSA; 99% purity), 2 mM MES, adjust to pH 5.5 with Tris-base. Filter sterilize, and store at 4°C for no more than 1 month.
2. Enzyme digest solution: For one set of protoplasts, add 1 mL enzyme buffer to 0.0001 g Pectolyase Y-23 from *Aspergillus*

japonicas (Kyowa Chemical Products; UK distributor is R.W Unwin & Co. Ltd http://www.rwunwin.co.uk/), 0.0015 g Cellulase Onozuka R-10 (can be obtained from Duchefa Biochemie http://www.dechefa.com/), 0.001 g Cellulysin Cellulase from *Trichoderma viride* (can be obtained from VWR International http://www.vwr.com/), 0.03 g D-Sorbitol (see Note 4).

3. Methods

3.1. Plant Growth

All plants are grown on half strength Murashige and Skoog (Duchefa), 1% (w/v) sucrose, 0.8% Bacto Agar (BD, Oxford, UK), pH 5.7. Seeds are sterilized by vortexing for 1 min in 70% (v/v) ethanol, followed by 10 min incubation in 30% (v/v) sodium hypochlorite, 0.01% (v/v) Triton X-100. Seeds are then washed in five changes of sterile deionized water before sowing. Stratification for 48 h in the dark at 4°C is followed by growth (vertical plates) at 25°C, 16 h light: 8 h dark, (78 μmol/m^2/s irradiance).

3.2. Verification of Plant Material by Discharge Test

1. Reconstitute an aliquot of coelentrazine in 10 μL HPLC grade methanol.
2. Transfer into 20 mL assay buffer (this should give a final concentration of 5 μM, but 10 μM can also be used). This coelentrazine solution should be kept in the dark as much as possible. From here on, it is advisable to work under green or very low light if possible.
3. Pipette 150 μL coelentrazine solution into a well of a flat-bottomed white plate. Repeat for as many seedlings as you will test. Volume of solution will vary with well size; descriptions here are for a 96-well plate.
4. Using sterile, fine-tipped forceps, place a 6-day-old whole seedling into a well, make sure it is immersed. Leaf tissue from older plants can also be used. Use untransformed plant material as a control. Cover the plate with foil and leave overnight at room temperature.
5. Transfer the plate into the luminometer and record photons once every second. After the first 10 s or when a steady baseline is evident, 150 μL discharge solution can be injected into the samples.
6. There should be a rapid but transient increase in photon count as the coelentrazine reacts. Continue to measure photons for another 60 s.
7. The total photon count for a seedling should be at least 10^6.

3.3. Response of Individual Roots to Stimuli

1. Reconstitute an aliquot of coelentrazine in 10 μL HPLC grade methanol.
2. Transfer into 20 mL assay buffer (brought to room temperature) and mix. Protect from light and work in green or very low light.
3. Pipette 5 mL of coelentrazine solution into a 4 cm diameter sterile petri dish.
4. Using sterile (fine-blade) scissors, excise roots from 6-day-old seedlings. Transfer with sterile forceps (holding root gently near the cut end) and immerse in coelentrazine solution for 3–4 h in the dark (see Note 5). Untransformed plants act as a control. Scalpel blades can be used instead of scissors.
5. Wash roots by transferring into assay buffer without coelentrazine (this should have been brought to room temperature before use).
6. Using sterile forceps, place a single excised root into the well of a white 96-well plate with 100 μL assay buffer without coelentrazine. Ensure that the whole root is immersed in buffer.
7. Transfer the plate into the plate reader and allow the plant to recover in the dark for 30 min.
8. To establish a baseline photon count, take measurements every second for 35 s (see Note 6).
9. Injection of buffer (100 μL, with or without a test substance) can be done at the 35th second mark (see Note 7).
10. Photon counts can then be recorded every second for 30 s, then every 15 s for a further 2,000 s to observe any longer term changes in $[Ca^{2+}]_{cyt}$.
11. Discharge solution can then be injected and photon counts recorded for a further 60 s.
12. Photon counts are summed up to obtain the total counts. This can be used to calculate the $[Ca^{2+}]_{cyt}$ (see data analysis, Subheading 3.6).

3.4. Isolation of Protoplasts from Mature Root Epidermis

1. One protoplast preparation requires approximately 60–80 roots from 6-day-old plants.
2. For each seedling, excise the root from the shoot and cut approximately 1 cm from root apex using sterile scissors/scalpel blade (see Note 8). Discard apices.
3. Transfer root segments onto a glass slide using sterile forceps. To avoid desiccation, place a small volume of sterile deionized water onto the slide.
4. Once all the root segments are on the glass slide, soak up the water with tissue ensuring that you do not touch the root material.

5. Cut the root samples using a sterile razor blade. Cut them into approximately 0.5–1 mm segments (see Note 9).
6. Transfer the mature root pieces to 1 mL enzyme digest solution in a sterile container (such as a 15 mL capped centrifuge tube: Greiner).
7. Incubate the root pieces at 30°C, shaking at 60 rpm for 90 min (see Note 10).
8. Filter the solution through 50 μm pore mesh and collect protoplasts with a 1 min centrifugation at 7,000 × *g*. Discard the supernatant and resuspend the protoplasts in 500 μL assay buffer with D-sorbitol if you want to use the protoplasts for non-luminometric work. The yield can vary from 45 protoplasts per μL to 70 protoplasts per 1 μL. For luminometry, resuspend the protoplasts in 500 μL coelentrazine solution with D-sorbitol (see Subheading 3.5, step 3).
9. Protoplasts should be used immediately for experiments.

3.5. Responses of Mature Root Epidermal Protoplasts Expressing Apoaequorin to Stimuli

1. Reconstitute an aliquot of coelentrazine in 5 μL HPLC grade methanol.
2. Transfer into 10 mL assay buffer with D-sorbitol (this should give a final concentration of coelentrazine of 10 μM). The coelentrazine solution should be kept in the dark as much as possible.
3. Use 500 μL coelentrazine solution from step 2 to resuspend the protoplasts isolated in step 8 of Subheading 3.4.
4. Incubate the protoplasts for 3–4 h in the dark.
5. Centrifuge the protoplasts at 4,500 × *g* for 1 min and discard the supernatant.
6. Wash the protoplasts by resuspending in the dark in 1 mL assay buffer **with** sorbitol but **without** coelentrazine.
7. Centrifuge the protoplasts again at 4,500 × *g* for 1 min and discard 500 μL of supernatant.
8. Resuspend the protoplasts in the dark and add 100 μL of the protoplast suspension into a well of a white 96-well plate.
9. Transfer the plate into the luminometer and allow the protoplasts to recover in the dark for 30 min.
10. Proceed as Subheading 3.3, step 8.

3.6. Data Analysis

Counts can be used to estimate $[Ca^{2+}]_{cyt}$ by using the following equation, as described in ref. (14):

$$-\log[\mathrm{Ca}^{2+}]_{\mathrm{cyt}} = a \cdot \left[-\log\left(\frac{L_{\mathrm{c}}}{L_{\mathrm{t}}}\right)\right] + b$$

where L_c = luminescence counts per second, L_t = total luminescence counts, a = 0.3326, b = 5.559. Consulting ref. (14) on the use of this equation is strongly advised, particularly if you cannot control the temperature in the plate reader.

4. Notes

1. Choose a luminomter with temperature control and programmable delivery of test solutions. Luminescence output should be readily transferred into data handling software. Once you have the plate reader luminometer under control, stick with it—no two are the same. Alternatively, a cuvette-based luminometer can be used. With that, test solutions are injected with a syringe and care has to be taken to exclude light and ensure reproducible injection rates. It is important to be able to adjust the osmolarity of solutions to prevent protoplast bursting and to generate accurate osmotic controls for salinity tests.
2. Composition of this basal assay medium can be varied. Here we use high Ca^{2+} to mimic the high Ca^{2+} of sodic soils but you could repeat experiments with lower levels, or omit or chelate it for evidence of Ca^{2+} entry from the apoplast. The level of sorbitol we use for protoplasts applies to the growth conditions we use; you may find you need to vary this to optimize yield.
3. For every test concentration of NaCl, you will need a D-sorbitol osmotic control, so you can elucidate any NaCl-specific effects. The most commonly used oxidative stress is H_2O_2. Keep the stock refrigerated and adhere to the use-before date. Hydroxyl radicals can be generated using $CuCl_2$ and ascorbate, with or without H_2O_2. We have used a maximum of 1 mM $CuCl_2$ with 1 mM ascorbate (2) but also 0.2 mM $CuCl_2$, 0.2 mM ascorbate with 2 mM H_2O_2 (18). The copper and ascorbate solutions should be prepared freshly and delivered simultaneously from separate reservoirs if possible. Check that the pH of your assay holds steady. Superoxide anions can be generated with 1 mM hypoxanthine plus xanthine oxidase (0.01–100 munits mL^{-1}), but this can also produce H_2O_2, necessitating incorporation of catalase (11).
4. Wear gloves and face mask, the enzymes are fine powders. Vortex to dissolve. Incubate the mixture at 30°C for about 10 min. Centrifuge at 11,900 × g for 1 min. Collect the supernatant and use the same day. Some researchers make enzyme stocks, aliquot and freeze. You will find that you may need to adjust concentrations and incubation times for your plants; batches of enzymes from the same supplier can vary in their

activity, so you'll need to check your protocol for each new batch. Keep checking your protoplast yields; don't assume they will be the same each time you do an isolation. If yield suddenly drops, then check solution pH, make a new enzyme mix, and be prepared to re-order enzymes.

5. There is good evidence that time of day can influence the $[Ca^{2+}]_{cyt}$ response to cold stress (17), so conducting your experiments at a set time of day is advisable.

6. The movement of the plate into position could evoke a touch response. Incorporating a delay of around 30 s into the measurement programme should account for this.

7. Start by injecting control solution only, so you can determine the effect of flow rate on the touch response. This should be a very rapid $[Ca^{2+}]_{cyt}$ transient and needs to be minimized, so you can better resolve the response to test solution. Note that when making assay buffer containing test substance, use double the concentration you want. For example, if you want to test the effect of 200 mM NaCl on roots, the test buffer should contain 400 mM NaCl since it will be diluted by the control buffer that is already in the well. To investigate the effect of inhibitors or Ca^{2+} channel blockers (e.g., Gd^{3+}, La^{3+}), try incorporating them into the assay buffer for step 7 when the plants are in the plate reader and in the dark.

8. This method is adapted from refs. (2, 7, 18). You'll need to judge which areas to select from plants grown under your conditions. The apices can be used for isolating epidermal protoplasts from the elongation zone, but you will need more material and possibly a shorter incubation time.

9. This chopping step is important as too little chopping or weak chopping action may result in low protoplast yield. Don't chop for too long or segments turn brown, because yield and quality will decline.

10. When starting out, view the protoplast release, so you can judge the point of origin. Ideally, work up your protocol with plants expressing GFP in the epidermis (1). Protoplasts from the mature epidermis are around 20 μm diameter. Checking viability (with a stain such as Evans Blue) is a good idea, particularly after stress/inhibitor exposure.

Acknowledgements

We thank Vadim Demidchik, Jenny Mortimer, Zhonglin Shang for all their efforts in establishing these protocols and Alex Webb for the aequorin construct. This work was funded by the BBSRC, the Frank Smart Trust and the Brookes Trust.

References

1. Kiegle E, Moore C, Haseloff J et al (2000) Cell-type specific calcium responses to drought, NaCl, and cold in *Arabidopsis* root: a role for endodermis and pericycle in stress signal transduction. Plant J 23:267–278
2. Demidchik V, Shabala SN, Coutts KB et al (2003) Free oxygen radicals regulate plasma membrane Ca^{2+}- and K^+-permeable channels in plant root cells. J Cell Sci 116:81–88
3. Rentel MC, Knight MR (2004) Oxidative stress-induced calcium signalling in *Arabidopsis*. Plant Physiol 135:1471–1479
4. Demidchik V, Shabala SN, Davies JM (2007) Spatial variation in H_2O_2 response of *Arabidopsis thaliana* root epidermal Ca^{2+} flux and plasma membrane Ca^{2+} channels. Plant J 49:377–386
5. Tracy FE, Gilliham M, Dodd AN et al (2008) NaCl-induced changes in cytosolic free Ca^{2+} in *Arabidopsis thaliana* are heterogeneous and modified by external ionic composition. Plant Cell Environ 31:1063–1073
6. Demidchik V, Cuin TA, Svistunenko D et al (2010) Arabidopsis root K^+-efflux conductance activated by hydroxyl radicals: single-channel properties, genetic basis and involvement in stress-induced death. J Cell Sci 123:1468–1479
7. Mahajan S, Pandey GK, Tuteja N (2008) Calcium- and salt-stress signaling in plants: shedding light on SOS pathway. Arch Biochem Biophys 471:146–158
8. Chung J-S, Zhu J-K, Bressan RA et al (2008) Reactive oxygen species mediate Na^+-induced *SOS1* stability in Arabidopsis. Plant J 53:554–565
9. Halperin SJ, Gilroy S, Lynch JP (2003) Sodium chloride reduces growth and cytosolic calcium, but does not affect cytosolic pH, in root hairs of *Arabidopsis thaliana* L. J Expt Bot 54:1269–1280
10. Mittler R, Vanderauwera S, Suzuki N et al (2011) ROS signalling: the new wave? Trends Plant Sci 16:300–309
11. Taniguchi N, Gutteridge JMC (2000) Experimental protocols for reactive oxygen and nitrogen species. Oxford University Press, Oxford
12. Swanson SJ, Choi W-G, Chanoca A et al (2011) In vivo imaging of Ca^{2+}, pH and reactive oxygen species using fluoresecent probes in plants. Annu Rev Plant Biol 62:273–297
13. Knight MR, Campbell AK, Smith SM et al (1991) Transgenic plant aequorin reports the effects of touch and cold-shock and elicitors on cytoplasmic calcium. Nature 352:5 24–526
14. Fricker MD, Plieth C, Knight H et al (1999) Fluorescence and luminescence techniques to probe ion activities in living plant cells. In: Mason WT (ed) Fluorescent and luminescent probes for biological activity. Academic, San Diego
15. Mithöfer A, Mazars C (2002) Aequorin-based measurements of intracellular Ca^{2+}-signatures in plant cells. Biol Proced Online 4:105–118
16. Sai J, Johnson CH (2002) Dark-stimulated calcium ion fluxes in the chloroplast stroma and cytosol. Plant Cell 14:1279–1291
17. Dodd AN, Jakobsen MK, Baker AJ et al (2006) Time of day modulates low-temperature Ca^{2+} signals in Arabidopsis. Plant J 4:962–973
18. Foreman J, Demidchik V, Bothwell JMH et al (2003) Reactive oxygen species produced by NADPH oxidase regulate plant cell growth. Nature 422:442446

Chapter 11

Infrared Thermography in Plant Phenotyping for Salinity Tolerance

Richard A. James and Xavier R.R. Sirault

Abstract

The recent advances made in the use of infrared thermal imaging (thermography) as a non-invasive, high-throughput technique for the screening of salinity tolerance in plants is reviewed. Taking wheat seedlings as an example, the methods and protocols used to impose a homogeneous salt stress to a large number of genotypes, as well as capturing infrared images of these genotypes and automatically processing the images are described in detail in this chapter. We also present the source code of the Matlab program applied to automatically identify plants and batch process IR images.

Key words: Salt, Wheat, Barley, Transpiration rate, Stomatal conductance, Osmotic stress

1. Introduction

The growth of crop plants in response to salinity is initially reduced by the decrease in soil water potential (1). This is essentially a water stress that results from the osmotic effect of salts in the soil solution. This makes it harder for the roots to extract water from the soil, so is similar to that imposed by soil drying (2). Growth can also decrease as a result of the salts taken up by the plant to toxic concentrations. This "salt-specific effect" usually occurs later—over a period of weeks to months—as salts accumulate in the older leaves. This build-up of salt results in leaf injury and death and a subsequent reduction in the supply of assimilate to the growing regions (1, 3).

Salt tolerance mechanisms in crop plants are generally described under three categories (4): (1) tolerance of high external salt concentrations in the soil, called "osmotic stress tolerance," (2) control

Sergey Shabala and Tracey Ann Cuin (eds.), *Plant Salt Tolerance: Methods and Protocols*, Methods in Molecular Biology, vol. 913, DOI 10.1007/978-1-61779-986-0_11, © Springer Science+Business Media, LLC 2012

of salt uptake, and (3) tolerance of high internal salt concentrations in leaves, called "tissue tolerance." Osmotic stress is the major cause of the reduction in growth and yield of cereals in saline soils.

Genotypic variation for osmotic stress tolerance can be assessed by measuring stomatal conductance because it is likely that the factors controlling growth of plants under osmotic stress also regulate stomatal conductance. Stomatal conductance is reduced immediately with the onset of salinity (5), and follows a virtually identical response profile to that of growth (6). Additionally, over a time-scale of days, the reduction in growth rate is often matched by a similar reduction in stomatal conductance (7–9).

However, phenotyping for osmotic stress tolerance by measuring stomatal conductance is problematic. This is due to poor repeatability that largely results from stomatal sensitivity to small changes in environmental conditions. The precision and speed of porometry also affect the accuracy and reproducibility of measurements. An alternative approach for measuring stomatal conductance uses infrared (IR) thermography. Leaf temperature varies with transpiration rate (10), which is largely a function of stomatal conductance. As infrared measurements of leaf temperature correlate well with estimates of stomatal conductance (11, 12), this technology can be used to assess osmotic stress tolerance in plants (11). Tolerance to osmotic stress is the only salt tolerance mechanism that is amenable for accurate and specific phenotyping using IR thermography.

Infrared thermography can graphically display the surface temperature distribution of a plant by focusing the long-wave radiation emitted and reflected by that plant onto a temperature-sensitive detector called a micro-bolometer. The plant's temperature determines how much radiation it will emit. Therefore, the measurement of thermal radiation emitted by the plant can be used as an indicator of its temperature.

Screening for phenotypic variation in osmotic stress tolerance is contingent on the ability to measure the specific effects of osmotic stress on seedling temperature without the complication of other salinity-induced changes that may also impact on seedling temperature. This could include changes in leaf morphology, and the technique requires that measurements are taken before salt levels in the leaves build up to potentially toxic concentrations and before the age-associated decline in stomatal conductance (13). Taking these factors into consideration, IR thermography measurements should ideally be completed on young seedlings (leaf 2–3 stage for cereals), shortly after the final desired concentration of salinity is attained (3–5 day). Measurements on young seedlings have additional benefits such as more accurate IR measurements and greater high-throughput screening potential.

Tolerance to osmotic stress is best assessed by simultaneously measuring the temperature of seedlings of the same variety treated with salt relative to non-salt controls. This overcomes complications due to any potential intrinsic variation in leaf temperature determinants such as stomatal conductance, waxiness, or color. It also overcomes the need for a temperature reference to determine the absolute temperature of the seedlings. Determining absolute leaf temperatures can be problematic due to a number of confounding factors such as changes in ambient temperature and humidity or reflected background radiation that could introduce errors into this measurement.

In this chapter, we outline the process of growing plants in salt-stressed conditions suitable for early IR thermography measurements. The actual IR thermography measurement process is described for wheat and barley plants grown hydroponically along with details of the subsequent automated image capture and analysis protocols.

2. Materials

2.1. Seed Preparation

1. Petri dishes (90 mm diameter).
2. Filter paper (Whatman Grade 1 circles, 90 mm).
3. 1% hypochlorite solution.
4. Thiram 800 (or other suitable antifungal agent).

2.2. Growing Media

1. 2 M $CaCl_2$ stock solution.
2. 5 M NaCl stock solution.
3. Hoagland's nutrient solution.
4. 50:50 coarse river sand:perlite mix.
5. Quartz gravel (granulation 8–12 mm).
6. Aquarium pumps (>400 L/h).
7. Hydroponic containers.

2.3. Acquiring Thermal Images

1. Bronze tinted Polymethyl methacrylate plate (e.g., Perspex™ ca. 40 × 60 cm, PlastiX, Sydney).
2. Infrared camera (ThermaCAM SC660 IR Camera) (FLIR Systems Inc., Boston, MA, USA, or equivalent, i.e., thermal resolution better than 0.050 K).
3. Firewire cable to connect camera to computer.
4. Tripod on which to mount the infrared camera.
5. Computer with a Firewire port and ThermaCam Researcher Professional software version 2.9 or above (FLIR Systems Inc.).

6. Controlled environment chamber with temperature and relative humidity control (e.g., Conviron chamber model PGC20 with CMP6050 Control System, Controlled Environments Ltd., Winnipeg, Canada).

2.4. Processing Thermal Images

1. Matlab software version 2009a or above (The Mathworks, Natick, MA, USA).
2. The Imaging Processing Toolbox version 6.3 or above (The Mathworks, Natick,. MA, USA).
3. Microsoft Excel version 2007 or above (Microsoft. Microsoft Excel. Redmond, Washington: Microsoft, 2007).

3. Methods

3.1. Seed Preparation

1. Select seeds for uniform shape and weight (for wheat and barley, select uniform seeds within a 5 mg weight range).
2. Surface sterilize seeds with 1% hypochlorite or a fungicide such as Thiram (1.4 g/L). Seed fungal infections on young seedlings can impair plant performance. For example, Na^+ uptake to the shoot can increase two to threefold with a fungal infection on the seed and crown of a young salt-treated wheat seedling.
3. Imbibe (1–2 h de-ionized H_2O) and germinate on moist filter paper (infused with antifungal agent such as Thiram 800) in Petri dishes for 2 days at 4°C. Plant only uniformly germinated seeds.

3.2. Growing Media

There are a number of different ways to grow and treat plants with salinity. The key considerations to determine which method to use relate to whether the plants are nutritionally well supplied, are well drained, aerated and not waterlogged, and importantly, the constancy and homogeneity of the salinity treatment to which they are exposed. Below are described two different growing methods that when used carefully, account for these key issues.

3.2.1. Supported Hydroponics

1. Supported hydroponics is preferable over hydroponics because it provides a supportive matrix around the roots. This prevents root breakage when solutions are changed or with bubbling for aeration. This matrix can vary. We use quartz gravel (14, 15), whereas other groups have used polycarbonate pellets (16).
2. The supported hydroponics system we use consists of a number of 40-L containers, each holding 40 square pots of 6.5 cm width and 15.8 cm depth. These contain quartz gravel and sit on a perforated stage (2 cm high) within the container. This set

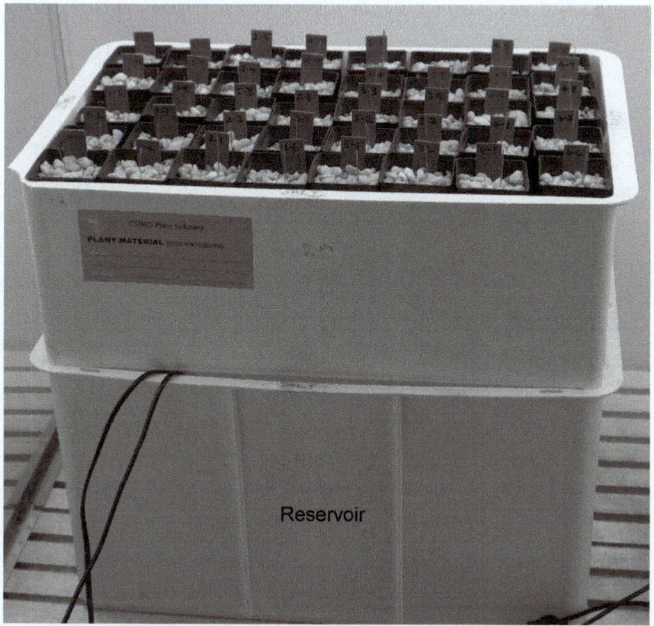

Fig. 1. Supported hydroponics system used for imposing a controlled and homogeneous salt-stressed to *Triticum aestivum* L. plants.

up sits upon a 60-L container of the identical width and length. This contains a 50-L reservoir of a nutrient and salt solution that is periodically pumped into the upper chamber and drained back into the reservoir for continual reuse. The frequency and duration of sub-irrigation events is controlled by a timer (Fig. 1).

3. Plant germinated seeds (one per pot) to a depth of about 1 cm. Carefully cover with fine quartz gravel, avoiding damage to either the roots or the small coleoptile. Water gently and cover pots with wet towels and insulation covers to maintain moist conditions during seedling establishment. Remove covers when coleoptiles emerge (24–48 h) and commence sub-irrigation with ¼ strength Hoagland's nutrient solution at a frequency of 30 min.

4. Increase to ½ modified Hoagland's nutrient solution (P reduced from 1 mM to 100 μM) 7 days after emergence (DAE). Solutions should be changed weekly and pH monitored and adjusted regularly because the nutrient solution is weakly buffered. For the salt treatment, at 8 DAE, add 25 mM NaCl to the irrigation solution once or twice daily until the required salt concentration is reached. We typically use 100–150 mM NaCl (see Note 1) for screening wheat and barley seedlings.

5. Supplemental Ca^{2+} ($CaCl_2$) should be added with the commencement of the salt treatment to maintain a Na^{+}:Ca^{2+} ratio of 15:1, based on the final concentration of the salinity (see Note 2).

6. For the control plants, maintain a duplicate hydroponics system containing only ½ modified Hoagland's nutrient solution (P reduced from 1 mM to 100 μM).

3.2.2. Sand/Perlite Media

1. Pack pots with a 50:50 mix of coarse river sand:perlite, leaving a 2 cm space at the top of the pots for irrigation events. The pots should be much taller than they are wide to assist with drainage and help prevent waterlogging (17). Fine mesh is required at the base of the pot to prevent loss of sand. It may be necessary to provide a larger matrix layer (e.g., large quartz gravel) in the base of the pot to assist with drainage.
2. Flush pots several times with tap water to wash out any fine silt or dirt. Drainage to approximate field capacity is important after every flush or irrigation event to prevent waterlogging. After the initial free drainage, extra water can be removed by tilting pots at 45° for 5–10 min. Additionally, wicks can be inserted in the base of the pots or pots can be placed on layers of towels, depending on how open the base of the pot is for good contact of soil to towel layer.
3. Plant germinated seeds 2 cm deep, one per pot. Sprinkle with water and cover pots to reduce evaporation and help maintain a humid environment. Remove covers when the coleoptiles begin to emerge. Keep the top layer of soil moist over the next few days to ensure root growth and seedling establishment.
4. Commence gentle watering (~5 mL) with ¼ Hoagland's nutrient solution at 3 (DAE) and increase to ½ Hoagland's nutrient solution at 7 DAE.
5. At 8 DAE, flush the pots with 25 mM NaCl, followed by adequate drainage to approximate field capacity. Once or twice daily, incrementally increase the concentration of the salt solution used for flushing by no more than 25 mM NaCl, until the desired NaCl concentration is achieved. Supplemental Ca^{2+} ($CaCl_2$) to maintain a Na^+:Ca^{2+} ratio of 15:1 based on the final concentration of the salinity is also advised when using the sand/perlite media.
6. Pots should be watered daily to excess and then adequately drained (as above) with the final salt solution (also containing supplementary Ca^{2+}). This will minimize the build-up of salts in the rhizosphere as the plants continue to take up water.
7. A duplicate set of pots should be set up and watered daily with a ½ Hoagland's nutrient solution for the control.

3.3. Growing Conditions

1. The conditions in the controlled environment chamber should be set to maximize the IR signal. Set light intensity between 500 and 1,000 μmol m^{-2} s^{-1}; optimal for cereals such as wheat and barley. Similarly, by acquiring thermographs of seedlings in

an atmosphere with a VPD ranging from 1.2 to 1.6 kPa, one can maximize transpiration rate while maintaining lower stomatal conductance. This keeps the seedling slightly cooler than the background, while maintaining higher effective sensitivity even at high salinity levels.

2. Care is needed to prevent a build-up of CO_2 (see Note 3) when taking IR measurements in a confined space such as a controlled environment chamber. To overcome this issue, the IR camera should be controlled remotely and the operator needs to minimize the time spent in the controlled environment chamber between readings.
3. Age and growth stage of seedlings is important for the accuracy of IR measurements. Wheat and barley seedlings with 2.5–3 leaves is an ideal growth stage so that the leaves fall in the same orientation, perpendicular to the IR camera. On more advanced plants, some of the leaves go out of focus as they fall away from the perpendicular plane with the IR camera. Consequently, they give a different IR signal to those leaves in focus.
4. Once the desired salt concentration is achieved, allow 2–3 day for the plants to equilibrate with the new conditions before commencing IR readings.

3.4. Acquiring Thermal Images

1. Thermal images of seedlings should be acquired between 11h00 and 14h00 in the controlled environment chamber described above; both stomatal conductance and transpiration tend to be relatively constant over this time period.
2. The background is particularly important when taking IR thermographs of plants. We used a bronze tinted acrylic (Perspex™) plate as a background (see Note 4).
3. For acquisition, a ThermaCAM SC660 infrared camera (FLIR systems Inc, MA, USA) is used (see Note 5). The camera is placed at a distance of ca. 90 cm from the plant and is left in the growth chamber for about 2 h before starting the IR measurements series. This allows the optics of the IR camera to reach thermal equilibrium with the air temperature.
4. A default emissivity value of 0.95 is entered in the camera settings. Emissivity of most vegetation is usually deemed to be ca. 0.95 (see Note 6).
5. Similarly, the distance of the camera to the subject needs to be measured and entered into the camera settings, together with air temperature and relative humidity (see Note 7).
6. Place pots side by side in front of the Perspex™ plate (Fig. 2) and avoid overlapping leaves. Pots should be ~10 cm apart. Allow a set amount of time of about 1–2 min before taking an IR image. During that time, focus on the image using the cameras controls for a clear IR thermograph (Fig. 3), and make

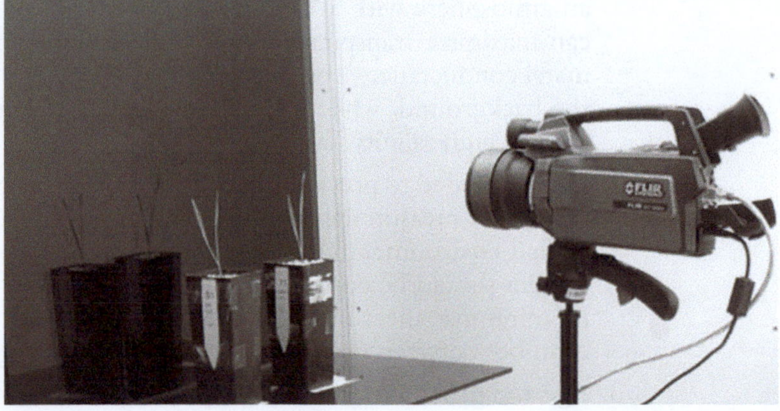

Fig. 2. Thermographic set up for assessing leaf temperature differences simultaneously of a salt-treated seedling on the *left* relative to a non-salt (control) seedling on the *right*. The infrared thermograph is taken within a Conviron growth chamber (model PGC 20) with controlled temperature, humidity, and light.

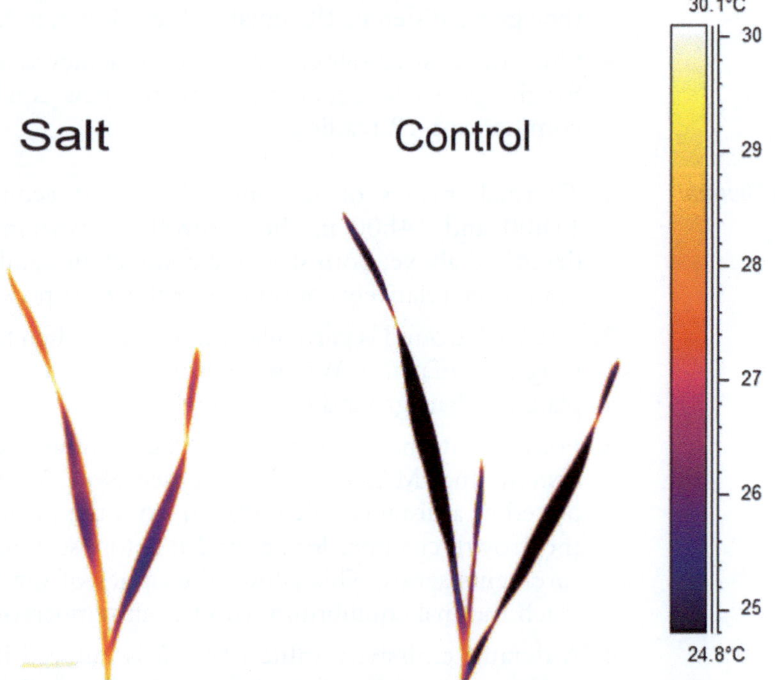

Fig. 3. False color, thermal infrared image of wheat seedlings (cultivar Axe) in 100 mM NaCl (*left*) vs. non-salt "control" (*right*) visualized as an 8-bit grey image (iron bow-palette look-up table). The background temperature is out of scale, thus appearing "*white*." Seedlings are 2 weeks old.

small adjustments to the orientation and location of pots as necessary. In this set up the relative difference in leaf temperature due to salinity is measured. We do not try to determine absolute leaf temperature (see Note 8).

7. IR Images are stored directly in a folder on the computer hard drive.

3.5. Processing Thermal Imaging

1. Convert IR images into Matlab format files (*.mat extension) using ThermaCAM Researcher Pro or FLIR Quickreport software.
2. Load files into Matlab software release 2011a (The MathWorks, Natick, MA, USA) and check that the matrices are read as matrices of raw temperature in units of Kelvin.
3. Transform matrices into 8-bit resolution, grey level images (Fig. 4) using the custom-built function "temp2img.m" (see Note 9).
4. Use the Matlab script (which labels the custom-built function, "temp2img.m") to automatically identify the seedlings in the 8-bit grey image (see Note 10). The code was written to batch process hundreds of IR images at a time. The detection algorithm relies on Otsu's automatic thresholding method (11). This assumes only two populations of pixels in an image, i.e., background and foreground (see Note 11). A binary matrix is then computed to create a mask or silhouette of the plant (Fig. 5) in the image. The mask is used to derive the temperature of the two seedlings in the thermograph by multiplying the original image by the mask values according to arithmetic array rules, i.e., element by element multiplication (Fig. 6). The Matlab script then computes the average temperature for each seedling and reports the difference in temperature between the control and salt-treated seedling within a MS Excel spreadsheet (Table 1).

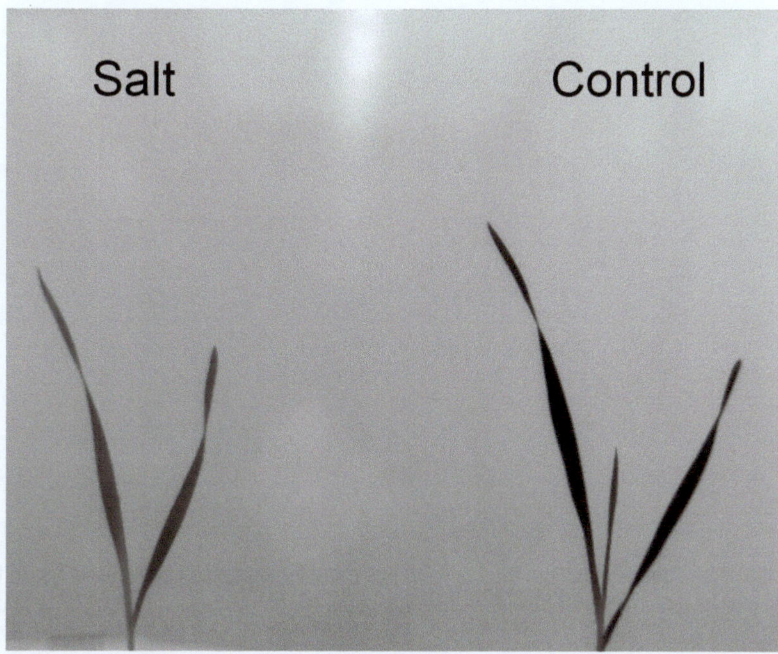

Fig. 4. Thermal infrared image of wheat cultivar Axe seedlings after applying the Matlab "temp2img.m" function.

Fig. 5. Masked image of wheat cultivar Axe seedlings after applying Otsu' thresholding method. Selected pixels are colored in "*white*." Please note that an erosion of one pixel has also been carried on the mask.

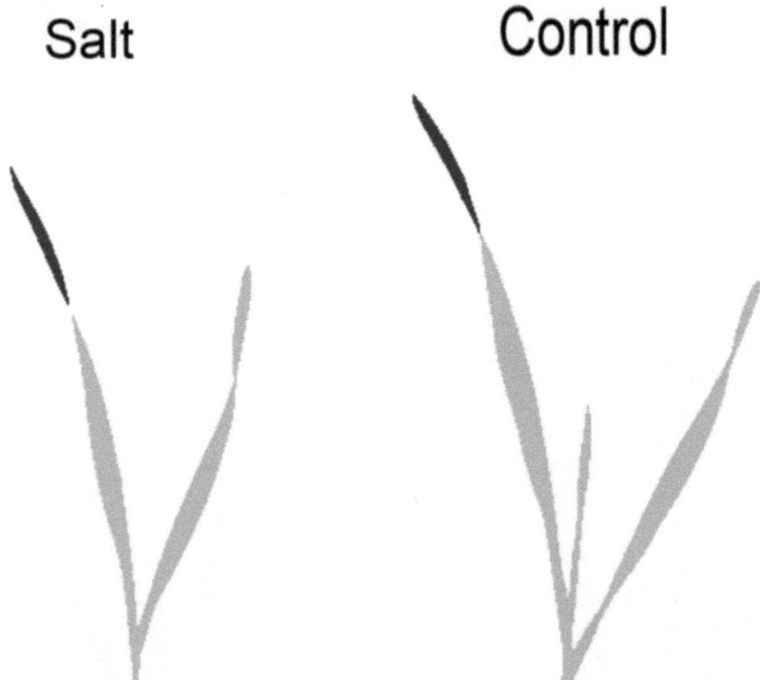

Fig. 6. Results of the segmentation algorithm for analyzing infrared thermographic image; seedlings in the binary matrix are labeled with an integer value using the "bwlabel" function of the image-processing toolbox 9The Mathworks.

Table 1
Tabulated excel format output

Filenames	Temps_left (Kelvin)	Temps_right (Kelvin)	Area_left	Area_right	Difference
D171_0001.MAT.png	300.20	298.48	3,464	4,835	-1.72
D171_0002.MAT.png	299.56	298.58	3,788	3,921	-0.98
D171_0003.MAT.png	299.64	298.52	2,723	3,696	-1.13
D171_0004.MAT.png	299.67	298.52	3,540	4,176	-1.15
D171_0005.MAT.png	299.71	298.57	3,298	3,709	-1.14
D171_0006.MAT.png	300.16	298.81	3,322	3,591	-1.35
D171_0007.MAT.png	299.96	298.90	3,068	3,559	-1.05

4. Notes

1. The slow incremental addition of salt concentrations is critical to avoid plasmolysis of root cells and to allow the plant time to osmotically adjust so as not to lose turgor and wilt (18).
2. If supplemental Ca^{2+} is not added, other cations such as Na^{+} out-compete for uptake sites, resulting in Ca^{2+} deficiency.
3. We have found that the CO_2 concentration can increase by at least 50% over a period of 30 min, due to expired air from an operator. This led to a decrease in stomatal conductance and transpiration rate by about 30%, resulting in "hotter" plants.
4. The benefit of Perspex™ is that it is readily available, it provides a homogenous background as it does not transmit IR radiation (it appears opaque in the IR) and it has a lower emissivity that results in an apparent temperature ca. 2°C "hotter" than the air temperature. Consequently, it provides a good contrast against the cooler transpiring seedlings. Note that at a high salt concentration, from and above 150 mM NaCl, the temperature difference between the plant and the background starts to shrink, eventually to less than 0.5°C, so the image of the salt-treated plant will start blending into the image of the background. This means that the image-processing algorithm will have to be modified. Remember that the ability to automatically separate the pixels representing the plant from the background depends on a consistent difference in apparent temperature between the two. A highly reflecting surface, i.e.,

with very low emissivity, (e.g., a polished metal) is unsuitable as a background for thermography study.

5. The SC660 infrared camera uses a focal plane array, uncooled micro-bolometer with a 640 × 480-detector array that gives a very good spatial resolution of the plant—good spatial resolution is critical when one looks at small seedlings. This model has also a very good thermal resolution (0.045 K) and accuracy (±1%).
6. The main problem in radiation thermometry is that the emissivity of the target is usually not exactly known. For plant material, values ranging from 0.94 to 0.97 have been measured. More importantly, the emissivity of a surface will vary with the angle of emission and the temperature of this surface; this is why care is taken to account for variability of leaf angles of the various plants. By screening seedlings rather than more advanced plants, leaf angles are no longer randomly distributed, providing a comparable surface to the IR camera.
7. Anything between the surface being measured and the detector will remove some of the thermal radiation in some region of the spectrum. This error is called "Absorption error." It can be corrected by the software if RH is measured before image acquisition.
8. As indicated previously, several factors (e.g., surface emittance, ambient temperature variation, humidity, background thermal radiation) introduce errors in the measurement of absolute leaf temperature. By measuring relative temperature differences between salt stressed and control seedling simultaneously, most of these problems are overcome.
9. The function scales and normalizes the temperature data in a grey level intensity image; the Matlab code for the function "temp2img" is indicated below:

```
function I = temp2img(T)
% function I = temp2img(T)
%
% Function to convert an input matrix of temperatures (in Kelvin)
  to an 8 bit grey level image suitable for viewing.
%% Inputs: T = matrix of temperatures
% Outputs: I = 8 bit grey level image representing temperatures
Tmax = max(T(:)); % Calculate the maximum temp in matrix
Tmin = min(T(:)); % Calculate the minimum temp in matrix
I_double = (T—Tmin)/(Tmax-Tmin); % Scale and normalize
  temp matrix
I = uint8(floor(255*I_double)); % convert from double to uint8
```

10. Matlab Source code (Script)

```
%% Batch processing IR images (FLIR SC660)
% Copyright 2010 Dr Xavier Sirault, HRPPC
%% Close all windows, clean Workspace, clean Command
   Window
close all; clear all; clc;
%% Get the name and number of files to be processed
fileNames = uigetfile('*.mat', 'multiselect', 'on')';
nData = length(fileNames);
%% Transform all *.mat files into temperature images with
   extension *.png
IRData = cell(size(fileNames)); %create preallocated cell
   'IRData' containing all temperature matrices
for i = 1:length(fileNames)
Data = load(fileNames{i,1});
I1 = fieldnames(Data);
I2 = getfield(Data,I1{1});
%Document 'IRData' cell
IRData{i,1} = I2;
% Visualization matrices of temperature
temperature_image = temp2img(I2);
% Writing images in current directory
fnam = sprintf('%s.png',fileNames{i,1});
imwrite(temperature_image,fnam);
end
%% Define PlantMask for each image
directory = which('D170_0001.mat.png');
filelist = dir([fileparts(directory) filesep '*.png']);
imageNames = {filelist.name}';
nImages = length(imageNames);
Masks = cell(size(imageNames)); %create preallocated cell
   'Masks' containing all PlantMasks
for i = 1:nImages
mat_png = imread(imageNames{i,1});
PlantMask = temp2img(mat_png);
%make a binary file
level = graythresh(PlantMask);
PlantMask = ~im2bw(PlantMask,level);
%figure, imshow(PlantMask)
```

```
fnam = sprintf('PlantMask_%s',imageNames{i,1});
imwrite(PlantMask,fnam);
%Visualization of what has been detected on each image
Duplicate = mat_png;
Duplicate(PlantMask) = 255;
fnam2 = sprintf('MaskedImage_%s',imageNames{i,1});
imwrite(Duplicate,fnam2);
%Fill the cell 'Masks' with all PlantMasks data
Masks{i,1} = PlantMask;
end
%% Create left and right Mask—applicable to all images
[row, col] = size(mat_png);
x = [0;320;320;0;0];
y = [0;0;480;480;0];
Left = roipoly(mat_png,x,y);
Right = ~Left;
%imshow(Right), figure, imshow(Left)
%% Combine left/right masks with PlantMask
PlantLeftMask = cell(size(Masks)); %preallocated cell for
  PlantLeftMask
PlantRightMask = cell(size(Masks));%preallocated cell for
  PlantRightMask
for i = 1:length(Masks)
PlantLeftMask{i,1} = and(Left, Masks{i,1});
PlantRightMask{i,1} = and(Right, Masks{i,1});
%remove all objects containing fewer than 50 pixels
PlantLeftMask{i,1} = bwareaopen(PlantLeftMask{i,1},50);
PlantRightMask{i,1} = bwareaopen(PlantRightMask{i,1},50);
end
%imshow(PlantLeftMask{1,1}), figure, imshow(PlantRight
  Mask{1,1})
%% Label each plant in the combined binary matrix with
  an integer value
IDPlantLeft = repmat(PlantLeftMask,1,5); %preallocated cell
  for Plant identification
IDPlantRight = repmat(PlantRightMask,1,5); %preallocated
  cell for Plant identification
%Plant on the left
```

```
for i = 1:length(PlantLeftMask)
LabelsLeft,numplantsLeft] = bwlabel(PlantLeftMask{i,1},4);
PlantLeft = regionprops(LabelsLeft,'Area');
% Plant Statistics
WholePlantLeft = [PlantLeft.Area];
WholePlantAreaLeft = sum(WholePlantLeft);
% Recombine Temperature matrix and PlantMask
Avg_Temps_Left = sum(sum(PlantLeftMask{i,1}.*IRData
  {i,1}))/WholePlantAreaLeft;
% Document IDPlantRight cell
IDPlantLeft{i,1} = LabelsLeft;
IDPlantLeft{i,2} = numplantsLeft;
IDPlantLeft{i,3} = WholePlantAreaLeft;
IDPlantLeft{i,4} = Avg_Temps_Left;
IDPlantLeft{i,5} = imageNames{i,1};
RGBLeft = label2rgb(LabelsLeft); % color display of identified
  objects
fnam3 = sprintf('Identified_Left_%s',imageNames{i,1});
imwrite(RGBLeft,fnam3);
end
%Plant on the right
for j = 1:length(PlantRightMask)
LabelsRight,numplantsRight] = bwlabel(PlantRightMask{j,1},4);
PlantRight = regionprops(LabelsRight,'Area');
% Plant Statistics
WholePlantRight = [PlantRight.Area];
WholePlantAreaRight = sum(WholePlantRight);
% Recombine Temperature matrix and PlantMask
Avg_Temps_Right = sum(sum(PlantRightMask{j,1}.*IRData
  {j,1}))/WholePlantAreaRight;
% Document IDPlantRight cell
IDPlantRight{j,1} = LabelsRight;
IDPlantRight{j,2} = numplantsRight;
IDPlantRight{j,3} = WholePlantAreaRight;
IDPlantRight{j,4} = Avg_Temps_Right;
IDPlantRight{j,5} = imageNames{j,1};
RGBRight = label2rgb(LabelsRight); % color display of
  identified objects
```

```
fnam3 = sprintf('Identified_Right_%s',imageNames{j,1});
imwrite(RGBRight,fnam3);
end
%% Calculate difference in temperature (Left Plant—Right Plant)
Diff_Temps = cell(size(fileNames));
for k = 1:length(fileNames)
Temps = IDPlantLeft{k,4}—IDPlantRight{k,4};
Diff_Temps{k,1} = Temps;
end
%% Store results in EXCEL
A = IDPlantLeft(:,5);
B = IDPlantLeft(:,4);
C = IDPlantRight(:,4);
D = IDPlantLeft(:,3);
E = IDPlantRight(:,3);
results =
[{'FILENAMES','Temps_left(Kelvin)','Temps_right (Kelvin)','Area_left','Area_Right'};A,B,C,D,E];
xlswrite('IR_Results',results,'','A1');
```

11. If more than two populations of pixels are present, other methods are available to create a binary mask from the 8-bit grey level image. However, this will require the involvement of an image analyst. It is also possible to manually capture the average temperature of a leaf by "drawing" the contour of the leaf by hand using the FLIR Quickreport software (delivered with the IR camera at no cost). Obviously, this is a very tedious task, non-amenable to automation and subjective.

Acknowledgements

The authors wish to thank Dr Robert Furbank for helpful comments on the manuscript and Mrs Carol Blake and Mr Scott Berry for their technical skills. Part of this research was conducted at the High Resolution Plant Phenomics Center (Canberra node of the Australian Plant Phenomics Facility).

References

1. Munns R (1993) Physiological processes limiting plant growth in saline soil: some dogmas and hypotheses. Plant Cell Environ 16:15–24
2. Epstein E (1980) Responses of plant to saline environments. In: Rains DW, Valentine RC, Hollaender A (eds) Genetic engineering of osmoregulation. Plenum Press, New York
3. Munns R, Schachtman DP, Condon AG (1995) The significance of a two-phase growth response to salinity in wheat and barley. Aust J Plant Physiol 22:561–569
4. Munns R, James RA, Lauchli A (2006) Approaches to increasing the salt tolerance of wheat and other cereals. J Exp Bot 57: 1025–1043
5. Rahnama A et al (2010) Stomatal conductance as a screen for osmotic stress tolerance in durum wheat growing in saline soil. Funct Plant Biol 37:255–263
6. Passioura JB, Munns R (2000) Rapid environmental changes that affect leaf water status induce transient surges or pauses in leaf expansion rate. Aust J Plant Physiol 27: 941–948
7. Fricke W et al (2004) Rapid and tissue - specific changes in ABA and in growth rate in response to salinity in barley leaves. J Exp Bot 55:1115–1123
8. Yeo AR, Caporn SJM, Flowers TJ (1985) The effect of salinity upon photosynthesis in rice: gas exchange by individual leaves in relation to their salt content. J Exp Bot 36:124–148
9. Sibole JV et al (1998) Role of sodium in the ABA-mediated long-term growth response of bean to salt stress. Physiol Plant 104:299–305
10. Fuchs M (1990) Infrared measurement of canopy temperature and detection of plant water stress. Theor Appl Climatol 42: 253–261
11. Sirault XRR, James RA, Furbank RT (2009) A new screening method for osmotic component of salinity tolerance in cereals using infrared thermography. Funct Plant Biol 36:970–977
12. Jones HG (1999) Use of thermography for quantitative studies of spatial and temporal variation of stomatal conductance over leaf surfaces. Plant Cell Environ 22:1043–1055
13. James RA et al (2002) Factors affecting CO_2 assimilation, leaf injury and growth in salt-stressed durum wheat. Funct Plant Biol 29:1393–1403
14. Munns R, James RA (2003) Screening methods for salinity tolerance: a case study with tetraploid wheat. Plant Soil 253:201–218
15. James RA et al (2008) Genetic variation in tolerance to the osmotic stress component of salinity stress in durum wheat. Funct Plant Biol 35:111–123
16. Genc Y, McDonald GK, Tester M (2007) Reassessment of tissue Na^+ concentration as a criterion for salinity tolerance in bread wheat. Plant Cell Environ 30:1486–1498
17. Passioura JB (2006) The perils of pot experiments. Funct Plant Biol 33:1075–1079
18. Munns R (2002) Comparative physiology of salt and water stress. Plant Cell Environ 25:239–250

Chapter 12

In vivo Imaging of Nitric Oxide and Reactive Oxygen Species Using Laser Scanning Confocal Microscopy

Yan-Jie Xie and Wen-Biao Shen

Abstract

Both nitric oxide (NO) and reactive oxygen species (ROS) are versatile molecules that mediate a variety of cellular responses in plants. In this chapter, methods for imaging NO and ROS using laser scanning confocal microscopy (LSCM) are presented. Arabidopsis roots, dyed with DAF-FM or H_2DCF, are observed using the Leica TCS-SP2 LSCM. NO or ROS production are imaged and their kinetic changes monitored with the laser excitation and emission wavelengths at 488 nm and between 500 and 530 nm, respectively. In addition, Leica software is employed to visualize and calculate the fluorescence intensity data.

Key words: Nitric oxide, Reactive oxygen species, Laser scanning confocal microscopy

1. Introduction

Nitric oxide (NO) is a type of free radical reactive gas that was first described in the 1980s as an endothelium-derived relaxant of vascular smooth muscle. To date, NO has been recognized as an important biological mediator that plays significant roles in key physiological processes in mammals, such as participating in the maintenance of neuronal signaling, host response to infection, and insulin secretion (1–6). Alongside these animal studies, NO has also generated a lot of interest in various aspects of plant biology. There is an increasing body of evidence showing that NO is a signaling molecule in plants where it plays an important role in a wide range of physiological processes (7–13). As free NO radical is notably unstable, it is a transient species with a half-life of approximately 5–10 s, it has been difficult to perform direct NO detection with satisfactory sensitivity (14). However, NO is readily oxidized to

Sergey Shabala and Tracey Ann Cuin (eds.), *Plant Salt Tolerance: Methods and Protocols*, Methods in Molecular Biology, vol. 913, DOI 10.1007/978-1-61779-986-0_12, © Springer Science+Business Media, LLC 2012

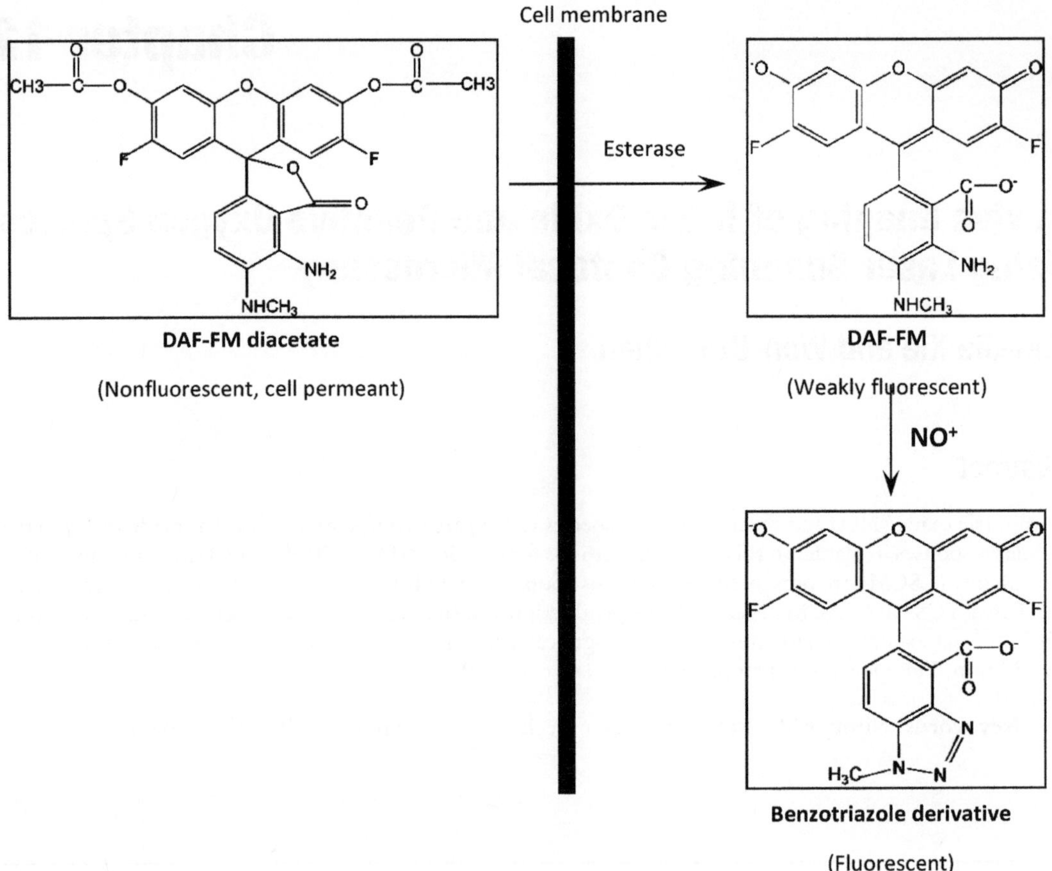

Fig. 1. Reaction scheme for the detection of nitric oxide (NO) by DAF-FM.

the nitrosonium cation (NO^+). This is moderately stable in aqueous solutions and reacts with the NO-sensitive fluorescent probes 4-amino-5-methylamino-2′7′-difluorofluorescein (DAF-FM) and 4,5-diaminofluorescein (DAF-2).

Both DAF-FM and DAF-2 have been widely used for measuring NO production in plant biological studies (15–19). DAF-FM diacetate (DAF-FM DA) and DAF-2 diacetate (DAF-2 DA) are both membrane permeant and can be deacetylated by intracellular esterases to DAF-FM and DAF-2, respectively. DAF-FM and DAF-2, however, remain essentially nonfluorescent until they react with the NO^+ to form a fluorescent heterocycle, which will be trapped in the cell cytoplasm (see Fig. 1). The DAF-FM reagent possesses a number of important advantages over DAF-2. First, the detection limit for DAF-FM is approximately 3 nM, whereas it is about 5 nM for DAF-2 (20, 21). Second, the fluorescence quantum yield of DAF-FM is reported to be 0.005 and it increases 160-fold to 0.81 after reacting with NO (20). Third, the spectra of the NO adduct of DAF-FM are independent of pH when it is above 5.5

(20). The fluorescence of DAF-FM and DAF-2 are measured using laser scanning confocal microscopy (LSCM). LSCM is a valuable tool for obtaining high-resolution images of a variety of biological specimens. Combined with a fluorescent probe, LSCM provides a practical approach for monitoring the dynamics of intracellular NO production.

Reactive oxygen species (ROS), including singlet oxygen, superoxide, and the hydroxyl radical, are versatile molecules that mediate a variety of cellular responses in plant cells. Nonetheless, although there is no equilibrium sensor that continuously monitors the level of ROS, several molecular probes have been used to measure the dynamic changes of ROS production *in vitro* and *in vivo*. The highly fluorescent 2′7′-dichloroflupréscein (DCF), which is oxidated from 2′7′-dichlorodihydrofluorescein diacetate (H_2DCF DA), is frequently used to detect the generation of ROS in plants. Although it is probably the most commonly used probe for monitoring intracellular ROS, the specificity of H_2DCF DA is still questioned. At the same time, it has also been reported that H_2DCF DA can be photooxided spontaneously, becoming fluorescent in the presence of a wide variety of ROS such as peroxyl radicals, hydroxyl radicals, and even NO and peroxynitrite anion.

2. Materials

2.1. Plant Material

Arabidopsis thaliana Col-0 ecotype (see Note 1) was used in this chapter. Seeds were surface-sterilized and washed three times with sterile water, then cultured in Petri dishes on solid Murashige and Skoog medium (MS, pH 5.8). Plates containing seeds were kept at 4°C for 2 days and then transferred into a growth chamber with 16/8 h (24/18°C) day/night regimes with 150 μmol m^{-2} s^{-1} irradiation.

2.2. Chemicals

1. HEPES buffer solution: 50 mM HEPES, adjusted to a pH of 7.5.
2. DAF-FM DA (purchased from Sigma-Aldrich; see Note 2)

3. Methods

LSCM is a well-established technique in biological sciences (see Note 3). It provides a two-dimensional (2-D) structure by collecting light from a single focal plane of the sample and excluding light that is out of focus (3-D images are also available through microscopic tomography).

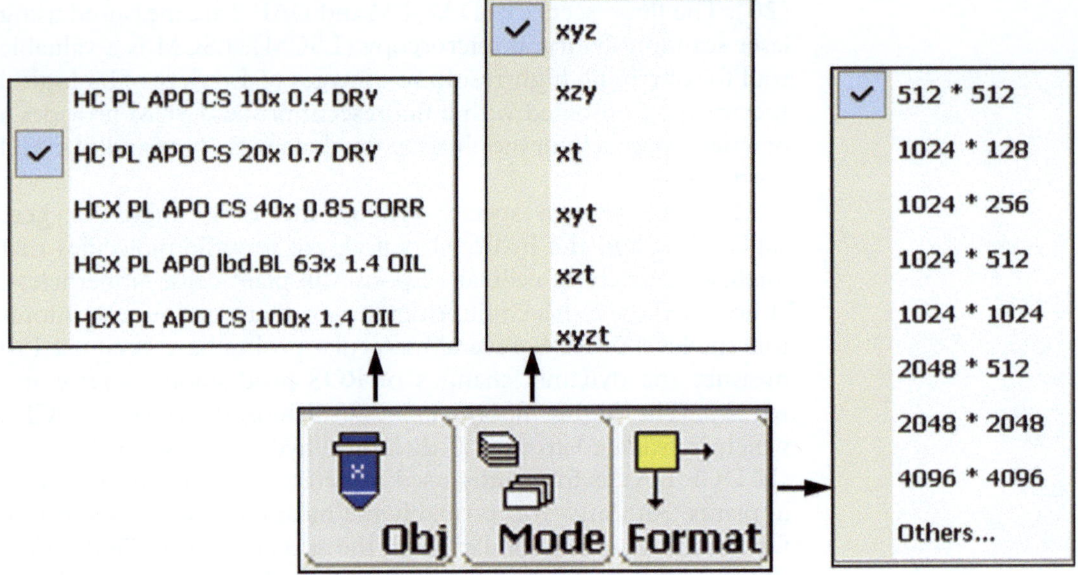

Fig. 2. Choosing the proper objective and resolution for scanning.

3.1. Specimen Culture

Treat *Arabidopsis thaliana* seedlings with 150 mM NaCl treatment (supplemented in MS medium) for 1 h (see Note 4).

3.2. Nitric Oxide Staining

1. Collect whole Arabidopsis seedlings after treatment with 150 mM NaCl (see Note 5), then immediately incubate seedlings with fresh HEPES buffer containing 10 μM DAF-FM DA for 30 min (see Note 6).
2. Wash three times with fresh HEPES buffer, 5 min per washing.

3.3. Detecting Fluorescence by LSCM

3.3.1. Bright Field Imaging

1. Transfer the Arabidopsis seedlings to a slide with a cover slip.
2. Bring the specimen (Arabidopsis root tips) into focus, and use a ×20 objective ("Obj" button), with a scan resolution of 512×512 ("Format" button) (see Fig. 2).

3.3.2. Fixing Parameters of Fluorescence Acquisition

Collect the fluorescence acquisition using Leica TCS-SP2 confocal laser scanning microscope equipped with an argon laser providing an illumination at 488 nm. DAF-FM has excitation peak with maximum at 495 nm and emits green light with a peak wavelength of 515 nm. Collect the emission fluorescence data in the range of 500 and 530 nm (see Fig. 3).

3.3.3. Optimizing the Acquisition Parameters

Optimize the image quality as soon as the scan runs, which needs a balance between high quality of the image and avoiding bleaching the specimen.

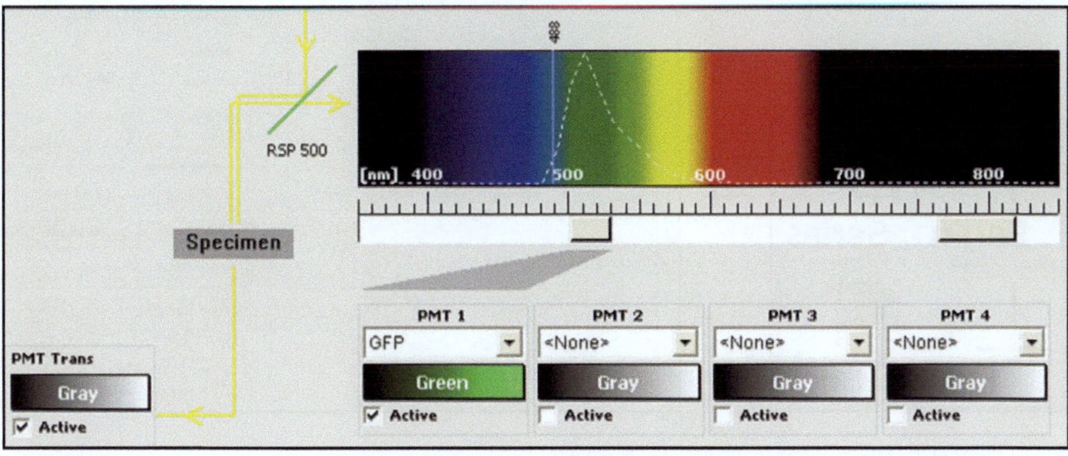

Fig. 3. Opting for the appropriate parameters for fluorescence acquisition.

1. The exact z-position within the specimen.
2. The amplification factor of the selected detector. This factor determines the range of light intensity.
3. The diameter of the detection pinhole. The amount of light reaching the detector is improved when the pinhole diameter increased; however, it will lead to the attenuated resolution and contrast.

3.4. Nitric Oxide Distribution Evaluation

1. Use the corresponding knob to adjust the z-position of the control panel.
2. Define the "Begin" and "End" position limit of the data set to be acquired (see Fig. 4).
3. Define the number of optical sections and sampling times (see Fig. 5).
4. Start to collect the Z-stack and save the image data.

3.5. 3D-Series Acquisition of Intracellular Nitric Oxide

After optimizing all image parameters, a series of 2D images is collected (see Fig. 6a) to form the 3D spatial data (see Fig. 6b).

3.6. Calculation of Relative Fluorescence Intensity

Leica provides fluorescence intensity calculations for a region of interest (ROI) (see Fig. 7a). After choosing the proper tool in "tool buttons" for selecting ROI, the fluorescence intensity data for each volume of the series can be calculated automatically by using Leica software (see Fig. 7b), the detailed intensity information on intensity is also displayed (see Fig. 7c).

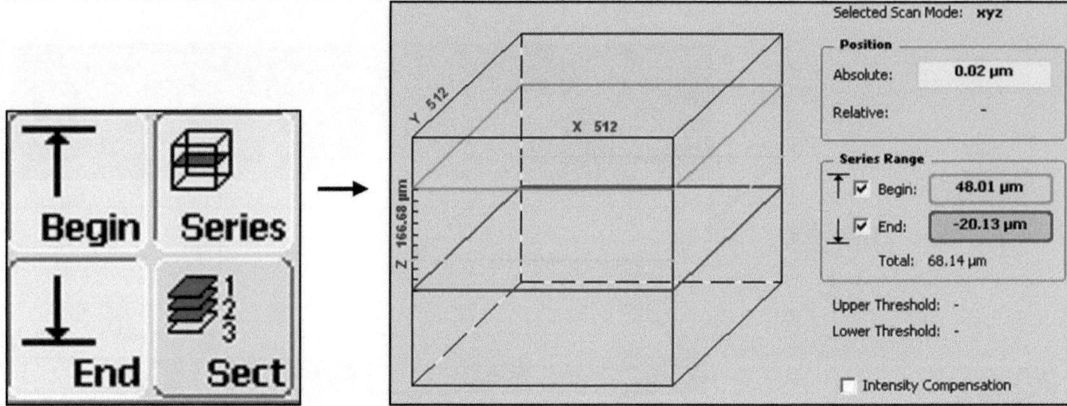

Fig. 4. Choosing the proper series range for laser scanning.

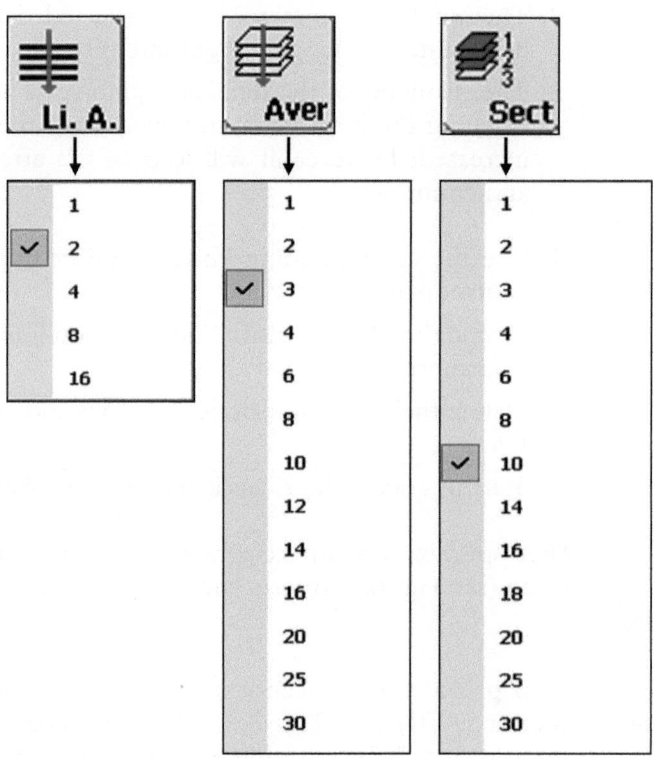

Fig. 5. Define the number of optical sections and sampling times.

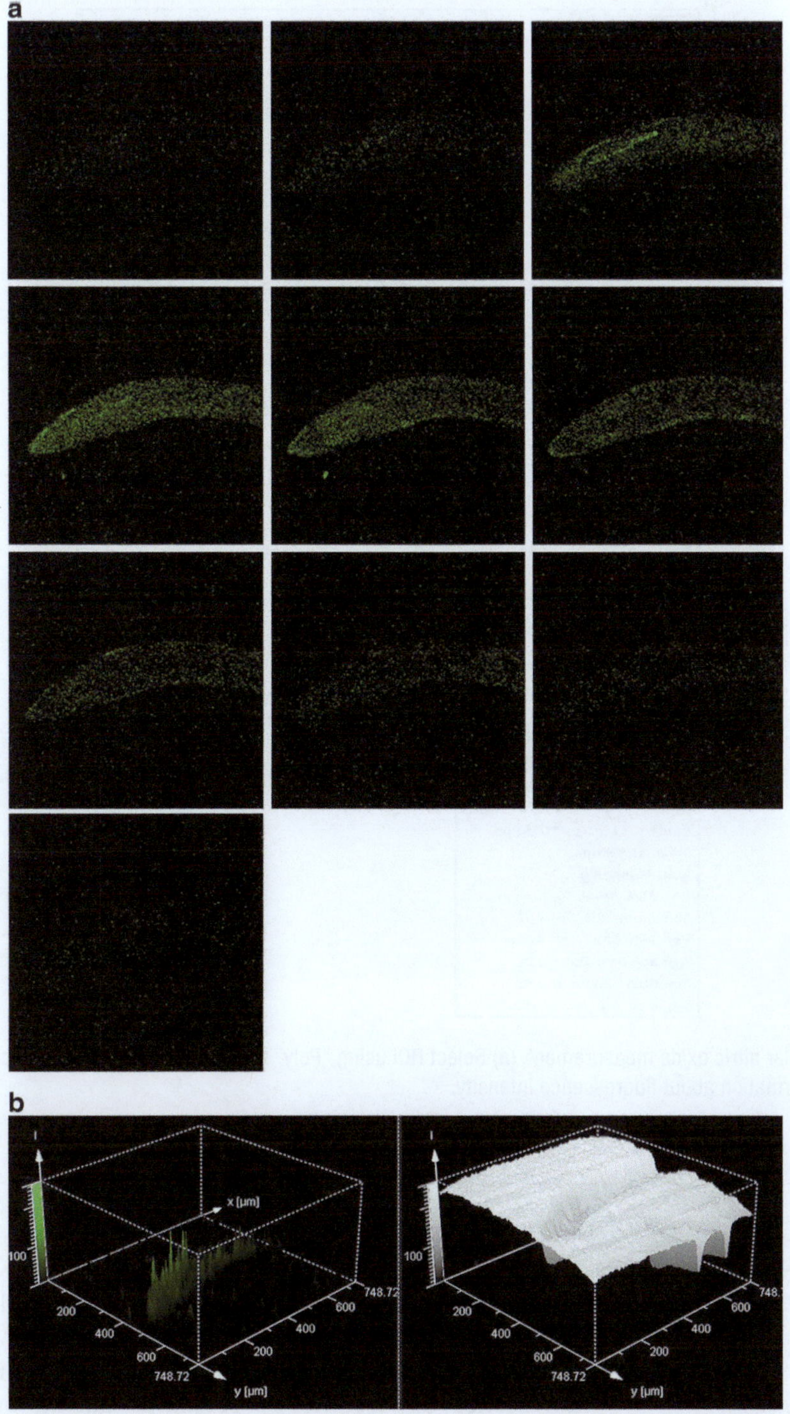

Fig. 6. Confocal series of optical sections of nitric oxide distribution. (**a**) An optical series of DAF-FM-labeled intracellular nitric oxide. (**b**) Three-dimensional representation for nitric oxide distribution of specimen. The volume data in (**a**) were projected by Leica software.

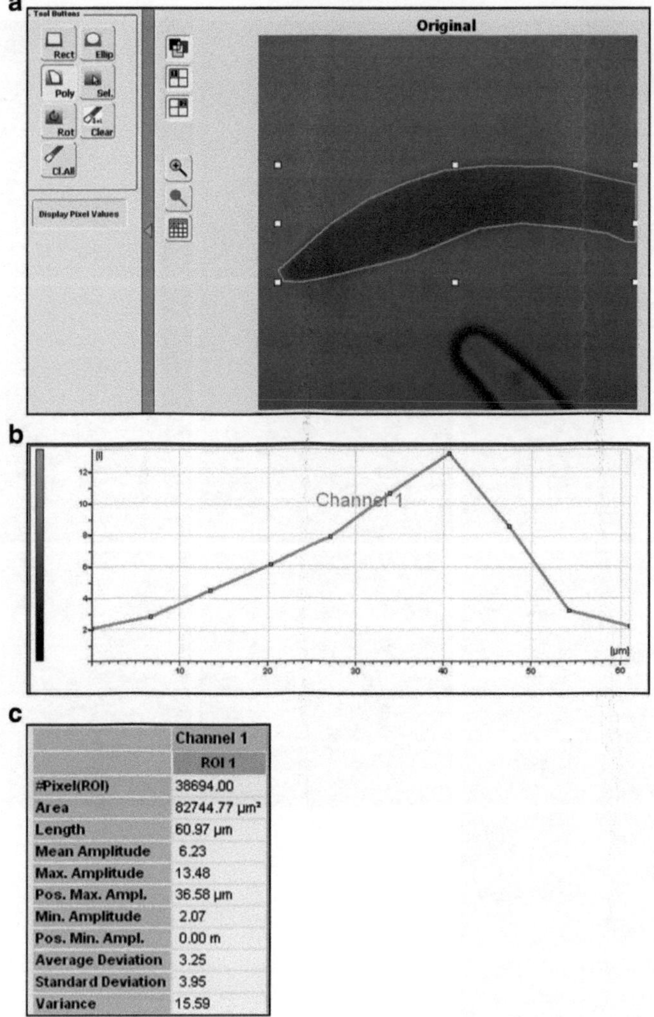

	Channel 1
	ROI 1
#Pixel(ROI)	38694.00
Area	82744.77 µm²
Length	60.97 µm
Mean Amplitude	6.23
Max. Amplitude	13.48
Pos. Max. Ampl.	36.58 µm
Min. Amplitude	2.07
Pos. Min. Ampl.	0.00 m
Average Deviation	3.25
Standard Deviation	3.95
Variance	15.59

Fig. 7. Intracellular nitric oxide measurement. (**a**) Select ROI using "Poly" tool. (**b**) Intensity data for each volume of ROI. (**c**) Detailed information about fluorescence intensity.

4. Notes

1. 5-days-old *Arabidopsis thaliana* seedlings (Col-0 ecotype) were used throughout this chapter. It should be noticed that both DAF-FM and DAF-2 could also be used to monitor NO fluorescence of Arabidopsis seedlings during other growth stages.
2. A DAF-FM DA stock solution (5 mM, dissolved in DMSO) was purchased from Sigma-Aldrich. This solution should be

stored at –20°C. Because the operating and monitoring procedure of H_2DCF DA is the same as DAF-FM DA, DAF-FM DA is used throughout this chapter.

3. The principle of LSCM can be described as follows. The microscope lenses focus the laser light on one point in the detection sample at a time called the focal point. The microscope and the optics of the scanner module focus the light emitted from the focal point to a second point, called the confocal point. The pinhole aperture, located at the confocal point, allows light from the focal point to pass through the detector. Light emitted from outside the focal point (i.e., all light coming from regions of the specimen above or below the plane of focus) is rejected by the aperture. The laser moves rapidly from point to point to produce the scanned image. The in-focus information of each specimen point is recorded by a light-sensitive detector positioned behind the confocal aperture. The analog output signal is digitized and fed into a computer. Analysis of original experimental images is also provided by Leica confocal software.
4. 150 mM NaCl treatment is able to induce DAF-FM fluorescence, because a significant fluorescence distribution image can be observed.
5. Though we only observed DAF-FM fluorescence in Arabidopsis seedling roots, whole seedlings were immersed into the buffer containing NO specific probe in order to avoid interference of NO production from the wounding responses associated with cutting the whole plants.
6. When incubating, it is advisable to shake slightly at room temperature in the dark until next step.

Acknowledgments

This work was supported by the National Natural Science Foundation of China (grant nos 31170241, 30971711 and 30671248), the Natural Science Foundation of Jiangsu Province of China (grant no. BK2009309), Fundamental Research Funds for the Central Universities (grants nos KYJ200912 and KYZ200905), the Priority Academic Program Development of Jiangsu Higher Education Institutions, and the Technology Support Program in Jiangsu Province (grant no. BE2010382).

References

1. Palmer RM, Ferrige AG, Moncada S (1987) Nitric oxide release accounts for the biological activity of endothelium-derived relaxing factor. Nature 327:524–526
2. Prast H, Philippu A (2001) Nitric oxide as modulator of neuronal function. Prog Neurobiol 64:51–68
3. Esplugues JV (2002) NO as a signaling molecule in the nervous system. Br J Pharmacol 135:1079–1095
4. Culotta E, Koshland DE Jr (1992) NO news is good news. Science 258:1862–1865
5. Snyder SH (1992) Nitric oxide: first in a new class of neurotransmitters. Science 257:494–496
6. Vincent SR (1992) Nitric oxide and arginine-evoked insulin secretion. Science 258:1376–1378
7. Durner J, Klessig DF (1999) Nitric oxide as a signal in plants. Curr Opin Plant Biol 2: 369–374
8. Wendehenne D, Pugin A, Klessig DF, Durner J (2001) Nitric oxide: comparative synthesis and signaling in animals and plant cells. Trends Plant Sci 6:177–183
9. Crawford NM, Guo FQ (2005) New insights into nitric oxide metabolism and regulatory functions. Trends Plant Sci 10:195–200
10. Lamattina L, García-Mata C, Graziano M, Pagnussat G (2003) Nitric oxide: the versatility of an extensive signal molecule. Annu Rev Plant Biol 54:109–136
11. Besson-Bard A, Pugin A, Wendehenne D (2008) New insights into nitric oxide signaling in plants. Annu Rev Plant Biol 59:21–39
12. Zhang H, Shen WB, Zhang W, Xu LL (2005) A rapid response of β-amylase to nitric oxide but not gibberellin in wheat seeds during the early stage of germination. Planta 220:708–716
13. Courtois C, Besson A, Dahan J, Bourque S, Dobrowolska G, Pugin A, Wendehenne D (2008) Nitric oxide signaling in plants: interplays with Ca^{2+} and protein kinases. J Exp Bot 59:155–163
14. Hetrick EM, Schoenfisch MH (2009) Analytical chemistry of nitric oxide. Annu Rev Anal Chem 2:409–433
15. Zhao MG, Tian QY, Zhang WH (2007) Nitric oxide synthase-dependent nitric oxide production is associated with salt tolerance in arabidopsis. Plant Physiol 144:206–217
16. Zhao MG, Chen L, Zhang LL, Zhang WH (2009) Nitric reductase-dependent nitric oxide production is involved in cold acclimation and freezing tolerance in Arabidopsis. Plant Physiol 151:755–767
17. Asai S, Ohta K, Yoshida H (2008) MAPK signaling regulates nitric oxide and NADPH oxidase-dependent oxidative bursts in *Nicotiana benthamiana*. Plant Cell 20: 1390–1406
18. Xie YJ, Ling TF, Han Y, Liu KL, Zheng QS, Huang LQ, Yuan XX, He Z, Hu B, Fang L, Shen ZG, Yang Q, Shen WB (2008) Carbon monoxide enhances salt tolerance by nitric oxide-mediated maintenance of ion homeostasis and up-regulation of antioxidant defense in wheat seedling roots. Plant Cell Environ 31:1864–1881
19. Lozano-Juste J, León J (2010) Enhanced abscisic acid-mediated responses in *nia1nia2noa1-2* triple mutant impaired in NIA/NR- and AtNOA1-dependent nitric oxide biosynthesis in Arabidopsis. Plant Physiol 152:891–903
20. Kojima H, Nakatsubo N, Kikuchi K, Kawahara S, Kirino Y, Nagoshi H, Nagano T (1998) Detection and imaging of nitric oxide with novelfluorescentindicators:diaminofluoresceins. Anal Chem 70:2446–2453
21. Kojima H, Urano Y, Kikuchi K, Higuchi T, Hirata Y, Nagano T (1999) Fluorescent indicators for imaging nitric oxide production. Angew Chem Int Ed Engl 38:3209–3212

Part III

Biochemical Assays

Chapter 13

Metabolomics for Salinity Research

Ute Roessner and Diane M. Beckles

Abstract

Soil salinity devastates agriculture. It reduces crop yields and makes arable land unsuitable for later use. Many species have evolved highly efficient strategies to sense, transduce, and build up tolerance to high salinity and even sensitive species have endogenous mechanism for coping with this stress. These underlying physiological and metabolic mechanisms can be unraveled using metabolomics. Here we describe detailed protocols of how to extract polar metabolites for analysis using GC-MS and LC-MS. We also touch briefly on considerations that should be taken into account when designing the experiment and how the resulting data may be analyzed and visualized in a biological context.

Key words: Metabolomics, Metabolite extraction, GC-MS, LC-MS, Polar metabolites, Data analysis

1. Introduction

As much as 20% of all currently irrigated crop land may be classified as salt stressed (1) and this percentage is predicted to increase the result of twenty-first-century global climate patterns (2) and anthropomorphic activity. While it is clear that there are system-wide responses to high salt that occur at the gene, transcript, and protein level, monitoring the response of the metabolome may be particularly informative. This is because the metabolite content of the cell is often a better indicator of its physiological state than transcript and protein complement. In addition, mechanisms of salinity tolerance involve specific changes in the levels and movement of metabolites and ions, both at the cellular and at the whole plant level (3, 4). Therefore, metabolomics, the art and science of

Sergey Shabala and Tracey Ann Cuin (eds.), *Plant Salt Tolerance: Methods and Protocols*, Methods in Molecular Biology, vol. 913, DOI 10.1007/978-1-61779-986-0_13, © Springer Science+Business Media, LLC 2012

determining and quantifying the metabolites in a biological system, is particularly well suited for understanding physiological responses to salinity.

Salt stress causes changes in the relative levels of metabolites and the preferential accumulation of a subset of small protective molecules in plants. When cells are exposed to abnormal salt levels, they first experience rapid osmotic changes. This affects root growth within minutes, which in turn, alters shoot development. Compatible solutes such as amino acids, sugars, polyols, and quaternary ammonium compounds accumulate and are associated with an osmotolerant response (5), and the specific compounds that are activated may be species-dependent (6). If the external salt concentration increases further, ions are then shunted to the leaves for storage. This can lead to metabolic dysfunction and toxicity (3, 4). The changes in metabolites and ions associated with both the adaptive response and the progression of toxicity can be effectively monitored using metabolite-profiling technologies.

There are numerous protocols available for plant metabolite analyses. These include those for untargeted profiling of as many metabolites as possible and those for more accurate and quantitative methods using complementary analytical platforms such as GC-MS, LC-MS, CE-MS or NMR (7). CE-MS is less commonly used but it has excellent potential for wider use because it presents so many advantages for both targeted and unbiased profiling (for review see ref. (7)). Its primary advantage is that only very small (picolitre) starting amounts of tissues are needed, making single-cell studies feasible. NMR is unrivaled in its comprehensiveness; it can also be used for real-time flux imaging in vivo and for the structural elucidation of novel compounds. However, it is not often used because there are some drawbacks especially when analyzing plant tissues (7). A powerful and valuable complementary tool to study plant salt stress is ionomics (8, 9). This reveals the changes in ions and elements such as Na, Ca, Fe, and Cu in tissues. Digested plant material is analyzed using Inductively Coupled Plasma Mass Spectroscopy (ICP-MS) or Inductively Coupled Plasma Optical Emission Spectroscopy (ICP-OES). These techniques and their uses are reviewed elsewhere (10).

The two most common methods for metabolite profiling are GC-MS and LC-MS, due to a combination of cost, historical use, and data content per run. GC-MS analysis provides a readout of primary metabolites, some of which are associated with the changes in osmotic, redox status, growth, and nutritional status of the plant under stress. LC-MS analysis will capture more complex compounds such as hormones and secondary metabolites that may be part of the stress signal transduction or protective pathways (for review see ref. (11)). These two techniques will be the subject of this chapter.

2. Considerations for Planning the Stress Metabolite Profiling Experiment

The results of any metabolite profiling of plant tissues exposed to salt stress will be defined by the species and tissues chosen (genotype), the intensity, length, and the type of salt (ion) applied (environment) and their interaction. Interpretation of the data will be limited by the experimental parameters used, so some effort must be placed in designing the experiment to achieve the desired scientific objectives. Regardless of the system chosen, the effect of salinity on growth, development, and on various processes, e.g., photosynthesis, leaf necrosis, dry weight, shoot, and root elongation, should be measured in the samples used for metabolite profiling. This will allow the data to be put in a physiological context (12, 13).

Growth media. Soil or hydroponic systems vary in their electrochemical conductance and physical interaction with roots and lead to drastically different physiological outcomes (14). Soil more closely mimics field conditions, but its consistency and complex composition inhibits the ability to properly analyze roots, as well as identifying the chemical state of the media. Hydroponic systems allow for more precision and reproducibility in controlling relative ion concentrations, but it may not be possible to extrapolate the results to soil-based media.

Salt type and concentrations. Judicious choice of salt concentration and length of exposure may permit the examination of acclimation to increasing salt concentrations, salt stress initiation/induction, and salt toxicity responses. The final concentration-time regimen chosen is best achieved by rigorous experimentation prior to metabolite profiling, using other independent approaches such as ion leakage, tissue conductance, etc. to monitor physiological changes to salt stress. As would be predicted different ions (Na^+, B^{3+}, Cu^{2+}, Cl^-, etc.) or ion combinations would activate distinct physiological, biochemical, and signaling pathways. These will produce different metabolite profiles (2).

Spatiotemporal resolution. Physiological changes to salt stress occur at the level of the organelle, cell, tissue, or organ. The choice of tissues analyzed will give different physiological snapshots of plant responses to stress. For example, Na^+ accumulation occurs in leaves, but the root is the organ that is directly exposed to salt and involved in its uptake. Even within the root, distinct tissues and cells respond differently to high salt and the changes vary in their timing and amplitude, sometimes in an interdependent manner (15, 16). The sequestering of excess Na^+ ions to the vacuole further highlights the non-uniform distribution of small molecules and elements in the cell in response to salt stress (15). Most experiments are conducted on whole organs. This damps metabolite signals, partially obscuring any mechanistic features of the response. Although there are techniques for high resolution spatial mapping of metabolites (17), most routine analyses use mixtures of different cells. This is generally acceptable, but data interpretation should take this into account.

Developmental stage. The effects of salinity vary with the age of the plant tissue. Stress occurring at the seedling stage may mean that the plant does not survive to maturation, while older established plants may fail to reproduce or otherwise undergo early senescence (15). It is also possible to dissect osmotic from ionic effects, based on the age of the tissue examined: actively growing leaves will be affected by the former while older leaves undergoing transpiration will be a better indicator of the latter (15).

Plants studied. The plant systems examined may include different cultivars, mutants, or transgenic plants that show variation in salt tolerance or sensitivity. If distantly related species varying in salt tolerance are compared, it becomes more difficult to identify the responses that are specifically due to salinity, from those that arise from phylogeny. Here, time course experiments, where profiling is done at multiple points during the progression of the stress in all of the species compared, will permit changes related to salt tolerance to be pinpointed (13).

Data analysis. Depending on the number of experiments performed, it is possible to identify biomarker metabolites that could be associated with salinity tolerance, sensitivity, etc. (17). In addition, metabolites that might be protective and toxic can be putatively identified if the experiment is carefully designed.

3. Materials

As mentioned previously, there is a wide array of analytical platforms for metabolite profiling. Here we present two methods, both commonly used, described from our laboratory for untargeted profiling of leaf and root polar metabolites using GC-MS (modified from (18)) and LC-MS (modified from (19)).

3.1. Tissue Harvesting and Quenching of Metabolism

1. Reaction tubes (e.g., 2 mL safe lock Eppendorf tubes) labeled with identifiable names for each sample to be harvested.
2. Liquid nitrogen in an appropriate storage container.

3.2. Storage of Samples

1. –80°C freezer.

3.3. Homogenization and Weighing of Tissue

1. Mortar and pestle (or any other device such as ball mills or ultra turrax).
2. New labeled reaction tubes (e.g., 2 mL safe lock Eppendorf tubes).
3. Accurate balance (to at least four decimal places).

3.4. Metabolite Extraction and Derivatization for GC-MS Profiling

1. Solvents: 100% (v/v) methanol, 100% (v/v) chloroform, distilled water.
2. Internal standards (e.g., $^{13}C_6$-Sorbitol and $^{13}C^{15}N$-Valine).
3. Autosampler vials with inserts and caps (e.g., from Agilent Technology or any other consumable vendor).
4. Derivatization: pyridine, methoxiamine hydrochloride, *N*-methyl-*N*-(trimethylsilyl)trifluoroacetamide (MSTFA) + 1% (v/v) trimethylchlorosilane (TMCS).
5. Retention time standard mixture (0.029% (v/v) *n*-dodecane, *n*-pentadecane, *n*-nonadecane, *n*-docosane, *n*-octacosane, *n*-dotriacontane, *n*-hexatriacontane dissolved in pyridine).
6. Vortexer.
7. Thermoshaker for reaction tubes (e.g., Eppendorf).
8. Bench top centrifuge for reaction tubes (up to 11337 × *g*).

3.5. GC-MS Analysis

1. GC-MS system: 7890A Agilent Gas Chromatograph and a 5975C Agilent quadrupole mass spectrometer (Agilent Technologies, Santa Clara, USA) equipped with a Gerstel 2.5.2 autosampler (Gerstel, Germany).
2. Hot inlet injector equipped with a focus liner (Agilent Technologies).
3. Analytical column: 30 m VF-5MS column with 0.25 μm film thickness with a 10 m Integra guard column (Agilent Technologies).
4. Helium as carrier gas.
5. Ethylacetate for syringe washing.

3.6. GC-MS Data Analysis

1. Chemstation or Mass Hunter for GC-MS (Agilent Technologies).
2. AMDIS (NIST).
3. AnalyzerPro (SpectralWorks).
4. GC-MS mass spectral libraries (e.g., NIST08, GMD, in-house libraries).
5. Statistical software package (e.g., R, Unscrambler, SIMCA, MatLab, etc.).
6. Pathway mapping software (e.g., VANTED, (20)).

3.7. Metabolite Extraction for LC-MS Profiling

1. Solvents: 2:3:3 v/v/v mixture of water/acetonitrile/isopropanol.
2. Internal standards (e.g., $^{13}C_6$-Sorbitol and $^{13}C^{15}N$-Valine).
3. Vortexer.
4. Sonicator.
5. Bench top centrifuge for reaction tubes (up to 11337 × *g*).

3.8. LC-MS Analysis

1. LC system: 1200 series LC (Agilent Technologies) comprising a vacuum degasser, binary pump, with a thermostated autosampler and column compartment.
2. MS system: 6520 QTOF MS (Agilent Technologies) used with a dual sprayer ESI source.
3. Analytical column: Cogent diamond Hydride 2.1 mm 6,100 mm, 4 mm particle size (MicroSolv Technology).
4. Solvents: (a) organic mobile phase solvent (B) composed of 90% (v/v) Acetonitrile (can) with 0.1% (w/v) ammonium acetate and 0.1% (v/v) acetic acid and (b) aqueous mobile phase (A) was composed of 100% deionized water with 0.1% (w/v) ammonium acetate and 0.1% (v/v) acetic acid (pH 3.4).

3.9. LC-MS Data Analysis

1. Mass Hunter Qual (Agilent Technologies).
2. Mass Profiler Professional (Agilent Technologies).

4. Methods

4.1. Tissue Preparation

Tissue harvesting is the most crucial step in any metabolomics experiment. Metabolic changes occurring after harvest are rapid. Metabolism needs to be stopped as soon as possible following harvesting in order to obtain an accurate picture of the in vivo physiological state of the tissue and not those due to wounding. Ideally, samples are shock frozen in liquid nitrogen. Depending on how much tissue is available, it is recommended that it is quickly frozen, then ground and weighed. However, if there are limited amounts of tissue available, it can be weighed before freezing, then homogenized to avoid any further losses (see Notes 1 and 2).

4.2. Sample Preparation for GC-MS

1. Weigh approximately 30 mg homogenized leaf in a 2 mL Eppendorf tube (round bottom-shaped). Add 500 μL 100% (v/v) Methanol, vortex (enzymatic activity stops here) (see Notes 2 and 3).
2. Add 20 μL stock ^{13}C Sorbitol/Valine (1 mg/mL stock solution) as an internal quantitative standard for the polar phase (see Note 4).
3. Vortex.
4. Shake 15 min at 70°C in Thermoshaker.
5. Centrifuge for 15 min at 11337 × *g*.
6. Transfer supernatant to a new reaction tube.
7. Add 500 μL H_2O to the pellet.
8. Vortex.

9. Centrifuge for 15 min at 11337 × *g*.
10. Combine both supernatants.
11. Add 400 μL $CHCl_3$.
12. Vortex for at least 15 s.
13. Centrifuge for 10 min at 11337 × *g*.
14. Transfer the upper polar phase (as much and clean as possible) into fresh Eppendorf tubes (see Note 5).
15. Add to the polar phase again 300 μL $CHCl_3$, vortex 15 s.
16. Centrifuge for 10 min at 11337 × *g*.
17. Add supernatant to previous supernatant (as clean as possible), shake.
18. Take aliquots for derivatization into an insert for the instrument autosampler vial (see Notes 6 and 7).
19. Dry aliquots under *vacuo* without heating by placing the insert with the extract into a reaction tube. Once dry, it can be placed into the autosampler vial prior to analysis. Place in bag with Silicagel at room temperature (see Note 8).
20. Derivatization: Add 10 μL Methoxyaminehydrochloride (30 mg/mL in Pyridine) using the Gerstel autosampler on the rack of the instrument (see Notes 9–11).
21. Transfer vial into the shaker with the temperature adjusted. Shake for 2 h at 37°C.
22. Add 20 μL MSTFA.
23. Add 2.5 μL Alkane Mix to the polar phase.
24. Transfer vial into the shaker with adjusted temperature. Shake for 30 min at 37°C.
25. Put vial back on the autosampler rack (see Note 12).
26. Let the derivatization solution settle for 2 h prior to injection (see Note 13).

4.3. GC-MS Analysis and Data Analysis

1. Inject 1 μL of derivatized sample into the injector set at 250°C.
2. GC oven start temperature set at 70°C.
3. Following injection, hold the temperature of the GC oven for 1 min and then increase the oven temperature by 7°C per minute up to 325°C.
4. Hold for 6 min.
5. Helium carrier gas flow rate is set at 0.8 mL/min.
6. Eluting compounds are transferred through the transfer line set at 280°C into the mass spectrometer.
7. Eluting compounds are ionized and fragmented using electron impact ionization (EI) using 70 eV.

8. Resulting ions are scanned using the quadrupole mass analyzer with two scans per min over a mass range of 50–600 *m/z* scanning range.
9. For untargeted analysis, resulting chromatograms are deconvoluted using the AnalyzerPro software program (Spectralworks). All deconvoluted mass spectra related to individual compounds are quantified and exported as a matrix for further analysis (see Note 14).
10. For targeted analysis, deconvoluted mass spectra using the AMDIS software program (NIST, US) are matched against the Golm Metabolome Database (21) using both mass spectral features and retention time index related to the alkane mixture.
11. Matched compounds are included into a target component list in AnalyserPro that searches and quantifies each compound in the target component list using specified identifier ions for more accurate quantification.
12. In both cases, the resulting areas of peaks are normalized using the area of the internal standard (e.g., $^{13}C_6$-sorbitol) and the fresh or dry weight obtained before extraction.
13. The resulting data matrix can then be further evaluated by calculating the averages for biological replicates, standard deviations, *x*-fold comparisons and statistics such as Student's *t*-test or ANOVA (see Subheading 3.6).

4.4. Sample Preparation for LC-MS

1. Weigh approximately 50 mg of ground tissue into 2 mL reaction tubes (e.g., Eppendorf).
2. Add 500 μL of a 2:3:3 (v/v/v) mixture of water/acetonitrile/isopropanol.
3. Sonicate for 15 min at room temperature.
4. Centrifuge at 11337 × *g*.
5. Transfer supernatant to a new reaction tube.
6. Re-extract pellet with the addition of another 500 μL of a 2:3:3 v/v/v mixture of water/acetonitrile/isopropanol.
7. Sonicate for 15 min at room temperature.
8. Centrifuge at 11337 × *g*.
9. Combine both supernatants.
10. Transfer to autosampler vial for analysis.

4.5. LC-MS Analysis and Data Analysis

1. Place sample vials onto the thermostad autosampler set at 5°C (see Note 12).
2. Inject 1 μL of extract onto the Diamond Hydrate column.
3. Column oven is set at 50°C.

4. The organic mobile phase solvent (B) is composed of 90% (v/v) ACN with 0.1% (w/v) ammonium acetate and 0.1% (v/v) acetic acid. The aqueous mobile phase (A) is composed of 100% deionized water with 0.1% (w/v) ammonium acetate and 0.1% (v/v) acetic acid (pH 3.4) (see Notes 15 and 16).
5. Set the following HPLC gradient:
   (a) The column flow rate is 0.4 mL/min.
   (b) The gradient starts at 100% B.
   (c) Then linearly decrease to 40% B over 10 min.
   (d) This is then followed by a 1-min hold at 40% B.
   (e) The column is then re-equilibrated at 100% B for 6 min.
6. Set the electrospray conditions as follows:
   (a) The nebuliser pressure is set at 45 psi.
   (b) The gas flow rate is set at 10 L/min.
   (c) The gas temperature is set at 300°C.
   (d) The capillary voltage is set at 4,000 V.
   (e) The fragmentor is set at 150 V.
   (f) The skimmer is set at 65 V.
   (g) Both positive and negative ionization are carried out. This requires two injections per sample (see Note 17).
7. Set the mass spectrometer conditions as follows:
   (a) Scanning function is set to extended dynamic range mode.
   (b) The data are collected in the *m/z* range 70–1,700 amu.
   (c) Data are collected in full scan mode. Additionally, a data-dependent scanning mode can be applied to obtain MS/MS spectra for abundant compounds for subsequent mass spectral interpretations.
8. Resulting chromatograms are deconvoluted using the Mass Hunter incorporated Molecular Feature Finding algorithm (Agilent Technologies) resulting in data matrices containing peaks presented by an accurate *m/z* value related to a retention time. These data matrices can then be further used for subsequent statistical analyses either within the Mass Profiler Professional software (Agilent Technologies) or using any Third Party statistical software tools (see Note 14).

4.6. Data Visualization and Interpretation

There are numerous ways to analyze, mine, and visualize metabolomics data. These all aid the interpretation in a biological context (22). Following deconvolution, different normalization and transformation techniques can be applied prior to univariate and multivariate statistical analyses. It would be outside the scope of this

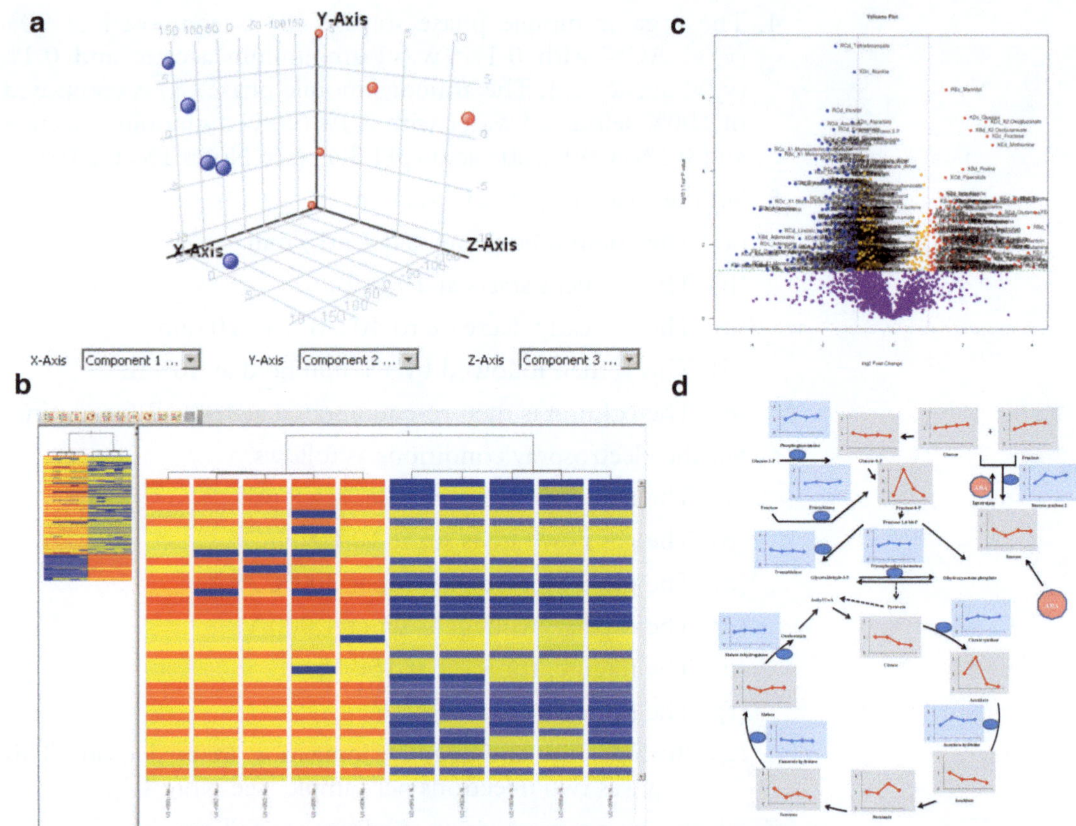

Fig. 1. Schematics of possible data visualization tools. (**a**) Principal component analysis plot created in Mass Profiler Professional (Agilent Technologies). (**b**) Hierarchical cluster analysis dendrogram combined with heatmap visualization of metabolite levels created in Mass Profiler Professional (Agilent Technologies). (**c**) Volcano plots created using R. (**d**) Mapping of metabolite level changes onto a metabolic pathway. Example taken from (23). Reprinted with permission from (23). Copyright 2010 American Chemical Society.

chapter to provide detailed protocols here, so the reader is referred to the extensive literature available (for review see ref. (22)). Figure 1 shows a few potential methods useful to visualize the data; Principal Component Analysis (A), Heatmap presentation of metabolite levels combined with Hierarchical Cluster Analysis (B), Volcano plots based on *x*-fold and *P*-value following Student's *t*-test as thresholds (C) and Mapping of metabolite changes onto metabolic pathways (D).

1. Deconvolute chromatograms using appropriate software (provided with the instrument, third party software programs, or open source codes).
2. Normalize the resulting data matrix consisting of peaks, with their respective area or height under the peak for each chromatogram. The most commonly used normalization techniques include an internal standard and fresh weight/cell number etc. normalization or normalization to the median of each compound analysed.

3. Transform the resulting normalized data matrix. The most frequently used transformation techniques are $\log_{10}$ transformation.
4. Now the matrix can be loaded into statistical software packages for analyses as demonstrated in Fig. 1. Graphs A and B were performed used Mass Profiler Professional (Agilent Technologies). Graph C was performed using a free script written in R and graph D was produced manually using SigmaPlot and Powerpoint (Microsoft).

5. Notes

1. When growing and treating plants for a metabolomics experiments, it is important to consider appropriate biological replication for the subsequent statistical and multivariate analyses.
2. It is important to ensure that tissue is **always frozen** while weighing or transferring between reaction tubes. At warmer temperatures, enzymatic activity can occur and cause spurious changes in metabolites that are unrelated to the treatment.
3. The weight has to be recorded for subsequent normalization of metabolite levels per gram fresh or dry weight.
4. Internal standards are compounds not present in the biological sample (e.g., stable isotope labeled compounds) and are included prior to or during metabolite extraction. Ideally, stable isotope labeled internal standards that have identical chemical properties as the metabolites under analysis should be used. However, the costs involved can be prohibitive.
5. The upper phase (water/methanol phase) contains polar/hydrophillic metabolites, the lower phase (chloroform phase) the apolar/hydrophobic compounds. This phase can be further used for analyses of those compounds using lipidomics technologies. This though is not described here.
6. The volume for each aliquot depends on the tissue type and amount extracted. Often there needs to be two aliquots of different volumes (e.g., 100 and 5 μL) analyzed to capture both low-abundant and high-abundant metabolites. This makes it possible to capture as many metabolites as possible within the dynamic range of the instrument.
7. Important: Always try to prepare more than one replicate (technical replicate) for backup. Once the samples have been derivatized, they cannot be stored. If there is equipment failure during a run, the experiment can be salvaged if identical samples are on hand.

8. If long-term storage is required, dried aliquots should be kept under argon to avoid oxidation and degradation of metabolites. Always keep in the dark.
9. To prepare the methoxiamine solution, weigh 30 mg of *O*-Methylhydroxylamine hydrochloride in a reaction tube. After addition of 1 mL pyridine, heat for 5 min at 50°C to better dissolve in solution.
10. Important: Always prepare MSTFA fresh (never store open ampoules; any contact with moisture will degrade the solution).
11. Safety note: Derivatization reagents (methoxiamine, pyridine, and MSTFA) are extremely toxic and should be handled in a fumehood while wearing gloves. Pregnant women should avoid handling them at all.
12. Make sure that the samples are randomized on the autosampler rack to reduce the impact of degradation onto subsequent statistical analysis.
13. Derivatized compounds are prone to degradation due to moisture and air. Therefore, online derivatization should be carried out on the instrument where each sample can be derivatized exactly at the same time prior to injection. This will reduce any impact of degradation and also unfinished derivatization reactions.
14. There are many commercial and open source software programs available for GC-MS and LC-MS data analysis and it is beyond the remit of this chapter to describe them in detail.
15. This moderate pH buffer is used to accommodate both positive and negative ionization conditions.
16. Ensure that the column is re-equlibrated sufficiently. This may take much longer than with conventional C18RP columns.
17. Some newer mass spectrometer offer fast switching between positive and negative ionization. However, this often leads to a loss of sensitivity that cannot be overcome by increasing the concentration of compounds in the sample. In our laboratory, we prefer to analyze each sample individually.

References

1. Yokoi S, Bressan RA, Hasegawa PM (2002) Salt stress tolerance of plants. JIRCAS Working Report. p 25–33
2. Mittler R, Blumwald E (2010) Genetic engineering for modern agriculture: challenges and perspectives. Annu Rev Plant Biol 61:443–462
3. Munns R (2002) Comparative physiology of salt and water stress. Plant Cell Environ 25:239–250
4. Munns R (2005) Genes and salt tolerance: bringing them together. New Phytol 167:645–663
5. Bowne J, Bacic A, Tester M, Roessner U (2011) Abiotic stress and metabolomics. Annu Plant Rev 43:61–85
6. Sanchez DH, Siahpoosh MR, Roessner U, Udvardi U, Kopka J (2008) Plant metabolomics reveals conserved and divergent meta-

bolic responses to salinity. Physiol Plant 132: 209–219

7. Roessner U, Beckles DM (2009) Metabolite measurements. In: Schwender J (ed) Plant metabolic networks. Springer, New York
8. Sanchez DH, Pieckenstain FL, Escaray F, Erban A, Kraemer U, Udvardi MK, Kopka J (2011) Comparative ionomics and metabolomics in extremophile and glycophytic Lotus species under salt stress challenge the metabolic pre-adaptation hypothesis. Plant Cell Environ 34:605–617
9. Sanchez DH, Pieckenstain FL, Szymanski J, Erban A, Bromke M, Hannah MA, Kraemer U, Kopka J, Udvardi MK (2011) Comparative functional genomics of salt stress in related model and cultivated plants identifies and overcomes limitations to translational Genomics. PLoS One 6:e17094
10. Salt DE, Baxter I, Lahner B (2008) Ionomics and the study of the plant ionome. Annu Rev Plant Biol 59:709–733
11. Kusano M, Tohge T, Fukushima A, Kobayashi M, Hayashi N, Otsuki H, Kondou Y, Goto H, Kawashima M, Matsuda F, Niida R, Matsui M, Saito K, Fernie AR (2011) Metabolomic approaches toward understanding nitrogen metabolism in plants. J Exp Bot 62:1439–1453
12. Roessner U, Patterson JH, Forbes MG, Fincher GB, Langridge P, Bacic A (2006) An investigation of boron toxicity in barley using metabolomics. Plant Physiol 142:1087–1101
13. Widodo PJH, Newbigin E, Tester M, Bacic A, Roessner U (2009) Metabolic responses to salt stress of barley (*Hordeum vulgare* L.) cultivars, Sahara and Clipper, which differ in salinity tolerance. J Exp Bot 60: 4089–4103
14. Tavakkoli E, Pichu R, McDonald GK (2010) The response of barley to salinity stress differs between hydroponic and soil systems. Funct Plant Biol 37:621–633
15. Munns R, Tester M (2008) Mechanisms of salinity tolerance. Annu Rev Plant Biol 59:651–681
16. Long TA (2011) Many needles in a haystack: cell-type specific abiotic stress responses. Curr Opin Plant Biol 14:325–331
17. Beckles DM, Roessner U (2011) Plant metabolomics—applications and opportunities for agricultural biotechnology. In: Altmann A., Hasagawa PM (eds) Plant biotechnology and agriculture: prospects for the 21st century. Elsevier, London
18. Jacobs A, Lunde C, Bacic A, Tester M, Roessner U (2007) The impact of constitutive expression of a moss Na^+ transporter on the metabolomes of rice and barley. Metabolomics 3:307–317
19. Callahan DL, De Souza D, Bacic A, Roessner U (2009) Profiling of polar metabolites in biological extracts using diamond hydride-based aqueous normal phase chromatography. J Sep Sci 32:2273–2280
20. Junker BH, Klukas C, Schreiber F (2006) VANTED: a system for advanced data analysis and visualization in the context of biological networks. BMC Bioinformatics 7:109
21. Kopka J, Schauer N, Krueger S, Birkemeyer C, Usadel B, Bergmüller E, Dörmann P, Weckwerth W, Gibon Y, Stitt M, Willmitzer L, Fernie AR, Steinhauser D (2005) GMD@CSB.DB:the Golm metabolome database. Bioinformatics 21:1635–1638
22. Roessner U, Nahid A, Hunter A, Bellgard M (2011) Metabolomics—the combination of analytical chemistry, biology and informatics. In: Moo-Young M, Butler M, Webb C, Moreira A, Grodzinski B, Cui ZF, Spiros A (eds) Comprehensive biotechnology, 2nd edn. Elsevier, New York
23. Rao SR, Ford KL, Cassin AM, Roessner U, Patterson JH, Bacic A (2010) Proteomic and metabolic profiling of rice suspension culture cells as a model to study abscisic acid signalling response pathways in plants. J Proteome Res 9:6623–6634

Chapter 14

Purification of Plant Plasma Membranes by Two-Phase Partitioning and Measurement of H^+ Pumping

Anette Lund and Anja Thoe Fuglsang

Abstract

Purification of plasma membranes by two-phase partitioning is based on the separation of microsomal membranes, dependent on their surface hydrophobicity. Here we explain the purification of plasma membranes from a relatively small amount of material (7–30 g). The fluorescent probe ACMA (9-amino-6-chloro-2-metoxyacridine) accumulates inside the vesicles upon protonation. Quenching of ACMA in the solution corresponds to the H^+ transport across the plasma membrane. Before running the assay, the plasma membranes are incubated with the detergent Brij-58 in order to create inside-out vesicles.

Key words: Two-phase partitioning, H^+ pumping, Plasma membranes, H^+-ATPase, Microsomes, ACMA, Inside-out vesicles

1. Introduction

Maintaining the membrane potential during salt stress is of major importance because it allows other transporters to remove the sodium ions from the cytosol. The plasma membrane H^+-ATPase pumps protons out of the cell against the proton gradient at the expense of ATP, thus they are a major contributor towards maintaining the membrane potential. Indeed, it has been demonstrated that barley varieties with a higher tolerance towards salt stress possesses a higher basal H^+-ATPase activity, so have a superior capacity to maintain a negative membrane potential (1).

In this chapter, we first describe the purification of plasma membranes by two-phase partitioning, followed by a method for measuring the H^+ pumping activity. Purification of plasma membranes by two-phase partitioning is based on the separation of

Sergey Shabala and Tracey Ann Cuin (eds.), *Plant Salt Tolerance: Methods and Protocols*, Methods in Molecular Biology, vol. 913, DOI 10.1007/978-1-61779-986-0_14, © Springer Science+Business Media, LLC 2012

microsomal membranes, dependent on their surface hydrophobicity. The addition of the detergent Brij-58 to the plasma membranes creates 100% inside-out vesicles that can be used, for H^+ pumping studies (2). The fluorescent probe ACMA (9-amino-6-chloro-2-metoxyacridine) accumulates inside the vesicles upon protonation and the assay measures the quenching of the ACMA in the solution (3).

2. Materials

2.1. Purification of Microsomes

1. For the purification steps: ice buckets and beakers. GSA centrifuge tubes, SS34 centrifuge tubes, ultracentrifuge tubes, and 2 mL homogenizers (glass with Teflon pestle).
2. Plant material: 7–30 g (or more) of leaves, roots, seedlings, depending on your experiment.
3. Blender with sharp knives.
4. Homogenization buffer: 50 mM MOPS, 5 mM EDTA, 0.33 M sucrose, adjust pH to 6–7 with KOH.
5. Additions to 200 mL homogenization buffer: 1.2 g PVPP (insoluble polyvinylpyrrolidione), 176 mg ascorbate (final concentration 5 mM), 670 μL of 0.5 M DTT (Dithiothreitol, final concentration 1.67 mM).
6. Nylon filter (240 μm mesh).
7. Protease inhibitor: PMSF 1.0 mL 0.1 M stock in Isopropanol, final concentration 0.5 mM.

2.2. Two-Phase Partitioning

1. Phase partitioning polymer solution: 3.72 g of 20% Dextran T500 solution, 1.86 g of 40% PEG4000 solution, 1.08 g sucrose, 225 μL potassium phosphate buffer 0.2 M pH 7.8, 18 μL KCl 2 M, add H_2O up to 9 g.
2. Buffer 330/5: 0.33 M sucrose, 5 mM potassium phosphate pH 7.8.

2.3. H^+ Pump Assay

1. For the H^+ pumping assay: fluorometer, quarts cuvettes with magnetic stirring bar.
2. Proton pumping buffer: 20 mM MOPS-KOH pH 7.0, 40 mM K_2SO_4, 25 mM KNO_3, 1 μM ACMA, 60 nM Valinomycin, 0.05% (w/v) Brij-58 (polyoxyethylene 20 acetyl ether).
3. Solutions of: $MgSO_4$, (500 mM) and ATP-KOH, (0.1 M, pH 7.0), Vanadate Na_3VO_4 (10 mM).

3. Methods

3.1. Preparation of the Polymer Solution for the Two-Phase Partitioning

Day 1:

Prepare the two-phase system and store at 4°C (see Note 1). Make two tubes per sample. These are prepared on a balance and are very critical for obtaining pure plasma membranes (see Notes 2 and 3).

3.2. Isolation of Microsomes

Day 2:

1. 7–30 g (or more) of plant material, cut into pieces and mixed with 200 mL ice-cold homogenization buffer in an ice-cold blender (see Notes 4 and 5).
2. Homogenizer 4 × 15 s (15 s break).
3. Filter the homogenized material through a nylon filter into a cold glass beaker.
4. Add PMSF (350 μL 0.1 M stock).
5. Transfer to GSA centrifuge tubes. Centrifuge the sample at 10,000 × *g* for 15 min.
6. Transfer supernatant to new SS34 centrifuge tubes and discard the pellet.
7. Centrifuge at 30,000 × *g* for 50 min (alternatively 50,000 × *g*, for 30 min in an ultracentrifuge).
8. Discard the supernatant (contains all soluble proteins).
9. Resuspend the pellet in buffer 330/5 using a glass homogenizer. The final volume must NOT be more than 3.5 mL. Add 1 mM DTT + 0.1 mM EDTA (see Note 6). This sample is called the microsomal fraction (MF).
10. Continue with the phase separation on a Two-phase system. Keep the rest of the microsomal fraction if more than 3 g.

3.3. Aqueous Polymer Two-Phase Partitioning

1. Add the microsomal fraction (3 g) on top of 9 g polymer solution (generating a 12 g system, 9 + 3 g). Mix thoroughly but with great care! (Turn slowly upside down 10–20 times, important to be carefully. Do NOT vortex!). Centrifuge the tubes at 1,000 × *g* for 5 min, at 4°C. Here two phases will appear. Make tube 2 at the same time, but add 3 g of 330/5 buffer instead of the microsomes (see Note 7).
2. Remove 90% of the upper phase from tube 2 (keep U_2 in a separate tube), thereby creating fresh lower phases (L_2). Take 90% of the upper phase (U_1), transfer it to lower phase 2 (L_2), and re-extract to increase purity. The upper phase should turn white/transparent as pure PM is obtained. To increase the yield of plasma membranes, re-extract the first lower phase L_1 with a fresh upper phase (U_2). See Fig. 1.

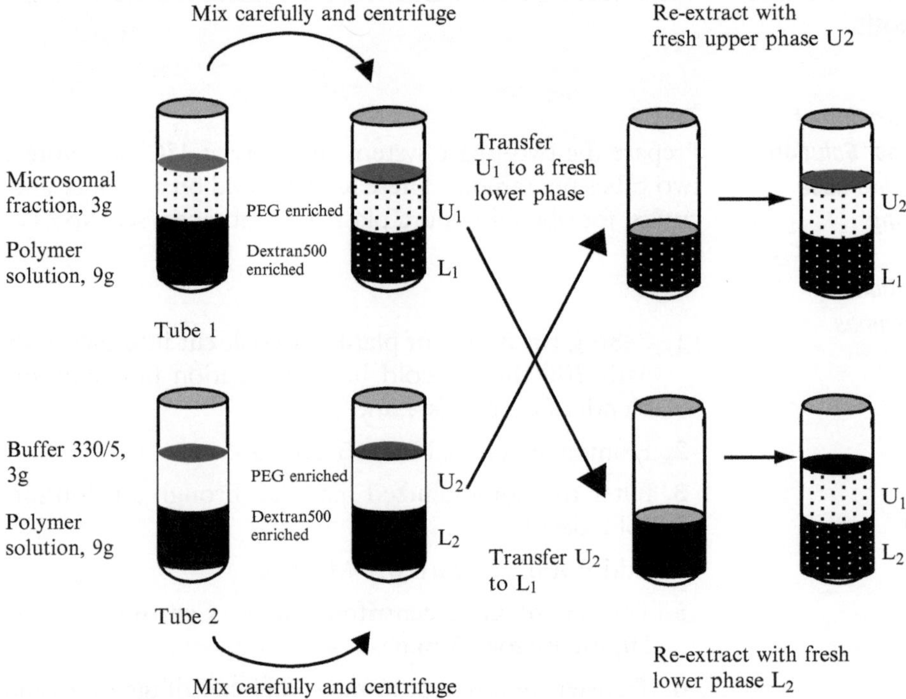

Fig. 1. Two phase partitioning. Generating the two-phase system and transfer of phases between the tubes. The microsomal fraction is added on top of the polymer solution in tube 1 and buffer 330/5 on top of the polymer solution in tube 2. After mixing and centrifugation, two new phases will appear. The upper phase is enriched in plasma membranes. The upper phase 2 (U_2) is removed and kept. The upper phase 1 (U_1) is then transferred to the lower phase 2 (L_2) and re-extracted in order to increase the purity of the obtained PM. Remaining PM is extracted from L_1 by the addition of U_2. The two fractions U_1 and U_2 can be pooled in the end of the experiment.

3. Dilute the first lower phase (L_1) ten times and save 10% for analysis of the purification (optional).
4. Remove PEG and dextran by a washing step: dilute MF, L_1, U_1, and U_2 with 330/5 and collect the membranes by ultracentrifugation (100,000 × *g* for 60 min). Resuspend the pellet in 330/5 + 5 mM KCL (add EDTA and DTT).
5. Divide every fraction into five portions and freeze them in liquid nitrogen. Store at −80°C. See Note 8.

3.4. Measurements of Proton Transport

Plant plasma membranes form sealed vesicles, so do not need to be reconstituted. The addition of Brij-58 to the assay buffer is essential in order to create inside-out vesicles.

1. Dilute the plasma membrane protein (10–20 μg/100 μL, see Note 9) in buffer 330/5. Keep on ice until used. You need 100 μL sample per assay reaction.

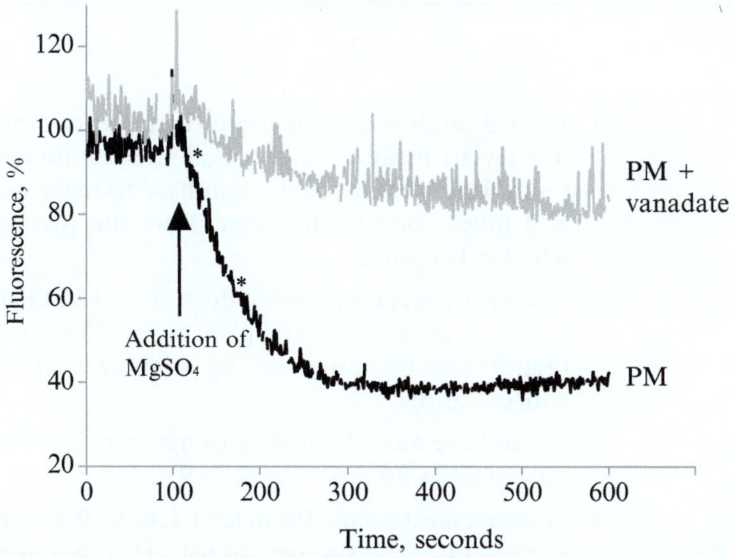

Fig. 2. H^+ pumping in plasma membranes assayed by ACMA quenching. The figure shows an example of H^+ pumping in an *Arabidopsis* plasma membrane fraction (*black trace*). The reaction is initiated by the addition of $MgSO_4$. Fluorescence is shown as the percentage of the fluorescence level just before addition of $MgSO_4$. When calculating the activity, the initial rate is used. In this example, it would be the slope (ΔF) between the *two asterisks* on the *black trace*, divided by the amount of protein (μg protein) and time (s) resulting in activity = ΔF/(μg protein × s). Proton transport not derived from PM H^+-ATPases is found by the addition of vanadate, a specific inhibitor of P-type ATPases to the assay reaction (*grey trace*).

2. Test settings on the spectrofluorometer using the ACMA containing assay buffer. See Note 10.
3. Transfer 100 μL PM solution to the cuvette.
4. Add 30 μL 0.1 M ATP-KOH to the cuvette. Fill up to a total volume of 1.5 mL with assay buffer.
5. Put the cuvette in fluorometer (remember to include the stirring bar) and begin to sample.
6. Wait until fluorescence is stable (60–120 s).
7. Add $MgSO_4$ to start the assay (6 μL 0.5 M).
8. Let the experiment run for a couple of minutes (see Note 11).
9. Optional: Dissipate membrane potential by adding 0.5 μM nigericin when the fluorescence has stabilized.
10. Make a sample with the addition of 100 μM vanadate as the background control. See Note 12.

An example of H^+ transport in *Arabidopsis* plasma membranes can be seen in Fig. 2. Also, the negative control where vanadate is added to the sample is shown.

4. Notes

1. Even though you need more than one two-phase system, do not try to make a batch of all the components and aliquot them. It is NOT possible. You have to make one phase system at a time. You can, however, store the two-phase systems at 4°C for 1 month.
2. The final concentrations of dextran and PEG are 6.2% and are found to work for several plant types. The purity of the membranes can be improved by optimization of the polymer concentrations.
3. If you have more than 30 g of material, it is beneficial to scale up and use a 36 g system (27 + 9).
4. Homogenization medium: for 1 L mix 10.46 g MOPS, 1.461 g EDTA, 112.96 g sucrose, adjust pH to 6–7 with KOH.
5. It is important to have a strong blender with exchangeable knives (strong household blender). A glass blender is recommended since it can be cooled with ice or in a cold room before use. If you have 10 g or less material, you can use a hand blender and reduce the volume to 100 mL.
6. It is important to get the microsomal fraction completely "dissolved" in the 330/5 buffer; this cannot be done by pipetting but requires a glass homogenizer. Add 2 μL 50 mM EDTA and 2 μL 0.5 M DTT/mL of 330/5 buffer used for solubilization of the pellet.
7. Hold on to the lid of the falcon tube when mixing in order to keep the samples cold.
8. Optional: Test the purity of your membrane fractions (MF, L_1, U_1, and U_2) by western blot with antibodies raised against proteins in different subcellular compartments.
9. When using PM from plants, 8–50 μg of protein can be used.
10. Always run an emission and excitation scan to ensure your setting and buffers are working properly. The properties of ACMA are as follows: Excitation 412 nm, make a scan between 350 and 450 nm, 0.1 point/s; Emission 480 nm, make a scan 450–500 nm. The signal should be around 10^4–10^6 cps (counts per second).The assay is time-based; we normally measure 1 point/s with a duration time of 600 s.
11. The initial (first minutes) rate of H^+ flux is used when calculating the rate of H^+ transport. After 5–10 min, the vesicles might become leaky to protons and you will observe a slight increase in fluorescence.

12. Vanadate inhibits P-type ATPases and will therefore block all proton pumping that is dependent on the plasma membrane H^+-ATPase. Vanadate can occur in several oxidation stages, so to obtain the correct stage, the vanadate solution must be boiled for 5 min and placed on ice before use.

References

1. Chen Z, Pottosin II, Cuin TA et al (2007) Root plasma membrane transporters controlling K^+/Na^+ homeostasis in salt-stressed barley. Plant Physiol 145:1714–1725
2. Johansson F, Olbe M, Sommarin M et al (1995) Brij 58, a polyoxyethylene acyl ether, creates membrane vesicles of uniform sidedness. A new tool to obtain inside-out (cytoplasmic side- out) plasma membrane vesicles. Plant J 7:165–173
3. Venema K PMG (1995) Metabolic modulation of transport coupling ratio in yeast plasma membrane H^+-ATPase. J Biol Chem 270:19659–19667

Chapter 15

Determination of Reactive Oxygen Species in Salt-Stressed Plant Tissues

Andrés Alberto Rodríguez and Edith L. Taleisnik

Abstract

Reactive oxygen species (ROS) participate in signaling events that regulate ion channel activity and gene expression. However, excess ROS exert adverse effects that stem from their interaction with macromolecules. Thus, the assessment of the effects of salinity on ROS changes are central to understanding how plants respond and cope with this stress. ROS determination in salt-stressed plants poses specific challenges. On the one hand, salinity comprises osmotic and ion-specific effects which may, in turn, have different effects on ROS production. On the other hand, changes in ROS production may happen when tissues from salinized plants are subject to water potential (Ψ) changes when incubated in non-isosmotic solutions. This chapter provides detailed accounts of methods for ROS detection in tissues from salt-stressed plants and includes suggestions for avoiding artifacts when dealing with such tissues.

Key words: Reactive oxygen species, Hydrogen peroxide, Superoxide, Hydroxyl radical, Salt stress, Fluorescence, Tetrazolium salts

1. Introduction

Reactive oxygen species (ROS) are highly reactive oxygen derivatives comprising both oxygen radicals and certain nonradicals (1). Oxygen radicals are independent species with unpaired electrons and include singlet oxygen (O_2^1), superoxide radicals ($^{\bullet}O_2^-$) and the hydroxyl radical ($HO^{\bullet}$). Nonradical ROS include H_2O_2 and ozone (O_3), among others.

ROS are generated in most plant cell compartments (2) and controlled ROS production appears to be a general characteristic of expanding plant cells and organs (3), where ROS participate in signaling events (4, 5) that regulate ion channel activity (6) and gene expression (7). However, ROS levels generally increase in many plant parts under biotic (8) and abiotic (9) stress conditions,

Sergey Shabala and Tracey Ann Cuin (eds.), *Plant Salt Tolerance: Methods and Protocols*, Methods in Molecular Biology, vol. 913, DOI 10.1007/978-1-61779-986-0_15, © Springer Science+Business Media, LLC 2012

including salinity (10). Excess ROS exert adverse effects (oxidative damage) that stem from their interaction with macromolecules such as lipids, proteins, and nucleic acids.

The ROS balance response to salt stress is expected to differ among tissues (11) and cell compartments (12, 13). Thus, the assessment of the effects of salinity on ROS changes is central to understanding how plants respond and cope with this stress. Techniques for measuring ROS have been extensively reviewed (14), and only a few specific references on plant systems are mentioned here (15–18).

ROS determination in salt-stressed plants poses specific challenges (19). On the one hand, osmotic and ion-specific effects of salinity on ROS production may be different (19, 20), while on the other hand, changes in ROS production may occur when tissues from salinized plants are subject to water potential (Ψ) changes (21) when incubated in non-isosmotic solutions (20). Essentially, when assessing ROS production under saline conditions, it is necessary to include realistic estimates of apoplastic ion concentration in the incubation medium, and to take into account the plant tissue Ψ.

This chapter provides accounts of methods for ROS detection in tissues from salt-stressed plants that we have personally practiced. Qualitative, histochemical methods are described for detecting ROS by epifluorescence, H_2O_2 by electron microscopy, and, $^{\bullet}O_2^-$ with nitro blue tetrazolium (NBT). Quantitative assessment methods for H_2O_2, $^{\bullet}O_2^-$, and $HO^{\bullet}$ production are also detailed. A method for the obtention of apoplastic fluid to estimate ion concentration in this compartment is also included.

ROS detection by epifluorescence is based on the formation of a fluorescent compound from 2′,7′-dichlorofluorescin diacetate (DCFH-DA) when the acetate group is cleaved and nonfluorescent 2′,7′-dichlorofluorescin (DCFH) is oxidated to the fluorescent DCF product in a peroxidase-dependent reaction (22). Endogenous peroxidase activity is considered to be sufficient to sustain this reaction. The non-polar unreactive DCFH-DA can be taken up by the cells and subsequently deacetylated by endogenous esterases to render the polar DCFH (23), to which the cell membrane is supposedly impermeable. Alternatively, DCFH-DA may be first deacelytated by apoplastic esterases and it will then detect extracellular ROS.

Subcellular H_2O_2 detection by electron microscopy is based on the formation of cerium perhydroxide crystals that are deposited after the reaction of $CeCl_3$ with endogenous H_2O_2 (24). Positive staining is detected in electron micrographs as the formation of electron-dense deposits. In vivo H_2O_2 production can be measured as the formation of a pink adduct that results from the reaction between H_2O_2, 4-aminoantipyrine (4-AAP) and 3,5-dichloro-2-hydroxybenzene sulfonic acid (DCHBS) in a peroxidase (POX)-dependent reaction (25).

The qualitative determination of $^{\bullet}O_2^-$ with NBT depends on the formation of a blue insoluble formazan precipitate that results from the reaction between $^{\bullet}O_2^-$ and NBT. Tetrazolium compounds that form soluble formazans can be used to quantify $^{\bullet}O_2^-$ production. The reaction between $^{\bullet}O_2^-$ and Na, 3′-[1-[(phenylamino)-carbonyl]-3, 4-tetrazolium](4-methoxy-6-nitro) benzene sulfonic acid hydrate (XTT) produces a soluble formazan that can be measured spectrophotometrically (26).

$HO^{\bullet}$ production can be quantified by spectrofluorometry as the formation of hydroxyl benzoate resulting from the reaction of $HO^{\bullet}$ and benzoate (BZ) (27).

As mentioned above, realistic estimates of apoplastic ion concentration should be included in the incubation media. The methods to obtain apoplastic fluid usually involve tissue centrifugation at low speed and a check for symplastic contamination by determining the glucose 6-P dehydrogenase (G6PDH) activity (3, 28).

2. Materials

The methods described below are for either intact seedlings or excised plant parts (2, 13, 19, 29). See Note 1 for tissue manipulation tips and Note 2 for the comments on the composition of incubation solutions.

All solutions should be prepared using distilled or deionized water and analytical grade reagents. Solutions can be prepared at room temperature and stored at 4°C (unless otherwise indicated).

2.1. Obtaining the Apoplastic Fluid and Estimating the Cytoplasmic Contamination and Na^+ Concentration

1. Net bags.
2. Falcon and Eppendorf tubes.
3. Spectrophotometer (UV lamp).
4. Atomic emission spectrophotometer or flame photometer.
5. Unrefrigerated centrifuge.
6. Reagents for determining G6PDH activity (30) and protein concentration (31).

2.2. Localized ROS Detection in Intact Seedlings Using 2′,7′-Dichlorofluorescin and Epifluorescence (See Note 3 on Specificity and Controls)

1. Intact plants or excised plant sections.
2. Petri dish.
3. Microwave oven.
4. Thermometer.
5. Epifluorescence or stereoscopic microscope with excitation filter BP 450–490 and emission filter LP 520 and attached camera. Alternative: confocal microscope with a 405 nm diode laser as source of excitation light and acquisition through a 515/30 BP filter (green channel).

6. Solutions and inclusion medium.
   (a) *2',7'-dichlorofluorescin diacetate (DCFH-DA).* For 1 mL of 25 mM DCFH-DA in ethanol, weigh 12 mg DCFH-DA into a 1.5 mL Eppendorf tube. Add 1 mL absolute ethanol and mix. Wrap the tube with aluminum foil. If not used immediately, this solution can be stored frozen at –20°C for up to 3 months.
   (b) *Potassium phosphate buffer.* For 50 mL of 20 mM potassium phosphate buffer, pH 6, weigh 23 mg KH_2PO_4 and 118 mg KH_2PO_4 and transfer to a glass beaker containing 40 mL water. Mix and adjust pH with HCl (see Note 4). Make up to 50 mL with water.
   (c) *KI control.* Weigh 1.66 g KI and add to the buffer solution above before adjusting to the final volume.
   (d) *Agar mixture.* Weigh 1 g agar. Add 90 mL of 20 mM potassium phosphate buffer to a 200 mL graduated glass beaker. Mix and place in a microwave oven and heat at maximum power setting for 1 min. Return the glass beaker to room temperature. Add 0.2 mL 25 mM DCFH-DA solution and wrap with aluminum foil. Make up to 100 mL. The final solution is 1% (w/v) agar, in 20 mM phosphate buffer pH 6, and 50 μM DCFH and should be prepared immediately before use (see Notes 5 and 6).

2.3. Subcellular H_2O_2 Detection by Electron Microscopy

Positive staining is detected in electron micrographs as the formation of electron-dense deposits. A treatment with KI is an appropriate control.

The solutions and procedures indicated do not include the processing of the plant material for transmission electron microscopy observation, detail is provided until the fixation step. From there on, use standard transmission electron microscopy techniques to embed, slice, mount, shade, and observe the tissue.

1. Small plant sections (approximately 3 mm or smaller), briefly rinsed after cutting to remove wound-induced ROS.
2. Electron microscopy embedding and mounting materials and equipment.
3. Transmission electron microscope.
4. Solutions.
   (a) *$CeCl_3$.* For a 5 mM $CeCl_3$ solution in 5 mM MOPS buffer, pH 7.2, weigh 62 mg $CeCl_3$ and 58 mg MOPS and transfer to 40 mL water in a glass beaker. Mix and adjust pH with NaOH (see Note 6). Make up to 50 mL with water. Transfer to a 50 mL Falcon tube wrapped with aluminum foil. Prepare and use the same day, do not store.
   (b) *KI control.* Weigh 1.66 g KI and add to the solution above before adjusting to final volume.

(c) *Sodium phosphate buffer*. For 100 mM sodium phosphate buffer pH 7.0 solution, weigh 0.584 g $NaH_2PO_4 \cdot H_2O$ and 1.547 g $Na_2HPO_4 \cdot 7H_2O$ and transfer to a glass beaker containing approximately 70 mL water. Mix and adjust pH with HCl (see Note 4). Make up to 100 mL with water.

(d) *Glutaraldehyde*. For 2.5 % glutaraldehyde in sodium phosphate buffer solution, add 40 mL sodium phosphate buffer solution to a 100 mL graduated glass beaker. Weigh 1.25 g glutaraldehyde and transfer to a glass beaker containing approximately 40 mL 100 mM sodium phosphate buffer pH 7.0. Mix, make up to 50 mL with sodium phosphate buffer solution, and store.

2.4. Determination of $^{\bullet}O_2^-$ with Nitro Blue Tetrazolium

Controls for this technique must include an effective $^{\bullet}O_2^-$ dismutating catalyst such as superoxide dismutase (SOD) or Mn salts.

1. Plant parts or tissue sections.
2. Falcon or Eppendorf tubes.
3. Vacuum pump.
4. Vacuum desiccator.
5. Glass slides or plates for mounting.
6. Camera.
7. Solutions.

(a) *Nitro blue tetrazolium (NBT)*. To prepare 50 mL of 122 μM NBT solution, weigh 5 mg NBT and transfer to a glass beaker containing approximately 40 mL water. Mix and make up to 50 mL with water. Transfer to a 50 mL Falcon tube wrapped with aluminum foil. For salinized plants, weigh an appropriate amount of NaCl or sorbitol (depending on whether the material to be tested is Na^+-including or not) and add before making up to final NBT solution volume (see Note 2). If not used immediately, solutions can be stored frozen at –20°C for up to 3 months.

(b) *$MnCl_2$*. For a 10 mM $MnCl_2$ solution, weigh 63 mg $MnCl_2$ and transfer to a glass beaker containing approximately 40 mL water. Mix and make up to 50 mL with water.

(c) *9:1 ethanol and glycerin*. Make up enough to completely submerge the stained tissue and sustain 30 min boiling.

2.5. Spectrophotometric Determination of In Vivo H_2O_2 Production

Use a KI treatment as a control for this technique.

1. Plant sections.
2. Falcon or Eppendorf tubes.

3. Vacuum pump.
4. Vacuum desiccator.
5. Spectrophotometer (visible light wavelength).
6. Solutions.
   (a) *4-aminoantipyrine (4-AAP)*. For a 5 mM 4-AAP solution, weigh 10 mg AAP and transfer to a 15 mL Falcon tube containing 5 mL water. Mix, make up to 10 mL with water, and store.
   (b) *Reaction mixture*. 100 μM 4-AAP, 1 mM 3, 5-dichloro-2-hydroxybenzene sulfonic acid (DCHBS), 0.06 mg^{-1} mL^{-1} horseradish POX (HPOX). Add 40 mL water to a 100 mL graduated glass beaker. Weigh 13 mg DCHBS, 3 mg HPOX, and transfer to the glass beaker. Add 1 mL 5 mM 4-AAP solution. Mix and make up to 50 mL with water. Transfer to a 50 mL Falcon tube wrapped with aluminum foil. For the NaCl treatment, weigh an appropriate amount of NaCl or sorbitol and add before mixing (see Note 2). For the H_2O_2 specificity control, add 1.66 g KI to the reaction mixture before adjusting to the final volume.
   (c) *Ethanol-glycerin*. For 9:1 ethanol-glycerin solution, add 45 mL ethanol and 5 mL glycerine to a glass beaker and mix to homogeneity. Transfer to a 50 mL Falcon tube.

2.6. Spectrophotometric Determination of In Vivo $^{\bullet}O_2^-$ Production

Suitable controls for this technique must include an effective $^{\bullet}O_2^-$ dismutating agent such as SOD or a Mn salt.

1. Falcon or Eppendorf tubes.
2. Vacuum pump.
3. Vacuum desiccator.
4. Spectrophotometer (visible light wavelength).
5. Solutions.
   (a) *$MnCl_2$*. For a 10 mM $MnCl_2$ solution, weigh 63 mg $MnCl_2$ and transfer to a glass beaker containing approximately 40 mL water. Mix and make up to 50 mL with water.
   (b) *3'-[1-[(phenylamino)-carbonyl]-3, 4-tetrazolium](4-methoxy-6-nitro) benzene sulfonic acid hydrate (XTT) solution.* For 0.5 mM XTT, weigh 17 mg XTT and transfer to a glass beaker containing approximately 40 mL water. Mix and make up to 50 mL with water. Transfer to a 50 mL Falcon tube wrapped with aluminum foil. For the NaCl treated plant material, weigh an appropriate amount of NaCl or sorbitol and add before mixing (see Note 2). For the $^{\bullet}O_2^-$ specificity control, add 0.5 mL of the $MnCl_2$ solution or SOD (50 μm mL^{-1}) before making up to the final volume. If not used immediately, solutions can be stored frozen at –20°C for up to 3 months.

2.7. Fluorometric Determination of In Vivo HO• Production

1. Falcon or Eppendorf tubes.
2. Vacuum pump.
3. Vacuum desiccator.
4. Spectrofluorometer.
5. *Benzoate (BZ) solution*. For a 2.5 mM BZ solution, weigh 18 mg BZ and transfer to a glass beaker containing water. Mix and make up to 50 mL with water. Transfer to a 50 mL Falcon tube wrapped with aluminum foil. For the salt-treated plants, weigh an appropriate amount of NaCl or sorbitol and add before mixing (see Note 2). Prepare and use the same day, do not store.

3. Methods

3.1. Obtaining the Apoplastic Fluid and Estimating the Cytoplasmic Contamination and Na^+ Concentration

1. Prepare a 15 mm long section from a plastic tube that can fit into a 50 mL Falcon tube, depending on the size of the plant material. Seal one end with plastic net. Introduce a net bag inside the tube. The net pores should be smaller than the surface of the plant segments (see next point).
2. Carefully introduce approximately 100 excised, rinsed leaf or root segments (10 mm in length) in the net bag. Make sure they are all positioned parallel to the tube side. Seal the net bag loosely to prevent the segments from moving.
3. Centrifuge the tubes for 1 min at 1,000 ×g to remove any rinsing solution.
4. Centrifuge again for 10 min at 2,000 ×g to collect the apoplastic fluid. Store the apoplastic fluid at –20°C.
5. Use the apoplastic fluid to check G6PDH activity, protein concentration, and Na^+ concentration. Several batches of apoplastic fluid will be necessary to carry out all assays. Compare apoplastic G6PDH activity on a protein concentration basis with that of whole tissue homogenate. It should not be higher than 5 % of whole tissue activity when the apoplastic fluid is not contaminated with cytosol.
6. Dilute uncontaminated apoplastic fluid samples with an appropriate water volume to determine Na^+ concentration by atomic emission spectrophotometry or flame photometry. See comments on estimation of apoplastic ion concentrations in Note 7.

3.2. Localized ROS Detection in Intact Seedlings Using 2′,7′-Dichlorofluorescin and Epifluorescence

1. Measure the temperature of the agar mixture. When it reaches 30°C, pour it on a Petri dish and submerge the plant material (see Notes 8 and 9). Incubate in the dark.
2. After 15–30 min, place the Petri dish under an epifluorescence or confocal microscope to observe DCF fluorescence (see Note 10). Briefly check the material and take a picture (see Notes 11 and 12).

3.3. Subcellular H_2O_2 Detection by Electron Microscopy

1. Place the plant sections in an Eppendorf tube and completely submerge in the $CeCl_3$ solution. Infiltrate by applying gentle vacuum pulses for 1 min. Incubate for 3 h.
2. Remove and fix the plant material in glutaraldehyde solution for 24 h at 4°C.
3. Process for electron microscopy.

3.4. Histochemical Determination of $^{\bullet}O_2^-$ with Nitro Blue Tetrazolium

1. Place the plant sections in an appropriate tube and add sufficient NBT solution to cover completely (see Notes 13 and 14). To infiltrate, place the open tubes in a desiccator, and apply vacuum pulses, gently releasing the vacuum between each pulse. Usually, 8–10 pulses, 10 s each, are sufficient for infiltrating the tissue.
2. Incubate for 2 h at 30°C in the dark. Comments about incubation time in Note 10 also apply for this technique.
3. Remove the plant material from the NBT solution and place in beaker or glass tube with an appropriate volume of the 9:1 ethanol-glycerin solution. Place the vessel in a water bath and boil for 10 min or until the green color has faded almost completely.
4. Mount the material on a glass slide or sheet and cover with glass slips.
5. Photograph.
6. See Note 15 for tips on quantifying staining intensity.

3.5. Spectrophotometric Determination of In Vivo H_2O_2 Production

1. Submerge the plant material completely (see Notes 14 and 15) in the reaction solution. Infiltrate by applying gentle vacuum for 1 min. Incubate for 2 h at 30°C in the dark.
2. Collect the incubation medium and centrifuge at $10{,}000 \times g$, for 5 min.
3. Measure the absorbance at 515 nm.
4. Transform 515 nm absorbance readings into the H_2O_2 concentration using the molar extinction coefficient 2.6×10^4 $M^{-1} cm^{-1}$.

3.6. Spectrophotometric Determination of In Vivo $^{\bullet}O_2^-$ Production

1. Submerge the plant material completely in the XTT solution (see Notes 14 and 15). Infiltrate by applying gentle vacuum for 1 min. Incubate for 5 h at 30°C in the dark.
2. Collect the incubation medium and centrifuge at $10{,}000 \times g$, for 5 min.
3. Measure the absorbance of the incubation medium at 470 nm.
4. Use the $^{\bullet}O_2^-$ molar extinction coefficient at 470 nm (2.16×10^4 $M^{-1} cm^{-1}$) to transform absorbance values into $^{\bullet}O_2^-$ concentration.

3.7. Fluorometric Determination of In Vivo HO• Production

1. Submerge the plant material completely in the BZ solution (see Notes 14 and 15). Infiltrate by applying gentle vacuum for 1 min. Incubate for 7 h at 30°C in the dark.
2. Collect the incubation medium and centrifuge at 10,000 × *g*, for 5 min.
3. Measure the fluorescence using a spectrofluorometer at 407 nm emission after excitation at 305 nm.
4. Run blanks without BZ in parallel to correct for unspecific fluorescence.
5. Prepare a BZ calibration curve by measuring a series of BZ dilutions (from 0 to 2.5 mM).
6. Transform the fluorescence values of the biological samples into HO• molar concentration using the calibration curve values.

4. Notes

1. For any ROS determination method, it is of the utmost importance to manipulate the tissue with great care, avoiding any pressure while handling it, as this may induce injury-related ROS production (19). When working with explants such as leaf segments, it is very important to rinse them briefly after excision to remove any ROS produced as a result of the injury.
2. It is recommended that the osmotic potential of infiltration and incubation solutions be as similar as possible to the tissues Ψ, which can be measured psychrometrically. The choice of a suitable organic osmotic agent for that purpose is contingent on its action as a ROS scavenger (32). Note that the organic solute sorbitol is a weak scavenger for $^{\bullet}O_2^-$ (19, 33) and HO• (34).

   For salt-treated plant material, it is important to estimate apoplastic Na^+ and include it in the incubation solutions. When studying roots obtained from hydroponically grown plants, the ion composition in the root apoplastic solution can be expected to approximately reflect that in the nutrient solution. However, this will not be the case in soil-grown roots. In other plant organs, the composition of the apoplastic solution will depend on the ion balance control exerted along the xylem and the transpiration stream (35). Therefore, an assessment of the apoplastic ion concentration should be performed and the estimated concentration included in the staining or assay solutions (36). Average tissue ion estimates will not necessarily reflect ion concentration in different compartments (10) and estimates of ion concentration from one plant part should obviously not be extrapolated to the subject tissue because significant differences in ion distribution within a plant are likely to exist (37).

3. Although DCFH is usually assumed to react mainly with H_2O_2, it can also become fluorescent in the presence of other oxidants (22), as was critically reviewed by (14). Appropriate controls, including a H_2O_2 scavenger such as KI (38), are essential for interpreting the results.
4. To adjust the pH, use a series of HCl solutions (e.g., 1 and 0.1 N), rather than concentrated HCl. This avoids sudden drops in pH.
5. The required volume of the agar solution depends on the size of the plant material; the solution should cover it evenly.
6. Use a series of NaOH solutions (e.g., 1 and 0.1 N) to avoid a sudden rise in pH.
7. Details on how to calculate apoplastic fluid dilution after infiltration can be found in (39). Although these methods can render only an approximate idea of the ion concentration in the apoplastic solution due to the very small volumes obtained and dilutions induced by infiltration, it is still preferable to obtain information from these procedures rather than to incubate the tissues without correcting for the estimated apoplastic ion concentration.
8. Note that the plant material is introduced before the medium has completely cooled. Keep the plant material submerged using dissection needles. When the agar begins to solidify, remove the needles.
9. If non-excised shoots are to be observed, roots can be kept in an appropriate nutrient medium, with or without NaCl, in a beaker or another vessel, alongside the Petri dish.
10. The incubation time depends on the type of plant material used. An estimated period is suggested above. For each material, try a series of increasing incubation periods until the right fluorescence intensity (sharp, not too high) can be observed.
11. It is important not expose the plant material to the microscope's light for too long. This exposure will increase the fluorescence intensity, produce a loss of detail at high light intensity, and will affect any comparisons between treatments. Expose for only a few minutes to check the material and then take a picture.
12. Fixed exposure times are essential if different treatments are being compared. Increasing the exposure time will also increase fluorescence [see, for example, ref. (20)].

13. Completely cover the plant material with the staining solution but use the lowest possible volume, increasing the ratio between solution and plant material will diminish staining quality.

14. Larger leaves can be rolled and placed in a Falcon tube, with due caution not to induce injuries or folds. Observe bubbles coming up to make sure air is drawn out and infiltration is effective.

15. Although this is essentially a qualitative technique, if the same incubation time is used for several simultaneous treatments, differences in blue color intensity will reflect differences in $^{\bullet}O_2^-$ production. Blue color intensity can be measured with an image processing software after splitting the images into blue, green, and red stacks. Color intensity in the blue channel can then be measured as luminance (Lu), (scale 0–255 from the darkest to the lightest) and transformed to optical density (OD), $OD = \log(Lu)^{-1}$.

Acknowledgments

The authors gratefully acknowledge support from Agencia Nacional de Promoción Científica y Tecnológica (ANPCYT), Consejo Nacional de Investigaciones Científicas y Técnicas (CONICET), Instituto Nacional de Tecnología Agropecuaria (INTA), and Universidad de San Martin (UNSAM), all from Argentina.

References

1. Halliwell B (2006) Reactive species and antioxidants. Redox biology is a fundamental theme of aerobic life. Plant Physiol 141: 312–322
2. Mittler R (2002) Oxidative stress, antioxidants and stress tolerance. Trends Plant Sci 9:405–410
3. Rodríguez AA, Grunberg K, Taleisnik E (2002) Reactive oxygen species in the elongation zone of maize leaves are necessary for leaf extension. Plant Physiol 129:1627–1632
4. Dat J, Vandenabeele S, Vranová E et al (2000) Dual action of the active oxygen species during plant stress responses. Cell Mol Life Sci 57:779–795
5. Miller G, Shulaev V, Mittler R (2008) Reactive oxygen signaling and abiotic stress. Physiol Plant 133:481–489
6. Demidchik V (2010) Reactive oxygen species, oxidative stress and plant ion channels. In: Demidchik V, Maathuis F (eds) Ion channels and plant stress responses, 1st edn. Springer, Berlin
7. Mittler R, Vanderauwera S, Gollery M et al (2004) Reactive oxygen gene network of plants. Trends Plant Sci 9:490–496
8. Lamb C, Dixon RA (1997) The oxidative burst in plant disease resistance. Annu Rev Plant Biol 48:25–275
9. Scandalios J (1993) Oxygen stress and superoxide dismutases. Plant Physiol 101:7–12
10. Munns R, Tester M (2008) Mechanisms of salinity tolerance. Annu Rev Plant Biol 59: 651–681
11. Bernstein N, Shoresh M, Xu Y et al (2010) Involvement of the plant antioxidative response in the differential growth sensitivity to salinity of leaves vs roots during cell development. Free Radic Biol Med 49:1161–1171
12. Mittova V, Tal M, Volokita M et al (2003) Up-regulation of the leaf mitochondrial and peroxisomal antioxidative systems in response to salt-induced oxidative stress in the wild salt-tolerant tomato species *Lycopersicon pennellii*. Plant Cell Environ 26:845–856

13. Rodríguez AA, Maiale SJ, Menéndez AB et al (2009) Polyamine oxidase activity contributes to sustain maize leaf elongation under saline stress. J Exp Bot 60:4249–4262
14. Tarpey MM, Wink DA, Grisham MB (2004) Methods for detection of reactive metabolites of oxygen and nitrogen: *in vitro* and *in vivo* considerations. Am J Physiol Regul Integr Comp Physiol 286:R431–R444
15. Sandalio LM, Rodríguez Serrano M, Romero Puertas MC et al (2008) Imaging of reactive oxygen species and nitric oxide *in vivo* in plant tissues. In: Colowick SP, Kaplan NO (eds) Methods in enzymology, 1st edn. Academic, London
16. Snyrychová I, Ayaydin F, Hideg É (2009) Detecting hydrogen peroxide in leaves *in vivo*—a comparison of methods. Physiol Plant 135:1–18
17. Kristiansen KA, Jensen PE, Møller IM et al (2009) Monitoring reactive oxygen species formation and localisation in living cells by use of the fluorescent probe CM-H_2DCFDA and confocal laser microscopy. Physiol Plant 136:369–383
18. Swanson SJ, Choi W-G, Chanoca A et al (2010) *In Vivo* Imaging of Ca^{2+}, pH, and reactive oxygen species using fluorescent probes in plants. Annu Rev Plant Biol 62:273–297
19. Rodríguez AA, Córdoba AR, Ortega L et al (2004) Decreased reactive oxygen species concentration in the elongation zone contributes to the reduction in maize leaf growth under salinity. J Exp Bot 55:1383–1390
20. Bustos D, Lascano R, Villasuso AL et al (2008) Reductions in maize root-tip elongation by salt and osmotic stress do not correlate with apoplastic $O_2^{\cdot -}$ Levels. Ann Bot 102:551–559
21. Felix G, Regenass M, Boller T (2000) Sensing of osmotic pressure changes in tomato cells. Plant Physiol 124:1169–1179
22. Cathcart R, Schwiers E, Ames BN (1983) Detection of picomole levels of hydroperoxides using a fluorescent dichlorofluorescein assay. Anal Biochem 134:111–116
23. Bass DA, Parce JW, Dechatelet LR et al (1983) Flow cytometric studies of oxidative product formation by neutrophils: a graded response to membrane stimulation. J Immunol 130:1910–1917
24. Bestwick CS, Brown IR, Bennet MHR et al (1997) Localization of hydrogen peroxide accumulation during the hypersensitive reaction of lettuce cells to *Pseudomonas syringae* pv phaseolicola. Plant Cell 9:209–221
25. Cona A, Rea G, Botta M et al (2006) Flavin-containing polyamine oxidase is a hydrogen peroxide source in the oxidative response to the protein phosphatase inhibitor cantharidin in *Zea mays* L. J Exp Bot 57:2277–2289
26. Frahry G, Schopfer P (2001) NADH-stimulated, cyanide-resistant superoxide production in maize coleoptiles analyzed with a tetrazolium-based assay. Planta 212:175–183
27. Schopfer P, Plachy C, Frahry G (2001) Release of reactive oxygen intermediates (superoxide radicals, hydrogen peroxide, and hydroxyl radicals) and peroxidase in germinating radish seeds controlled by light, gibberellin, and abscisic acid. Plant Physiol 125:1591–1602
28. Córdoba-Pedregosa MdC, Córdoba F, Villalba JM et al (2003) Differential distribution of ascorbic acid, peroxidase activity, and hydrogen peroxide along the root axis in *Allium cepa* L. and its possible relationship with cell growth and differentiation. Protoplasma 221:57–65
29. Campestre MP, Bordenave CD, Origone AC et al (2011) Polyamine catabolism is involved in response to salt stress in soybean hypocotyls. J Plant Physiol 168:1234–1240
30. Hong QZ, Copeland L (1991) Isoenzymes of glucose 6-phosphate dehydrogenase from the plant fraction of soybean nodules. Plant Physiol 96:862–867
31. Bradford M (1976) A rapid and sensitive method for the quantitation of microgram quantities of protein utilizing the principle of protein-dye binding. Anal Biochem 72:248–254
32. Smirnoff N, Cumbes QJ (1989) Hydroxyl radical scavenging activity of compatible solutes. Phytochemistry 28:1057–1060
33. Komagoe K, Takeuchi H, Inoue T, Katsu T (2010) Application of an oxygen electrode to evaluate superoxide anion-scavenging ability. Anal Sci 26:903–906
34. Morelli R, Russo-Volpe S, Bruno N, Lo Scalzo R (2003) Fenton-dependent damage to carbohydrates: free radical scavenging activity of some simple sugars. J Agric Food Chem 51:7418–7425
35. Sattelmacher B (2001) The apoplast and its significance for plant mineral nutrition. New Phytol 149:167–192
36. Rodríguez AA, Lascano H, Bustos L et al (2007) Salinity-induced reductions in NADPH oxidase activity in the maize leaf blade elongation zone. J Plant Physiol 164:223–230
37. Davenport R, James RA, Zakrisson-Plogander A et al (2005) Control of sodium transport in durum wheat. Plant Physiol 137:807–818
38. Frahry G, Schopfer P (1998) Hydrogen peroxide production by roots and its stimulation by exogenous NADH. Physiol Plant 103:395–404
39. Mühling KH, Läuchli A (2002) Effect of salt stress on growth and cation compartmentation in leaves of two plant species differing in salt tolerance. J Plant Physiol 159:137–146

Chapter 16

Quantification of the Antioxidant Activity in Salt-Stressed Tissues

Jelena J. Dragišić Maksimović and Branka D. Živanović

Abstract

Biochemical methods available for the measurement of antioxidant activity in salt-stressed tissues are reviewed, outlining the most important advantages and shortcomings of the methods. Here we consider commonly used methods for measuring total antioxidant capacity and phenolic content, ABTS and Folin-Ciocalteu's procedure, respectively. Moreover, we presented assays for determination of antioxidant enzymes activities: superoxide dismutase, catalase, and ascorbate peroxidase. This choice of methods enables us to elucidate a full profile of antioxidant activities, evaluating their effectiveness against various reactive oxygen species produced during salt stress.

Key words: Ascorbate peroxidase, Catalase, Protein concentration, Salinity, Superoxide dismutase, Total antioxidant capacity, Total phenolic content

1. Introduction

Exposure of plants to salinity is known to induce formation of reactive oxygen species (ROS) that cause cell damage, but ROS also are involved in cell growth processes (1). ROS are normally derived from essential aerobic metabolisms and it is necessary to keep ROS at a certain physiological level to enable normal plant growth and development. In plant cells, ROS production occurs in almost all cell compartments (cytosol, chloroplast, and mitochondria), including the apoplast (2, 3). Production of ROS appears to be a general characteristic of expanding plant cells and organs (4–6). The excess production of ROS during salt stress induced by different metabolic pathways with an ability to initiate cascade reactions results in the production of hydroxyl radical ($OH^{\bullet}$) and other reactive species, such as hydrogen peroxide (H_2O_2) and superoxide anion radical ($O_2^{\bullet -}$) (7). To diminish the excess of ROS,

Sergey Shabala and Tracey Ann Cuin (eds.), *Plant Salt Tolerance: Methods and Protocols*, Methods in Molecular Biology, vol. 913, DOI 10.1007/978-1-61779-986-0_16, © Springer Science+Business Media, LLC 2012

plants have developed an antioxidant defense system that comprises enzymatic and non-enzymatic components (7–9). Hence, ROS levels in cells need to be tightly balanced between ROS-scavenging and ROS-producing proteins (3). Multiple antioxidant enzymes that directly react with and scavenge ROS include superoxide dismutase (SOD), catalase (CAT), and various peroxidases (POD) such as ascorbate peroxidase (APX). SOD belongs to a family of metalloenzymes catalyzing the dismutation of $O_2^{\bullet -}$ into H_2O_2 and O_2. This is the first reaction of ROS detoxification and hence a key component of the ROS-scavenging system (10, 11). The H_2O_2 produced is then scavenged by CAT and several classes of peroxidases. CATs are heme proteins that catalyze the removal of H_2O_2 into H_2O and O_2 (12). PODs are distributed throughout the cell and also catalyze the reaction of H_2O_2 detoxification (13). APX has high substrate specificity for ascorbic acid as electron donor and higher affinity for H_2O_2 than CAT (14). APX is found to be the primary scavenger of H_2O_2 in the chloroplast and cytosol of plant cells (15, 16). It has been demonstrated that the increase in activities of antioxidant enzymes could be the basis for salt stress tolerance in a range of plant species (1, 17–22). Moreover, the antioxidative response is well correlated with growth response and salt tolerance of cultivars, indicating a possible cross-talk between these two responses. It therefore seems that under salinity, plants compromise the need of ROS for growth to overcome the possible oxidative damage induced by salinity (4, 23, 24). The apoplastic ROS have been suggested to play a role in cell expansion via effects on cell-wall loosening (25) and cell-wall stiffening (26–28). In addition, ROS can serve as second messenger molecules and induce signal transduction leading to salt adaptation (7, 29). Temporal and spatial regulation of each antioxidative enzyme determines the competence of the antioxidant capacity under saline stress. Their quantitative and qualitative aspects of changes could be related to the levels of resistance to salinity. In order to develop practical strategies for selecting salt tolerant variety of potential crops, there is a pressing need to gain detailed information on whether changes in physiological/biochemical parameters due to salt stress are attributable to detrimental effects of salt stress or are components of the adaptation mechanism.

On the other hand, a plant's free radical-scavenging ability is attributed also to plant polyphenols. These are essential to plant physiology for their contribution to growth, reproduction, and resistance to biotic and abiotic stresses (30). Besides removing free radicals, these compounds are capable of chelating metal, activating antioxidant enzymes, and inhibiting oxidases (31). The reported data indicate that the content of phenolic compounds in plants can be changed by salt stress. This is critically dependent on the salt sensitivity of plants (32) and the level of salt stress (33). It has been proposed that flavonoids and polyphenolic compounds act as reducing agents either as enzyme co-factors or as electron donors (34, 35). Accumulation of these compounds could be pronounced in tissues under stress conditions.

Thus, we present here combination of biochemical and analytical testing methods for quantification of the antioxidant activity in salt-stressed tissues. These methods could be additional tool for screening plants for salt tolerance.

2. Materials

In all protocol steps use ultrapure water (H_2O in the text, prepared by purifying deionized water to attain a sensitivity of 18 MΩ cm at 25°C) and analytical grade reagents. For the best results, all solution should be prepared directly before use and stored at 4°C during experiments.

2.1. Total Antioxidant Capacity and Phenolic Content Extraction Solution

Extraction solution consists of 70 mL of 80% methanol, 30 mL H_2O, and 0.1 mL of concentrated hydrochloric acid.

2.2. Total Antioxidant Capacity Test Components

The reaction mixture for total antioxidant capacity (TAC) determination contains 2 mM ABTS (2,2-azino-bis-3-ethylbenzothiazoline-6-sulfonic acid diammonium salt), 15 μM hydrogen peroxide (H_2O_2), and 0.25 μM horse radish peroxidase (HRP) in 50 mM potassium phosphate buffer pH 7.5.

1. *50 mM potassium phosphate buffer*. Dissolve 0.23 g KH_2PO_4 (MW 136.09 g/mol) and 0.58 g K_2HPO_4 (MW 174.18 g/mol) in 100 mL H_2O. Adjust pH to 7.5 using 1 M KOH solution (dissolve 0.56 g KOH [MW 56.1056 g/mol] in 10 mL H_2O). Phosphate buffer should be prepared directly before use and stored at 4°C.
2. *2 mM ABTS*. Dissolve 0.11 g ABTS (MW 548.68 g/mol) (see Note 1) in 100 mL of phosphate buffer (see Note 2).
3. *15 μM H_2O_2*. Prepare H_2O_2 solution, putting 1 μL of 30% H_2O_2 in 999 μL H_2O. Store solution at 4°C in the dark.
4. *0.25 μM HRP*. Dissolve 1.125 mg HRP (class II) in 1 mL H_2O. Store solution at 4°C.
5. *Preparing solution for the standard curve*. For the standard curve, use set (0.1–0.8 mM) of standard solutions of L-ascorbic acid (AA) prepared from stock solution of 0.8 mM AA. Dissolve 0.014 g of AA (MW 176.1 g/mol) in 100 mL H_2O to get 0.8 mM AA. Prepare the standard solutions from stock solution of AA, starting with 0.7 mM AA (use 8.75 mL of stock solution and fill it up to 10 mL with H_2O to get 0.7 mM

AA), and subsequently make dilutions to complete the set of standard solutions of AA (0.6, 0.5, 0.4, 0.3, 0.2, 0.1 mM).

2.3. Total Phenolic Content Procedure Components

1. *Preparing solution for the standard curve.* Use gallic acid (GA) as a standard for the calibration curve (0.1, 0.5, 1, 1.5, 2 mM). Dissolve 0.034 g GA (MW 170.12 g/mol) in 100 mL of 80% methanol to prepare a stock solution (2 mM). Use 7.5 mL of stock solution and fill it up to 10 mL with methanol to get 1.5 mM GA. Diluting a stock solution with methanol in ratio 1:2, 1:4, and 1:20, prepare 1, 0.5, and 0.1 mM GA, respectively.
2. *Folin-Ciocalteu reagent.* Dissolve 12.5 mL 2 N Folin-Ciocalteu reagent in 100 mL H_2O to get 0.25 N solution. Swirl to mix and incubate 5 min at room temperature (see Note 3).
3. *0.2 M sodium carbonate solution.* Dissolve 0.572 g $Na_2CO_3 \cdot 10H_2O$ (MW 286.14 g/mol) in 10 mL of H_2O.

2.4. Enzymatic Extracts Solution

Solution for tissue extraction consists of 50 mM potassium phosphate buffer (see recipe at Subheading 2.2) pH 7 containing following components (observing addition sequence):

1. *1 mM phenylmethanesulfonylfluoride (PMSF).* Dissolve 0.017 g PMSF (MW 174.19 g/mol) in 100 mL of phosphate buffer (see Note 4).
2. *0.2 mM EDTA.* Add 5.8 mg of EDTA (MW 292.24 g/mol) to the buffer (100 mL) (see Note 5).
3. *1% polyvinylpyrrolidone (PVP) (w/v).* Add 1 g of PVP 40 (average MW 40,000) to the buffer (100 mL) (see Note 6).
4. *0.1% Triton X 100 (v/v).* Add 100 μL to 100 mL buffer (see Note 7).

The major APX in green tissue (leaves) are localized in chloroplast and has a very short life time in an ascorbate-depleted medium. So, for APX activity measurement of photosynthetic tissues, 5 mM ascorbate has to be added in extract solutions (dissolve 88 mg of ascorbate [MW 176.1 g/mol] in 100 ml of buffer).

2.5. Protein Concentration of Enzymatic Extracts

Bradford total protein assay (36) is used to determine the concentration of protein in a sample.

1. *50 mM potassium phosphate buffer* pH 7. (see recipe in Subheading 2.2).
2. *BSA (bovine serum albumin) standard solution preparation.* Considering that the linear concentration range is 0.1–1.4 mg/ml of protein, prepare a standard solution using BSA as the standard protein in a serial dilution series (100, 250, 500, 750, 1,000 mg/mL) in the same buffer as the samples to be assayed (see Note 8). To prepare BSA stock solution, measure 3 mg BSA and dissolve with 3 mL of buffer (see Note 9).

Table 1
Preparing BSA standards for Bradford protein assay standard curve

Well designation	BSA stock solution (μL)	Buffer (μL)
Blank	0	1,000
Standard 1 (100 μg/mL)	100	900
Standard 2 (250 μg/mL)	250	750
Standard 3 (500 μg/mL)	500	500
Standard 4 (750 μg/mL)	750	250
Standard 5 (1,000 μg/mL)	1,000	0

The total volume of each standard sample is 1 mL

To get dilution series for a standard curve, prepared stock solution dilute with buffer as outlined in Table 1.

3. *Bradford reagent*. Dissolve 100 mg Coomassie Brilliant Blue G-250 in 50 mL 95% ethanol, add 100 mL 85% (w/v) phosphoric acid. Dilute the Bradford reagent fivefold in H_2O (1 part Bradford: 4 parts H_2O) (see Note 10) when the dye has completely dissolved, and filter through filter paper just before use (see Note 11). Set the completed Bradford reagent aside and allow it to come to room temperature before use.

2.6. Reaction Solutions for Determination of Enzymes Activities

1. *Reaction solutions for the total SOD activity*. Among the different methods used for the assay of SOD in plant tissue homogenate, the xanthine–xanthine oxidase-ferricytochrome *c* method is the first and the most widely used method (37).
   (a) *50 mM potassium phosphate buffer + 0.1 mM EDTA*. In 100 mL of potassium phosphate buffer (see recipe at Subheading 2.2) pH 7.8 add 2.9 mg EDTA (MW 292.24 g/mol).
   (b) *Solution of 5 × 10^{-5} M xanthine*. After preparing 1 mM NaOH (4 mg NaOH [MW 40 g/mol] in 100 mL of H_2O), dissolve 0.76 mg of xanthine in 10 mL of prepared NaOH solution.
   (c) *Solution of 6 × 10^{-9} M xanthine oxidase*. In 1 mL of phosphate buffer, add 11 μL of xanthine oxidase (EC 1.1.3.22) commercial suspension from buttermilk (Grade I, 0.68 units/mg protein).
   (d) *Solution of 10^{-5} M ferricytochrome c*. Dissolve 2.4 mg cytochrome *c* (from horse heart, MW approx. 12,400 Da) in 1 mL of H_2O.
   (e) *Stock solution of 3 mM KCN*. In 1 mL of prepared phosphate buffer, dissolve 2 mg of KCN (MW 65.12 g/mol).

2. *Reaction solutions for the CAT activity.* Among the numerous assays for catalase, the oldest and the simplest is the method in which the disappearance of peroxide is followed spectrophotometrically (38).
   (a) *50 mM potassium phosphate buffer.* Phosphate buffer (see recipe in Subheading 2.2), adjust to pH 7.
   (b) *30% H_2O_2.* Adjust H_2O_2 volume to total volume to get the desired response.
3. *Reaction solutions for the APX activity.* We optimized the standard assay (39) by slightly changes of components concentrations:
   (a) *50 mM potassium phosphate buffer.* Phosphate buffer (see recipe in Subheading 2.2) adjust to pH 7.
   (b) *0.1 M ascorbate.* Dissolve 17.6 mg of ascorbate (MW 176.1 g/mol) in 1 mL of buffer (2 μL of this solution will give a final concentration of 0.2 mM ascorbate in a 1 mL cuvette).
   (c) *10 mM H_2O_2.* Make a stock solution by addition of 1 μL of 30% H_2O_2 in 999 μL H_2O (20 μL of this solution are going to give final concentration of 0.2 mM H_2O_2 in 1 mL cuvette).

3. Methods

3.1. Plant Tissue Extraction for Total Antioxidant Capacity and Phenolic Content

Homogenize plant tissue in extraction solution (see Note 12) at ratio of 1:5 (fresh weight: extraction solution). Add 5 mL of extraction solution (see Subheading 2.1) to 1 g of homogenized tissue, stir and transfer to plastic tubes. After centrifugation at 13,000 × *g* for 10 min at 4°C, supernatants are transferred to the new tubes to be used for further analyses (see Note 13). A minimum of three extractions must be prepared for each sample.

3.2. Total Antioxidant Capacity Test

Determination of TAC is done following the ABTS method (40). This compound is chosen because the enzyme (HRP) facilitates the reaction with hydrogen peroxide, turning it into a green end-product:

$$\text{ABTS} + H_2O_2 \xrightarrow{\text{HRP}} \text{ABTS}^{\bullet+} + H_2O$$

The pre-formed radical monocation ($\text{ABTS}^{\bullet+}$) is reduced in the presence of hydrogen-donating antioxidants (ascorbate, phenolics). During this reaction, the green $\text{ABTS}^{\bullet+}$ is converted back to its colorless neutral form in the reaction. This can easily be monitored with a spectrophotometer, 730 nm at 25°C for 5 min.

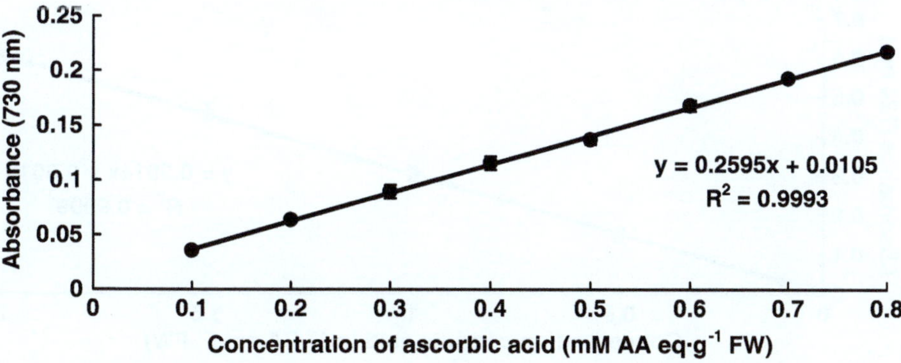

Fig. 1. Standard curve for total antioxidant capacity in millimolar of ascorbic acid equivalent per a gram of fresh weight (mM AA eq/g FW) using ABTS assay.

1. *Standard curve for TAC.* In the 1 mL cuvette, the reaction mixture (see Subheading 2.2) consists of 980 μL of ABTS-buffer solution, 1.55 μL H_2O_2, and 10 μL HRP until a stable absorbance is obtained due to ABTS radical formation. Afterwards, per 10 μL of different concentrations (0.1–0.8 mM, see Subheading 2.2) of ascorbic acid (AA) added for a standard curve set-up (Fig. 1).
2. *Spectrophotometric analysis.* Adding per 10 μL of methanolic extracts of plant tissue to the reaction mixture instead of ascorbic acid standards results in the absorbance decreasing as a consequence of ABTS radical depletion. Read absorbance alterations from the linear regression equation obtained from the standard curve at Fig. 1, where $x=(y-0.0105)/0.2595$ (see Note 14). Results are expressed as millimolar of AA equivalent per gram of fresh weight (mM AA eq/g FW) (see Note 15).

3.3. Total Phenolic Content Procedure

The widely used method for the analysis of total phenolic content (TPH) in plant extracts is the Folin-Ciocalteu's spectrophotometric procedure (41). Colorimetry is based on a chemical reduction of the reagent, a mixture of tungsten and molybdenum oxides. The products of the metal oxide reduction have a blue color that exhibits a broad light absorption with a maximum at 724 nm. The intensity of light absorption at that wavelength is proportional to the concentration of phenols.

1. *Standard curve for TPH.* Mix per 50 μL of GA calibration standards (see recipe in Subheading 2.3), 50 μL of samples, as well as blank (H_2O) with 475 μL of 0.25 N Folin-Ciocalteu reagent (see recipe in Subheading 2.3) and incubate for a 3 min at room temperature. Afterwards, add 475 μL of 0.2 M sodium carbonate solution (see recipe in Subheading 2.3) and incubate for 60 min at constant room temperature. Modern measuring equipment allow for microscaling the reaction to the volume of a UV-Vis cuvette, reducing the cost of the reagent

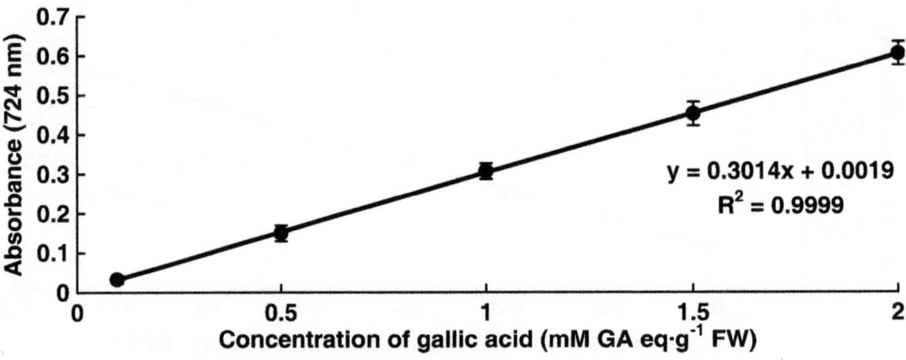

Fig. 2. Standard curve for total phenolic content in millimolar of gallic acid (GA) equivalent per gram of fresh weight (mM GA eq/g FW) obtained by Folin-Ciocalteu's method.

and waste disposal, so samples can be run in 96-wells microplate reader.

2. *Spectrophotometric analysis.* Put per 200 μL of each GA calibration standard, sample, or blank into a plate wells and read them choosing a filter at 724 nm. Read absorbance alterations using the linear regression equation obtained from standard curve, Fig. 2, where $x = (y - 0.0019)/0.3014$ (see Note 14). Results are expressed as millimolar of GA equivalent per gram of fresh weight (mM GA eq/g FW) (see Note 16).

3.4. Plant Tissue Extraction for Enzymes Activities

Preparation of enzymatic extracts includes tissue homogenization (see Note 12) in extraction buffer (see recipe in Subheading 2.4) at a ratio of 1:5 (fresh weight:extraction buffer). Add 5 mL of extraction buffer (see Subheading 2.1) to 1 g of homogenized tissue, stir, and transfer to plastic tubes. After centrifugation at 13,000 × *g* for 10 min at 4°C, supernatants are transferred to the new tubes in order to be used for further spectrophotometric determination of enzymes activities (see Note 17). A minimum of three extractions must be prepared for each sample.

3.5. Determination of Protein Concentration in Enzymatic Extracts

Perform the Bradford total protein assay in a 96-well Micro plate Reader. In each well (see Note 18), put 200 μL of Bradford reagents (see recipe in Subheading 2.5). Afterwards add per 10 μL of BSA standards (see recipe in Subheading 2.5), samples, and blank. Read absorbance choosing an optical filter at 595 nm. Prepare the standard curve with the dependent variable (protein concentration—mg/mL) on the X axis and the independent variable (absorbance—595 nm) on the Y axis. Calculate the protein concentration of the experimental samples using the standard curve equation (see Note 19).

3.6. Determination of Enzymes Activities

1. *Total SOD activity.* SOD activities are assayed by their capacity to compete with native or partially succinylated ferricytochrome *c* for $O_2^{\bullet-}$ radicals generated by the xanthine/xanthine

oxidase system. The reaction mixture contains sufficient xanthine oxidase (see recipe, Subheading 2.6) to produce an initial rate of ferricytochrome *c* reduction of $\Delta A_{550\ \mathrm{nm}} = 0.025 \pm 0.005$ per min. In a total volume of 3 mL, put the following components (see recipe at Subheading 2.6): 300 μL xanthine, 150 μL xanthine oxidase, 150 μL cytochrome *c*, and 20 μL of control (H_2O)/sample and fill it up to 3 mL with buffer + EDTA solution (see Note 20). Under these defined conditions, the amount of SOD required to inhibit the rate of reduction of cytochrome *c* by 50% in a coupled system at pH 7.8 at 25°C in a 3.0 mL reaction volume (i.e., to a rate of 0.0125 absorbance unit per min) is defined as 1 unit of activity.

Calculations:

$$\%\text{Inhibition} = \frac{\Delta K - \Delta A}{\Delta K} \times 100$$

$$\text{Volume activity (Units/mL)} = \frac{\%\text{Inhibition}}{(50\%) \cdot (0.02)}$$

$$\text{Enzyme activity (Units/mg)} = \frac{\text{Units/mL}}{\text{mg protein/mL}}$$

ΔK = Absorbance change in control

50% = Inhibition of the rate of cytochrome *c* reduction as per the Unit Definition

0.02 = Volume (in mL) of sample used

Activities of different SOD forms are identified and measured using KCN (42) given that three forms of the enzyme are distinguished from each other, based on their sensitivity to cyanide (see Note 21). Add 10 μL of KCN stock solution (see recipe at Subheading 2.6) to 3 mL of reaction mixture (see Note 22).

2. *CAT activity.* CAT activity is a measure of the absorbance change (decreasing) at 240 nm as a consequence of H_2O_2 degradation (43). Adjust H_2O_2 volume to get an absorbance of 0.85 ± 0.02 absorbance units (see Note 23). Regulate the sample volume, depending on absorbance. One unit is defined as the amount of enzyme that decomposes 1 μmol of H_2O_2 per minute at pH 7.0 at 25°C.

Calculations:

$$\text{Volume activity (Units/mL)} = \frac{\Delta A \cdot V_q}{0.0436 \cdot V_s}$$

$$\text{Enzyme activity (Units/mg)} = \frac{\text{Units/mL}}{\text{mg protein/mL}}$$

V_q = reaction volume in cuvette (in mL)

0.0436 = millimolar extinction coefficient of H_2O_2 at 240 nm ($cm^2/\mu mol$)

V_s = Volume (in mL) of sample used

3. *APX activity.* The hydrogen peroxide-dependent oxidation of ascorbate is followed by a decrease in the absorbance at 290 nm and 28°C. One milliliter of reaction mixture consists of 890 μL buffer, 2 μL ascorbate, and 30 μL sample (see recipe, Subheading 2.6). Record the reduction in ascorbate concentration by reading the absorbance continuously for 180 s after starting the reaction by adding 20 μL of H_2O_2 (see Note 24). One unit of APX activity is defined as the amount of enzyme that can oxidize 1 μmol of ascorbic acid per minute.

   Calculations:

$$\text{Volume activity (Units/mL)} = \frac{\Delta A \cdot 2 \cdot V_q}{2.8 \cdot V_s}$$

$$\text{Enzyme activity (Units/mg)} = \frac{\text{Units/mL}}{\text{mg protein/mL}}$$

V_q = reaction volume into cuvette (in mL)

2.8 = extinction coefficient of ascorbate at 290 nm (per mM cm)

V_s = volume (in mL) of sample used

4. Notes

1. ABTS causes serious skin, eye, and respiratory irritation! To avoid exposure, wear a dust mask, eye shields, and gloves.
2. Considering that ABTS solution is temperature and light sensitive, it should be prepared fresh each day, protect from the light (use tinted glass or aluminum-foil), and stored on ice until use. Under these conditions, the solution shows essentially no change in absorbance over a few hours.
3. Folin-Ciocalteu reagent is light sensitive, so it should be protected from light (use tinted glass or aluminum-foil) and prepared at least 5 min before the first sample incubation. The time of the colorimetric reading is critical, and the temperature will affect the extent of the reaction and degradation. Thus, the incubation must not be more than 8 min and without any changes in room temperature. During preparation, wear protective clothing, gloves, and eye protection; this reagent could cause skin corrosion/irritation.

4. PMSF is a serine protease inhibitor that is commonly used in protein solublization in order to deactivate proteases from digesting proteins of interest. That the half-life is short in aqueous solutions (110 min at pH 7 and 35 min at pH 8) is extremely important for experimental design. PMSF is considered a highly toxic inhibitor that could cause oral, dermal, and respiratory acute toxicity so requiring protective clothing, gloves, and a face mask.
5. EDTA is known to inhibit a range of metallopeptidases via the chelation of the metal ion required for catalytic activity. Since EDTA is an irritant (causes serious acute eye irritation), handle with gloves, safety glasses, and wash hands before and after use. Keep in suitable, closed containers for disposal. After EDTA addition, recheck pH!
6. PVP complexes with phenolics and alkaloids for their removal from plant samples, thus preventing the modification of proteins and any interference they may cause in spectrophotometric determinations of protein content. This is also reported to improve the stability of enzymes. May be harmful if inhaled, swallowed, or if absorbed through skin, and may cause eye irritation. Accordingly, use eye shields, gloves, and respirator filter, as well as a proper waste treatment method! PVP is readily soluble in cold water, so add PVP in cooled buffer slowly with slightly longer stirring.
7. Triton X100 is widely used non-ionic surfactant for solubilize membrane proteins under mild non-denaturing conditions. Harmful if swallowed; risk of serious damage to eyes. Wear protective gloves and glasses. Like any viscous material, pipette very slowly. Triton X100 is a high-foaming surfactant and therefore can produce many bubbles. Extremely slow stirring needs to be exercised during its use to avoid generating bubbles. Ordinarily, 0.5–1% Triton X is used, but taking into account that the Bradford assay is inhibited by the presence of detergents, we recommended 0.1%. This has no effect on assay results.
8. A standard curve is a plot of absorbance vs. a varying known protein concentration. A common method to prepare a standard curve is to prepare various known protein concentrations as standards. As long as the volume of the standard samples and the unknown samples are the same, the final concentration of the unknown is directly calculated from the least squares line of the standard curve. Always run samples in duplicate or triplicate and use the averages in the graph (to involve pipetting error). Finally, with every assay a "blank" must be included. The blank or the tube without a standard protein is usually made up in the same buffer as in the samples. The "blank" is used to set the instrument to the 100% transmittance or 0 absorbance.

9. Shake very gently!
10. Final volume depends on number of standards and samples (200 μL for each).
11. The Bradford reagent should be a light brown in color. Filtration may have to be repeated to remove the reagent of blue components. This "homemade" reagent works quite well, but is usually not as sensitive as the commercial one.
12. For effective disruption and homogenization, grind plant tissue in a mortar with pestle, using liquid nitrogen. Liquid nitrogen should be handled with care, following safety procedures. When fresh tissue is crushed into a paste, add an extraction medium. For rapidly and effortlessly tissue disruption, a tissue homogenizer, blender, or ultrasonic bath (Sonicator) could be used.
13. If the samples are not going to be used immediately after preparing, they should be stored at −20°C.
14. Be careful to enter the proper sign for the y intercept value as it may be a negative number! During fresh weight scaling, a dilution factor (5×) must be included in the calculation.
15. Results could be expressed as milligram of ascorbic acid equivalent per a gram of fresh weight (mg asc eq/g FW), when using masses of ascorbic acid (instead of molarities) for preparation of standard solutions.
16. Results could be expressed as milligram of gallic acid equivalent per a gram of fresh weight (mg GA eq/g FW), when using masses of gallic acid (instead of molarities) for preparation of standard solutions.
17. If the samples are not going to be used immediately after preparing, they should be stored at −80°C until analysis.
18. The numbers of wells depend on total number of blank, standards, and samples.
19. Results are expressed as mg of proteins per mL of extraction solution. Multiplying this number by the number of μL in the extraction solution, resulting in mg of proteins per gram of fresh weight.
20. If a higher amount of SOD (from samples) is required, increase the sample volume at the expense of buffer volume reducing.
21. While the Cu-Zn enzyme is sensitive to cyanide, the Fe and Mn enzymes are not. On the other hand, the Cu-Zn and Fe enzymes are inhibited by H_2O_2, whereas the Mn enzyme is H_2O_2-resistant. Therefore, only activity detected in KCN-treated samples is primarily Mn-SOD activity. Cu-Zn SOD activity can be determined by subtracting KCN-inhibited activity from the total SOD activity.

22. Addition of KCN stock solution requires buffer volume adjustment.
23. Use more H_2O_2 to increase or less to decrease the absorbance. Since the diluted H_2O_2 solution is stable for a maximum of 2 h, we recommend not to prepare a solution, but to mix with buffer directly into the cuvette immediately before measurement.
24. For calculation, use 30 s in linear range.

Acknowledgment

This work was supported by the Serbian Ministry of Education and Science (grants 173028 and 173040).

References

1. Bernstein N, Shoresh M, Xu Y et al (2010) Involvement of the plant antioxidative response in the differential growth sensitivity to salinity of leaves vs roots during cell development. Free Radic Biol Med 49:1161–1171
2. Mittler R (2002) Oxidative stress, antioxidants and stress tolerance. Trends Plant Sci 9:405–410
3. Mittler R, Vanderauwera S, Gollery M et al (2004) Reactive oxygen gene network of plants. Trends Plant Sci 9:490–498
4. Foreman J, Demidchik V, Bothwell JHF et al (2003) Reactive oxygen species produced by NADPH oxidase regulate plant cell growth. Nature 422:442–446
5. Peleg-Grossman S, Volpin H, Levine A (2007) Root hair curling and Rhizobium infection in Medicago truncatula are mediated by phosphatidylinositide-regulated endocytosis and reactive oxygen species. J Exp Bot 58:1637–1649
6. Potocký M, Jones MA, Bezvoda R et al (2007) Reactive oxygen species produced by NADPH oxidase are involved in pollen tube growth. New Phytol 174:742–751
7. Dat J, Vandenabeele S, Vranová E et al (2000) Dual action of the active oxygen species during plant stress responses. Cell Mol Life Sci 57:779–795
8. Munns R (2005) Genes and salt tolerance: bringing them together. New Phytol 167:645–663
9. Munns R, Tester M (2008) Mechanisms of salinity tolerance. Annu Rev Plant Biol 59:651–681
10. Fridovich I (1986) Superoxide dismutases. Adv Enzymol Relat Areas Mol Biol 58:62–97
11. Fridovich I (1995) Superoxide radical and superoxide dismutases. Annu Rev Biochem 64:97–112
12. Willekens H, Inzé D, Van Montagu M et al (1995) Catalases in plants. Mol Breed 1:207–228
13. Welinder KG (1992) Superfamily of plant, fungal and bacterial peroxidases. Curr Opin Struct Biol 2:388–393
14. Mittler R, Poulos TL (2005) Ascorbate peroxidase. In: Smirnoff N (ed) Antioxidants and reactive oxygen species in plants. Blackwell, Oxford
15. Jespersen HM, Kjaersgard IV, Ostergaard L et al (1997) From sequence analysis of three novel ascorbate peroxidases from *Arabidopsis thaliana* to structure, function and evolution of seven types of ascorbate peroxidase. Biochem J 326:305–310
16. Zamocky M, Janecek S, Koller F (2000) Common phylogeny of catalase-peroxidases and ascorbate peroxidases. Gene 256: 169–282
17. Hernandez JA, Ferrer MA, Jimenez A et al (2001) Antioxidant systems and $O_2^{\cdot-}/H_2O_2$ production in the apoplast of pea leaves. Its relation with salt-induced necrotic lesions in minor veins. Plant Physiol 127:817–831
18. Gómez JM, Hernández JA, Jiménez A et al (1999) Differential response of antioxidative enzymes of chloroplasts and mitochondria to long-term NaCl stress of pea plants. Free Radic Res 31:s11–s18

19. de Azevedo Neto AD, Prisco JT, Eneas-Filho J et al (2006) Effect of salt stress on antioxidative enzymes and lipid peroxidation in leaves and roots of salt-tolerant and salt-sensitive maize genotypes. Environ Exp Bot 56:87–94
20. Singh MP, Singh DK, Rai M (2007) Assessment of growth, physiological and biochemical parameters and activities of antioxidative enzymes in salinity tolerant and sensitive basmati rice varieties. J Agron Crop Sci 193: 398–412
21. Kim SY, Lim JH, Park MR et al (2005) Enhanced antioxidant enzymes are associated with reduced hydrogen peroxide in barley roots under saline stress. J Biochem Mol Biol 38:218–224
22. Bustos D, Lascano R, Villasuso AL et al (2008) Reductions in maize root-tip elongation by salt and osmotic stress do not correlate with apoplastic $O_2^{\bullet-}$ levels. Ann Bot 102:551–559
23. Rodr guez AA, Grunberg KA, Taleisnik EL (2002) Reactive oxygen species in the elongation zone of maize leaves are necessary for leaf extension. Plant Physiol 129:1627–1632
24. Walz C, Juenger M, Schad M et al (2002) Evidence for the presence and activity of a complete antioxidant defence system in mature sieve tubes. Plant J 31:189–197
25. Liszkay A, van der Zalm E, Schopfer P (2004) Production of reactive oxygen intermediates ($O^{2\bullet-}$, H_2O_2, and,OH) by maize roots and their role in wall loosening and elongation growth. Plant Physiol 136:3114–3123
26. Ogawa K, Kanematsu S, Asada K (1997) Generation of superoxide anion and localization of Cu Zn-superoxide dismutase in the vascular tissue of spinach hypocotyls: their association with lignification. Plant Cell Physiol 38:1118–1126
27. Córdoba-Pedregosa MC, Córdoba F, Villalba JM et al (2003) Differential distribution of ascorbic acid, peroxidase activity, and hydrogen peroxide along the root axis in *Allium cepa* L. and its possible relationship with cell growth and differentiation. Protoplasma 221:57–65
28. Córdoba-Pedregosa MC, Villalba JM, Córdoba F et al (2005) Changes in intracellular and apoplastic peroxidase activity, ascorbate redox status, and root elongation induced by enhanced ascorbate content in *Allium cepa* L. J Exp Bot 56:685–694
29. Miller G, Shulaev V, Mittler R (2008) Reactive oxygen signaling and abiotic stress. Physiol Plant 133:481–489
30. Bravo L (1998) Polyphenols: chemistry, dietary sources, metabolism, and nutritional significance. Nutr Rev 56:317–333
31. Heim KE, Tagliaferro AR, Bobilya DJ (2002) Flavonoid antioxidants: chemistry, metabolism and structure-activity relationships. J Nutr Biochem 13:572–584
32. Kim HJ, Fonseca JM, Choi JH et al (2008) Salt in irrigation water affects the nutritional and visual properties of romaine lettuce (*Lactuca sativa* L.). J Agric Food Chem 56:3772–3776
33. Yuan G, Wang X, Guo R et al (2010) Effect of salt stress on phenolic compounds, glucosinolates, myrosinase and antioxidant activity in radish sprouts. Food Chem 121:1014–1019
34. Yamasaki H, Sakihama Y, Ikehara N (1997) Flavonoid-peroxidase reaction as a detoxification mechanism of plant cells against H_2O_2. Plant Physiol 115:1405–1412
35. Pérez FJ, Villegas D, Mejia N (2002) Ascorbic acid and flavonoid-peroxidase reaction as a detoxifying system of H_2O_2 in grapevine leaves. Phytochemistry 60:573–580
36. Bradford MM (1976) A rapid and sensitive method for the quantitation of microgram quantities of protein utilizing the principle of protein-dye binding. Anal Biochem 72: 248–254
37. McCord JM, Fridovich I (1969) Superoxide dismutase: an enzymic function for erythrocuprein (hemocuprein). J Biol Chem 244: 6049–6055
38. Beers RF, Sizer IW (1952) A spectrophotometric method for measuring the breakdown of hydrogen peroxide by catalase. J Biol Chem 195:133–140
39. Nakano Y, Asada K (1981) Hydrogen peroxide is scavenged by ascorbate specific peroxidase in spinach chloroplast. Plant Cell Physiol 22:867–880
40. Arnao MB, Cano A, Acosta M (1999) Methods to measure the antioxidant activity in plant material. A comparative discussion. Free Radic Res 31:89–96
41. Singleton VL, Rossi JA Jr (1965) Colorimetry of total phenolics with phosphomolybdic-phosphotungstic acid reagents. Am J Enol Viticult 16:144–158
42. Bridges SM, Salin ML (1981) Distribution of iron-containing superoxide dismutase in vascular plant. Plant Physiol 68:275–278
43. Aebi H (1983) Catalase. In: Bergmeyer HU (ed) Methods of enzymatic analysis, 2nd edn. Verlag Chemie, Weinheim

Chapter 17

Quantification of Abscisic Acid, Cytokinin, and Auxin Content in Salt-Stressed Plant Tissues

Petre I. Dobrev and Radomira Vankova

Abstract

Plant hormones cytokinins, auxin (indole-3-acetic acid), and abscisic acid are central to regulation of plant growth and defence to abiotic stresses such as salinity. Quantification of the hormone levels and determination of their ratios can reveal different plant strategies to cope with the stress, e.g., suppression of growth or mobilization of plant metabolism. This chapter describes a procedure enabling such quantification. Due to the high variability of these hormones in plant tissues, it is advantageous to determine their content in the same sample. Reverse phase and ion exchange chromatography allows separation of the individual hormone fractions. Hormones as well as their metabolites can be identified and quantified by LC/MS.

Key words: Abscisic acid, Auxin, Cytokinin, Salinity, Hormone purification, Hormone analysis

1. Introduction

Plant interactions with the often adverse environment are, at least partially, regulated by phytohormones. The hormone analysis can provide important information about the physiological state of plants and can contribute to the prediction of their further behavior.

The key hormone in the defence to abiotic stresses, including salinity, is abscisic acid (ABA). ABA mediates both fast responses (modulation of ion flows resulting in stomata closure) and relatively longer-term changes in the expression of many stress-associated genes. Regulation of stomata aperture is crucial to cope with water deficit. Re-arrangement of transcriptome results in stimulation of the formation of a range of defence molecules, e.g., dehydrins.

Sergey Shabala and Tracey Ann Cuin (eds.), *Plant Salt Tolerance: Methods and Protocols*, Methods in Molecular Biology, vol. 913, DOI 10.1007/978-1-61779-986-0_17, © Springer Science+Business Media, LLC 2012

ABA-regulated genes represent over 10% of the genome in *Arabidopsis* seedlings (1). ABA is a sesquiterpenoid (15-carbon). According to the position of carboxyl group at side chain, *cis*- and *trans*-isomer can be distinguished, the former one exhibiting the biological activity. Due to the optically active C1′, bearing a hydroxyl group, ABA may occur either as (S)/(+) isomer, which is the most physiologically active, or as much less active (R)/(−) isomer. ABA can be deactivated by conjugation to glucose. ABA glucosylester and glucoside are important ABA storage forms. ABA can be irreversibly inactivated to phaseic, dihydrophaseic, or neo-phaseic acid.

Recent reports suggest that apart from the stimulation of the defence mechanisms, an important part of the stress response is a modulation of plant growth and development. Cell division is governed predominantly by two hormones—auxins and cytokinins. The active endogenous auxin is indole-3-acetic acid (IAA), which stimulates cell division and elongation, as well as promotes apical dominance, positively affects root growth, mediates tropic responses, and prevents abscision (2). Vast amount of IAA (90–95%) is present in plants in a conjugated form, mostly with amino acids (aspartate, leucine, alanine, or glutamine) or sugars. These conjugates represent either storage or deactivation forms. IAA can be also metabolized to indole-butyric acid.

Cytokinins are defined as substances stimulating (at the presence of auxin) cell division (cytokinesis) (3). They are indispensable for cell cycle transition at both check points—G_2/M and G_1/S (4). They positively affect photosynthesis (5), delay senescence, and enhance the sink strength (6). They play a critical role in balancing acquisition and distribution of macronutrients (7). Natural cytokinins are adenine derivatives with either isoprenoid or aromatic side chain. The physiologically active forms are cytokinin bases (predominantly *trans*-zeatin) and, to lesser extent, cytokinin ribosides. Cytokinins are deactivated either by cleavage of the side chain with cytokinin oxidase/dehydrogenase or by conjugation with glucose. Glucosylation of hydroxyl at the side chain results in reversible cytokinin *O*-glucosides (storage forms), glucosylation at purine ring in position N9– and N7–, is irreversible. Cytokinin phosphates (nucleotides) are immediate precursors of bioactive cytokinins. In some cases, their level correlates with the rate of cytokinin biosynthesis. They can be, however, also products of cytokinin back conversion, especially after sudden increase of the level of bioactive cytokinins (e.g., after application of exogenous cytokinin).

Cytokinins and ABA exhibit antagonistic effects on regulation of many processes involved in stress responses (cell division, stomata aperture, photosynthetic activity). Their ratio may reflect plant strategies to cope with stress. As the defence is highly energy demanding, suppression of growth, coinciding with high ABA and low cytokinin levels, might allow redirection of the (limited) energy sources to the defence. On the other hand, stabilization of photosynthesis, as the energy source, by maintenance of relatively high cytokinin levels could lead to mobilization of plant metabolism. This may be advantageous, especially in case of milder or short stresses.

Estimation of phytohormone content in plants is challenging due to their very low concentrations (in the range of picomoles per gram of fresh weight (FW)). At the same time, plant tissues are very rich in diverse classes of potential interferences, such as primary and secondary metabolites that are present at much higher concentrations (usually micro- to millimoles per gram FW). The first techniques used for the estimation of hormone content in plant tissues were biotests. Subsequent development of analytical methods enabled more specific determination of phytohormones. In the early days of hormone analysis, the quantification techniques had rather low selectivity and sensitivity. Thus, hormone analysis required large amounts of material (usually several grams) and involved several comprehensive purification steps. The whole procedure was rather demanding, very labor-, time-, and material-consuming.

Progress in hormone analyses is associated with the development of antibodies specific to individual hormones (8). Quantification of the hormone concentration in the sample was based on the competition for a limited amount of antibodies, either with radiolabeled standard (radioimmunoassay) or with standards coupled to the enzyme (e.g., phosphatase or peroxidase, enzyme immunoassay). The limitations of immunotechniques have been specificity of antibodies, as well as the presence of interfering substances in the sample. Due to relatively small size of hormone molecules, antibodies can be raised only to the conjugate of the hormone and high molecular "carrier," usually BSA. In consequence, antibodies have rather low recognition for the part of the molecule used for binding to the protein. Immunological analysis is still commonly used in case of ABA, as the structure of deactivation products differ sufficiently from the active molecule, but it could be quite misleading in case of cytokinins. The highly abundant cytokinin deactivation products, cytokinin *N*-glucosides, can exhibit quite significant cross-reactivity. Therefore, cytokinin fractionation by high performance liquid chromatography (HPLC) and subsequent immunological determination of cytokinins in the individual fractions is used. This approach is complicated by potential variation in retention times, caused by other compounds present in the sample (amounts of which highly prevail that of hormones) or by their effect on hormone-antibody interaction (which could be both negative as well as positive).

Recent advancements of analytical instrumentation allowed significant increase of efficiency, selectivity, sensitivity, and throughput of phytohormone analyses. The method of choice for phytohormone analysis is HPLC coupled to mass spectrometric (MS) detector. Modern HPLC can separate efficiently and quickly complex mixtures of compounds with a wide range of polarity without need of derivatization. Modern mass spectrometers have three features that make them almost ideal detector: sensitivity, selectivity, and speed. MS is one of the most sensitive types of detectors capable of detecting a few femtomoles of an analyte. The sensitivity of MS allows one to decrease significantly the amount of material needed for analysis to as little as tens to a hundred milligrams FW. This simplifies sample collection procedures and allows analysis of materials in limited amount, such as individual plant parts or tissues of small plants. Selectivity of MS is based on its intrinsic capability to measure the very specific parameter of a compound, its mass. Selecting and monitoring of a particular mass filters out all other masses. The recorded chromatogram contains only compounds with the same mass, isomers, or isobars (i.e., nonrelated compounds with the same weight). Even more selective is the tandem MS, also known as MS/MS. It contains two MS with a collision cell in between. The first MS allows the passage of one mass (precursor) that enters the collision cell, where it is broken into compound-specific fragments. The fragments then enter the second MS where they are filtered and only pre-selected fragments are allowed to pass through to the detector. Only compounds with pre-selected precursor-to-fragment mass transitions are detected. Tandem MS is so selective, that often chromatogram of crude extract can contain just one peak of the target compound. The superior selectivity of MS increases the confidence in compound identity and allows simplification of purification procedure to just one or a few steps. Recent progress in electronics of MS has made their responses very short, in the range of few milliseconds, allowing rapid measurement and switching between many pre-set masses. The MS speed permits simultaneous analysis of tens to hundreds of compounds in a single sample, leading to recent expansion of multi-metabolite profiling, also known as metabolome analysis.

In order to obtain reliable, physiologically relevant data on hormone content, severe precautions need to be taken for the sample collection. All phytohormones exhibit substantial diurnal variations. Thus, approximately the same time of a day should be used for sampling. It should be taken into the consideration that levels of hormones change during the year. Samples taken in the winter would substantially differ from the spring and summer ones (even in the case of plants grown in cultivation chambers without any contact with daylight). Also, the intensity of light (and its spectrum) significantly affects hormone pool, which complicates repetitions of the experiments in different labs. The individual tissues (leaf, root, stem) differ in hormone levels. Quite big differences are also among individual leaves and even within the leaf blade; more actively growing parts have higher cytokinin levels, e.g., the basal part of monocotyledonous leaves.

The procedure for phytohormone analysis includes: (a) sampling, (b) extraction, (c) purification, and (d) quantitative determination using HPLC-MS/MS. The procedure described by Dobrev and Kaminek (9) allows determination of three groups of plant hormones, cytokinins, auxins, and ABA, in a single sample.

2. Materials

2.1. Extraction

1. Mortar and pestle.
2. Liquid nitrogen (LQN).
3. Pipettes (50 μL, 5 mL).
4. Measuring cylinders (1 L, 100 mL).
5. Analytical balances.
6. Methanol, p.a. grade ("per analysis"—purity for chemical analyses).
7. Double-distilled water.
8. Formic acid, p.a. grade.
9. Appropriate stable isotope labeled internal standards ($^{2}H_{5}$-tZ, $^{2}H_{5}$-tZR, $^{2}H_{5}$-tZRMP, $^{2}H_{5}$-tZ7G, $^{2}H_{5}$-tZ9G, $^{2}H_{5}$-tZOG, $^{2}H_{5}$-tZROG, $^{2}H_{3}$-DZ, $^{2}H_{3}$-DZR, $^{2}H_{3}$-DZRMP, $^{2}H_{3}$-DZ9G, $^{2}H_{7}$-DZOG, $^{2}H_{6}$-iP, $^{2}H_{6}$-iPR, $^{2}H_{6}$-iPRMP, $^{2}H_{6}$-iP7G, $^{2}H_{6}$-iP9G; $^{2}H_{6}$-ABA, $^{2}H_{3}$-PA, $^{2}H_{3}$-DPA, $^{2}H_{4}$-7OH-ABA, $^{2}H_{5}$-ABA-GE, $^{13}C_{6}$-IAA).
10. 2 mL Eppendorf tubes for samples <100 mg FW, 50 mL centrifugation tubes for samples >100 mg FW.
11. Centrifuge.
12. Freezers (–20, –80°C).

2.2. Purification

1. Methanol, p.a. grade.
2. Formic acid, p.a. grade.
3. Acetic acid, p.a. grade.
4. Ammonia (Ammonium hydroxide 26%), p.a. grade.
5. MilliQ water.
6. SPE C18 column, e.g., Sep-Pak Plus, Waters, part # WAT036810.
7. SPE Oasis MCX column, 6 cc/150 mg, Waters, part # 186000256.
8. SPE Vacuum Manifold, 12 or 24-port, Supelco, part # 57250-U (to mount SPE columns for simultaneous purification of up to 12 or 24 samples).

9. SpeedVac, i.e., vaccum evaporator with vacuum pump (in case of oil pump also freezing trap (to −100°C) to collect the evaporating organic solvents).
10. Calf-intestine alkaline phosphatase (13 U/mg, Sigma, part # P7640).

2.3. MS Quantification

1. Acetic acid, LCMS grade.
2. Acetonitrile, LCMS grade.
3. Water, MilliQ grade.
4. Autosampler vials.
5. HPLC column, e.g., Luna C18(2), 3 μm, 150×2 mm, Phenomenex.
6. HPLC system, e.g., Ultimate 3000, Dionex.
7. MS detector, e.g., 3200 QTRAP LC/MS/MS, ABSciex.

2.4. Solvent and Standard Preparation

Extraction solvent is composed of methanol/double-distilled water/formic acid = 15/4/1, v/v/v, keep at −20°C.

Stable isotope labeled internal standards (100 pmol/50 μL per 1 g FW sample) dilute in 50% methanol in water.

Load solvent: 1 M formic acid: dilute 37.7 mL of 99% formic acid with double-distilled water to 1,000 mL; pH ~1.4.

Elute 1 solvent: 100% methanol.

Elute 2 solvent: 0.35 M NH_4OH: dilute 2.5 mL of 26% ammonia with 97.5 mL double-distilled water, pH ~11.

Elute 3 solvent: to 60 mL MeOH add 2.5 mL of 26% ammonia and adjust to 100 mL with double-distilled water.

Incubation buffer for alkaline phosphatase reaction: 0.1 M ammonium acetate, pH 10.

Calf-intestine alkaline phosphatase dissolved in incubation buffer at concentration 0.2 U/20 μL for 1 g FW sample (it should be prepared fresh).

3. Methods

3.1. Sampling

1. The plant material for hormone analysis should be cut, weigh precisely (also see Notes 1 and 2), and frozen in liquid nitrogen as quickly as possible; because degradative processes are initiated immediately after plant wounding. The fresh weight of plant material could be within a relatively broad range (0.05–1 g FW). Nonetheless, the precise weight (±1%) for each sample

must be recorded. The ideal amount of sample depends on the actual amount of hormone in the sample, which needs to be well above the detection limit. The samples can be stored at −80°C.

2. When leaf blades are collected from dicotyledonous plants, the main vein should be removed. This is because many hormones are transported via the vascular system, thus the presence of the main (or other big) vein would substantially affect the result of analysis. When monocotyledonous plants are sampled, the apical third of the leaf should be removed, as the already senescent part of the leaf will differ substantially from the basal part. In all cases, the leaves at the same developmental stage should be compared, when evaluating the effect of salinity. Parallel samples for determination of the ratio between fresh and dry weight should be taken when the results are going to be presented per dry weight basis.
3. When roots are sampled from plants grown in soil, the soil particles should be quickly removed by brief rinsing with cold tap water and gentle blotting with the tissue. When a hydroponic system is used, usually the roots can be cut, weigh, and freeze immediately.

3.2. Extraction

The aim of the extraction is quantitative release of target compounds from the plant tissue into an extraction solution, and preservation of their chemical integrity. For quantitative recovery, 5 to 1 volumetric ratio of extraction solvent to sample is used, and the sample is extracted twice. The compound degradation is avoided by working at low temperature with an extraction buffer containing a high proportion of organic solvent and a low pH. Extraction is followed by centrifugation. This removes large undissolved and precipitated biopolymers such as cellulose, large nucleic acids, and proteins.

1. Homogenize the frozen material with a mortar and a pestle in liquid nitrogen to a fine powder. Care should be taken to avoid any thawing of the tissue.
2. Transfer immediately the homogenized, frozen sample (with liquid nitrogen) into an appropriate centrifugation tube, precooled in liquid nitrogen.
3. Wash twice the mortar and pestle with five volumes of extraction buffer (cooled to −20°C) and add the wash to the sample in order to transfer quantitatively the sample to the tube. For example, for 100 mg FW sample, wash with two consecutive 0.25 mL of extraction buffer.

4. Add internal standards (100 pmol/1 g FW sample or 50 pmol per smaller samples, see Note 3). Mix.
5. Place tubes with samples, extraction solvent, and internal standards in freezer at –20°C for 1 h.
6. Centrifuge at 15,000 × *g* for 30 min (at 4°C).
7. Transfer supernatant into clean tube.
8. Add five volumes of extraction solvent to the pellet and mix.
9. Incubate for 30 min at –20°C.
10. Centrifuge again and combine the supernatants.

3.3. Purification

The aim of purification is to remove from the extract as much as possible of interfering substances without losing significant amount of target compounds. Two solid phase extraction (SPE) columns are used. The first one, C18, is used as a filter for removal of most of the lipophilic substances, with phytohormones passing through. The second SPE column (MCX) retains hormones that are sequentially eluted with appropriate elute solvents into three fractions containing different types of hormones. The fraction containing cytokinin phosphates is incubated with alkaline phosphatase to convert phosphates into ribosides. This conversion is performed due to the lower sensitivity of LC-MS/MS to cytokinin phosphates.

1. Using SPE Vacuum Manifold, condition SPE C18 column by washing it with 5 mL methanol, followed by 5 mL extraction solvent (see Subheading 2.4). Do not run the column dry (see Note 4).
2. Pass sample extract through SPE C18 column. Collect flow through and evaporate in SpeedVac at 40°C to about 1/10 of volume.
3. Dissolve residue into 1 mL of load solvent (see Subheading 2.4).
4. Condition SPE Oasis MCX column by washing it with 5 mL methanol followed by 5 mL load solvent. This column can run dry.
5. Apply sample onto SPE Oasis MCX column. Discard flow through.
6. Wash column with 5 mL load solvent and discard flow through.
7. Apply 5 mL elute 1 solvent to MCX column. Collect flow through. This is fraction 1. It contains phytohormones of neutral and acidic character: auxins and ABA.
8. Wash column with 5 mL double-distilled water and discard the flow through.

9. Apply 5 mL elute 2 solvent to MCX column. Collect flow through. This is fraction 2. It contains cytokinin phosphates (cytokinin nucleotides).
10. Apply 5 mL elute 3 solvent to MCX column. Collect flow through. This is fraction 3 and contains cytokinin bases, ribosides, and glucosides.
11. Evaporate all three fractions to dryness in SpeedVac at 40°C. Fractions 1 and 3 are ready for quantitative analysis by HPLC-MS/MS.
12. Dissolve fraction 2 in 1 mL incubation buffer for alkaline phosphatase. Add alkaline phosphatase and incubate at 37°C for 2 h. Stop the reaction by adding 20 μL glacial acetic acid. This is de-phosphorylated fraction 2.
13. Condition SPE C18 column by washing it with 5 mL methanol, followed by 5 mL water.
14. Pass de-phosphorylated fraction 2 through SPE C18 column. Discard flow through.
15. Wash column with 5 mL water.
16. Apply 5 mL 100% methanol and collect flow through. It contains cytokinin ribosides, the de-phosphorylation products of cytokinin phosphates. Evaporate in SpeedVac at 40°C. The de-phosphorylated fraction 2 is ready for analysis.

3.4. Quantification by HPLC-MS/MS

The purified fractions are applied to HPLC-MS/MS, where the individual compounds are separated. The individual hormones and their metabolites are quantified by comparison of the measured response ratio of endogenous hormone to its internal standard and the ratio of hormone of known concentration to internal standard.

The procedure and parameter settings for HPLC-MS/MS are very much instrument-specific. The optimal parameters should be found for the particular instrument (see Note 5). Thus, only general guidance for HPLC-MS/MS follows.

1. Dissolve dried sample into 50 μL 10% acetonitrile in water.
2. Centrifuge at 15,000 × *g* for 10 min at 4°C.
3. Transfer supernatant into autosampler vial.
4. Inject the sample. Depending on the amount of extracted sample, inject into HPLC an aliquot, i.e., 1/2 from 100 mg FW extract or 1/10 from 1 g FW extract.
5. HPLC conditions: run gradient of A: 5 mM acetic acid in water and B: 5 mM acetic acid in acetonitrile; from 10 to 50% B in 20 min at flow rate 0.25 mL/min. Flush column at 100% B for 5 min and equilibrate to initial conditions for 10 min.

4. Notes

1. Hormone content could be expressed either per fresh weight basis (which is directly measured) or by dry weight basis (which may diminish differences caused by different water content among the leaves). It should be taken into account that the individual leaves differ in their water content, so the corresponding ratio of fresh and dry weight should be used.
2. Hormone levels reflect very precisely the physiological state of individual plants. A considerable difference in hormone levels can be found between the plants in the experiment and even bigger between the independent experiments. Thus, several experiments (at least three) with replicate samples should be analysed to achieve physiologically relevant data.
3. Critical points are precise weighing of sample and addition of exactly known amounts of internal standards (which should be at room temperature), since these data will be used in the final calculation of the hormone content.
4. SPE column conditioning is important for their proper operation. Loss of a small portion of the sample is not that critical because added internal standards allow to evaluate the losses. However, the losses must not diminish the final amounts of hormones in the samples below the detection limit.
5. MS conditions: electrospray ionization at positive mode is advantageous for cytokinin analysis, at negative mode for the other (acidic) hormones. The most intensive ion is usually used for quantification and the others, for identity confirmation. However, in case of high interference, the other (less intensive) ion without substantial background should be used for quantification. Substances for which the labeled standards are not available are determined using the retention times and the mass spectra of unlabeled standards and the response ratio (labeled/unlabeled) of their closest derivative. The multilevel calibration curve is necessary to estimate the linear range of responses as well as detection limits.

   MS parameters need to be optimized for each hormone, especially, ion spray voltage, declustering potential, collision energy, and precursor to fragment transition.

   Since MS is very sensitive instrument, it is very important to keep it clean. Recommended is usage of solvents with the highest purity, e.g., LC/MS grade. Furthermore, the ion source and entrance of MS, gas filters, and oil of the vacuum pump become dirty with use. This leads to gradual decrease of sensitivity. Therefore, cleaning of the ion source and entrance to MS, and replacement of gas filters and pump oil should be carried out at regular intervals.

Acknowledgement

This work was supported by the MEYS CR, project no. LD11073.

References

1. Cutler SR, Rodriguez PL, Finkelstein RR et al (2010) Abscisic acid: emergence of a core signaling network. Annu Rev Plant Biol 61:651–679
2. Woodward AW, Bartel B (2005) Auxin: regulation, action, and interaction. Ann Bot 95: 707–735
3. Miller CO, Skoog F, Von Saltza MH et al (1955) Kinetin, a cell division factor from deoxyribonucleic acid. J Am Chem Soc 78:1345–1350
4. Dobrev P, Motyka V, Gaudinova A et al (2002) Transient accumulation of *cis*- and *trans*-zeatin type cytokinins and its relation to cytokinin oxidase activity during cell cycle of synchronized tobacco BY-2 cells. Plant Physiol Biochem 40: 333–337
5. Chernyadev II (2009) The protective action of cytokinins on the photosynthetic machinery and productivity of plants under stress. Appl Biochem Microbiol 45:351–362
6. Skoog F, Strong FM, Miller CO (1965) Cytokinins. Science 148:532–533
7. Hirose N, Takei K, Kuroha T et al (2008) Regulation of cytokinin biosynthesis, compartmentation and translocation. J Exp Bot 59: 75–83
8. Mertens R, Deusneumann B, Weiler EW (1983) Monoclonal-antibodies for the detection and quantitation of the endogenous plant-growth regulator, abscisic acid. FEBS Lett 160: 269–272
9. Dobrev P, Kaminek M (2002) Fast and efficient separation of cytokinins from auxin and abscisic acid and their purification using mixed-mode solid-phase extraction. J Chromatogr A 950:21–29

Part IV

Molecular Techniques

Chapter 18

Fluorescence-Activated Cell Sorting for Analysis of Cell Type-Specific Responses to Salinity Stress in *Arabidopsis* and Rice

Aurelie Evrard, Bastiaan O.R. Bargmann, Kenneth D. Birnbaum, Mark Tester, Ute Baumann, and Alexander A.T. Johnson

Abstract

Fluorescence-activated cell sorting (FACS) provides a rapid means of isolating large numbers of fluorescently tagged cells from a heterogeneous mixture of cells. Collections of transgenic plants with cell type-specific expression of fluorescent marker genes such as green fluorescent protein (*GFP*) are ideally suited for FACS-assisted studies of individual cell types. Here we describe the use of *Arabidopsis* and rice enhancer trap lines with tissue-specific *GFP* expression patterns in the root to isolate specific cell types of root tissues using FACS. Additionally, protocols are provided to impose a ramped salinity stress for 48 h prior to cell sorting.

Key words: FACS, Cell type-specific, Salinity, Rice, *Arabidopsis*, Protoplast, GAL4

1. Introduction

Fluorescence-activated cell sorting (FACS) of plant protoplasts expressing green fluorescent protein (*GFP*) has enabled numerous cell type-specific gene expression studies in the model dicotyledonous species *Arabidopsis* (*Arabidopsis thaliana*). The use of *Arabidopsis* lines with cell type-specific patterns of *GFP* expression (typically reporter lines with tissue-specific promoters driving *GFP* or GAL4 enhancer trap lines) as source material for FACS experiments permits thousands of protoplasts from a specific cell type to be harvested within minutes. Following the initial publication of a gene expression map spanning five cell populations and three developmental zones of the *Arabidopsis* root (1), FACS was used

Sergey Shabala and Tracey Ann Cuin (eds.), *Plant Salt Tolerance: Methods and Protocols*, Methods in Molecular Biology, vol. 913, DOI 10.1007/978-1-61779-986-0_18, © Springer Science+Business Media, LLC 2012

to examine developmental and spatiotemporal gene expression as well as cell type-specific responses to abiotic stresses and nitrogen uptake in the *Arabidopsis* root (2–4). FACS has also been utilized to collect numerically small cell types such as the root quiescent center and sperm cells for transcriptomic analyses (5, 6). Additionally, FACS of nuclei with nuclear-targeted GFP has been employed as an alternative means of harvesting genetic material from specific cell types of *Arabidopsis* (7). These studies demonstrate that FACS is an efficient means of isolating and analyzing a variety of GFP-labeled *Arabidopsis* cell types and provides an alternative to manual isolation techniques such as laser-capture microdissection.

While methods for FACS of *Arabidopsis* protoplasts are well established (8, 9), the use of FACS to examine cell type-specific gene expression in the model monocotyledonous species rice (*Oryza sativa*) has not been reported. Here, we demonstrate the use of GAL4 enhancer trap lines of rice and *Arabidopsis* to isolate specific cell types of the root from both species under control and saline conditions. We describe a supported hydroponics setup for *Arabidopsis* and rice that allows for growth of large numbers of plants under control and saline conditions, followed by high-throughput protoplasting from root tissues of both species. The FACS procedure to collect fluorescent protoplasts is also described in detail. The enhancer trap lines utilized in these experiments originate from large GAL4 collections of *Arabidopsis* (10) and rice (11) that display a wide variety of highly specific patterns of GFP expression in root tissues and other organs (Fig. 1). In addition to their use as GFP marker lines, the *Arabidopsis* and rice GAL4 collections have been used in cell type-specific overexpression studies to increase sodium exclusion and salinity tolerance in both model species (12, 13). The following methods show that it is possible to collect specific cell types of rice and *Arabidopsis* after exposure to salinity stress through use of FACS. While this protocol is geared towards transcriptomic analyses of the harvested cells, the protocol can be easily adapted for metabolic analyses of the cells through use of a suitable collection buffer.

2. Materials

2.1. GAL4-GFP Enhancer Trap Lines

1. Selected *Arabidopsis* enhancer trap lines in C24 background from the GAL4 catalogue http://www.plantsci.cam.ac.uk/Haseloff/assembly/page167/index.html.
2. Selected rice enhancer trap lines in cv. Nipponbare background from the *O. sativa* GAL4-GFP database http://129.127.183.5.

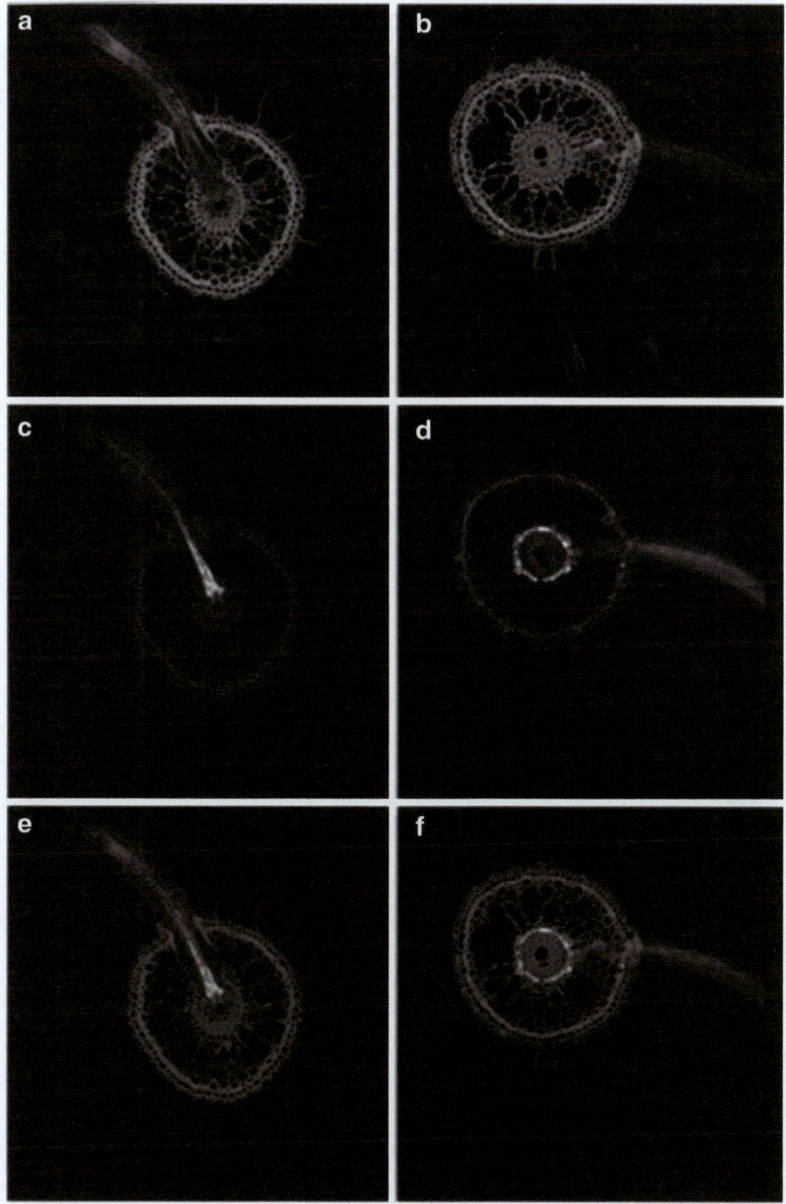

Fig. 1. Cell type-specific expression of GFP in roots of two GAL4-GFP rice enhancer trap lines as visualized by confocal laser scanning microscopy. Images of root cross sections are presented as fluorescence of the cell wall stain propidium iodide (**a**, **b**), GFP (**c**, **d**) and overlay of the propidium iodide and GFP fluorescence patterns (**e**, **f**). (**a**, **c**, **e**) Enhancer trap line ASG F03 displays bright GFP fluorescence in xylem parenchyma cells and the pattern is most evident in emerging lateral roots. (**b**, **d**, **f**) Enhancer trap line ASU A03 displays bright GFP fluorescence in cortical cells just outside of the endodermis (note that most cortical cells have been replaced by aerenchyma).

2.2. Arabidopsis Supported Hydroponics

1. Phytatray (Sigma).
2. 1 mL tip rack (96 tips per rack) cut to fit inside of the Phytatray (95 × 80 mm). Nitex 03-100/47 mesh (Sefar) cut to fit on top of the tip rack. Use 4 × 1 mL wide-bore tips (cut-off 1 mL tips)

to support the tip rack within the Phytatray. Autoclave tip rack and Nitex mesh prior to assembly within the Phytatray.

3. Growth cabinet set to 14 h light at 22°C (200 μm s^{-1} m^{-1}) and 10 h dark at 18°C.
4. 100 mL of 0.1% agarose. Autoclaved.
5. 100% Ethanol.
6. 100% bleach solution with 1 drop Tween-20 detergent. Made fresh on day of experiment.
7. Sterile water (see Note 1).
8. *Arabidopsis* growth solution: 1/4× MS basal salt mixture, 1× Gamborg's B5 vitamin solution, 0.5% sucrose, 0.05% MES, pH 5.7 (with 10 M KOH). Prepared in 1 L batches. Autoclaved.
9. *Arabidopsis* 25 mM NaCl growth solution: Same composition as growth solution supplemented with 25 mM NaCl and 0.22 mM $CaCl_2$ (see Note 2), pH 5.7 (with 10 M KOH). Autoclaved.
10. *Arabidopsis* 50 mM NaCl growth solution: Same composition as growth solution supplemented with 50 mM NaCl and 0.375 mM $CaCl_2$, pH 5.7 (with 10 M KOH). Autoclaved.

2.3. Rice Supported Hydroponics

1. 95×95-mm petri dish (BD Falcon).
2. Filter paper (Whatman) wrapped in foil. Autoclaved.
3. 70% Ethanol
4. 30% Bleach solution with 1 drop Tween-20 detergent. Made fresh on day of experiment.
5. Elfa® mesh basket (L527×W327×H85 nm). Soak in a 50% bleach overnight, rinse well.
6. 20 L Starmaid® plastic container (L558×W390×H144 nm) or equivalent.
7. Growth cabinet set to 12 h light at 28°C (700 μm s^{-1} m^{-2}) and 12 h dark at 25°C.
8. Rice growth solution: 1/4× MS basal salt mixture, pH 5.7 (with 10 M KOH). Prepared in 10 L batches (1 L should be autoclaved for rice germination in petri dishes).
9. Rice 25 mM NaCl growth solution: Same composition as growth solution supplemented with 25 mM NaCl and 0.22 mM $CaCl_2$ (see Note 2), pH 5.7 (with 10 M KOH). Prepared in 10 L batches.
10. Rice 50 mM NaCl growth solution: Same composition as growth solution supplemented with 50 mM NaCl and 0.375 mM $CaCl_2$, pH 5.7 (with 10 M KOH). Prepared in 10 L batches.

2.4. Protoplasting and FACS Materials and Solutions

1. Water bath at 60°C.
2. Clinical centrifuge.
3. Vacuum chamber.
4. BD FACSAria cell sorter or equivalent.
5. RNeasy plant mini kit (Qiagen)
6. 40 μm cell strainer (BD falcon).
7. 50 mL centrifuge tubes.
8. No. 10 surgical blades (Feather)
9. 1 M KCl. Autoclaved.
10. 1 M $CaCl_2$. Autoclaved.
11. 1 M Tris pH 8. Autoclaved.
12. *Arabidopsis* protoplasting solution (10 mL): directly before collecting roots prepare 400 mM mannitol (0.73 g), 20 mM MES hydrate (39 mg), 20 mM KCl (200 μL of 1 M KCl), 1.25% Cellulase R10 (125 mg), 0.3% Macerozyme R10 (30 mg) in 10 mL water. Adjust pH to 5.7 with 1 M Tris pH 8. Heat 10 min in 60°C water bath (solution should clear), cool to room temperature while stirring. Add 10 mM $CaCl_2$ (100 μL of 1 M $CaCl_2$), 0.1% BSA (10 mg), and 1.79 μL of β-mercapoethanol. Make fresh on day of protoplasting.
13. Rice protoplasting solution (30 mL): directly before collecting roots prepare 400 mM mannitol (2.2 g), 20 mM MES hydrate (117 mg), 20 mM KCl (600 μL of 1 M KCl), 1.25% Cellulase R10 (375 mg), 1.25% Cellulase RS (375 mg), 0.3% Macerozyme R10 (90 mg), 0.12% Pectolyase (36 mg) in 30 mL water. Adjust pH to 5.7 with 1 M Tris pH 8. Heat 10 min in 60°C water bath (solution should clear), cool to room temperature while stirring. Add 10 mM $CaCl_2$ (300 μL of 1 M $CaCl_2$), 0.1% BSA (30 mg), and 5.38 μL of β-mercapoethanol. Make fresh on day of protoplasting.

3. Methods

3.1. Arabidopsis Germination, Growth, and Application of Salinity Stress

All work should be carried out under sterile conditions. Perform each procedure as early as possible in the morning.

1. Weigh 40 mg of *Arabidopsis* seed (approximately 2,000 seed) into a 2 mL tube (40 mg is sufficient for two Phytatrays, scale up as required; also see Note 3).
2. Add 1.5 mL of 100% ethanol and mix for 1 min. Pipette off the ethanol.

3. Add 1.5 mL of 100% bleach solution and mix on an orbital shaker for 5 min. Pipette off the bleach.
4. Rinse multiple times with sterile water until clean. Pipette off final rinse.
5. Add 1.5 mL of 0.1% agarose solution and mix well by pipetting up and down. Store in the dark for 3 days at 4°C to ensure uniform germination.
6. Place a sterile tip rack covered with Nitex mesh into two Phytatrays; add 240 mL of sterile *Arabidopsis* growth solution (Fig. 2a). The growth solution should just touch the base of the Nitex mesh.
7. Pipette 750 μL of the agarose seed mixture (approximately 1,000 seed) onto the Nitex mesh of each Phytatray in an even distribution.
8. Close the lid of the Phytatrays and leave for 4 days in the growth cabinet.
9. On the morning of day 5, replace the growth solution in the Phytatrays with 240 mL of sterile *Arabidopsis* 25 mM NaCl growth solution (replace the growth solution of any control plants with 240 mL of sterile *Arabidopsis* growth solution; also see Note 4). Return Phytatrays to the growth cabinet for 24 h.
10. On the morning of day 6, replace the growth solution in the Phytatrays with 240 mL of sterile *Arabidopsis* 50 mM NaCl growth solution (replace the growth solution of any control plants with 240 mL of sterile *Arabidopsis* growth solution; also see Note 4). Return Phytatrays to the growth cabinet for 24 h.

3.2. Arabidopsis Protoplasting Protocol

This procedure has been modified from the original publication (7) to increase protoplast yield.

1. On the morning of day 7, prepare 10 mL of the *Arabidopsis* protoplasting solution and pour into a 50 mL flask (10 mL is sufficient for digestion of roots harvested from two Phytratrays, scale up as required).
2. Open Phytatrays and peel the Nitex mesh from the tip rack support, turning over to expose the roots. Quickly harvest the roots by running a surgical blade across the mesh. The roots should collect as a clump at the tip of the blade.
3. Tap the roots into 10 mL of *Arabidopsis* protoplasting solution and incubate for 1–2 h at 26°C on a benchtop orbital shaker set at 100 rpm.
4. Filter the solution two times through a 40 μm cell strainer fitted on top of a 50 mL centrifuge tube to remove debris.

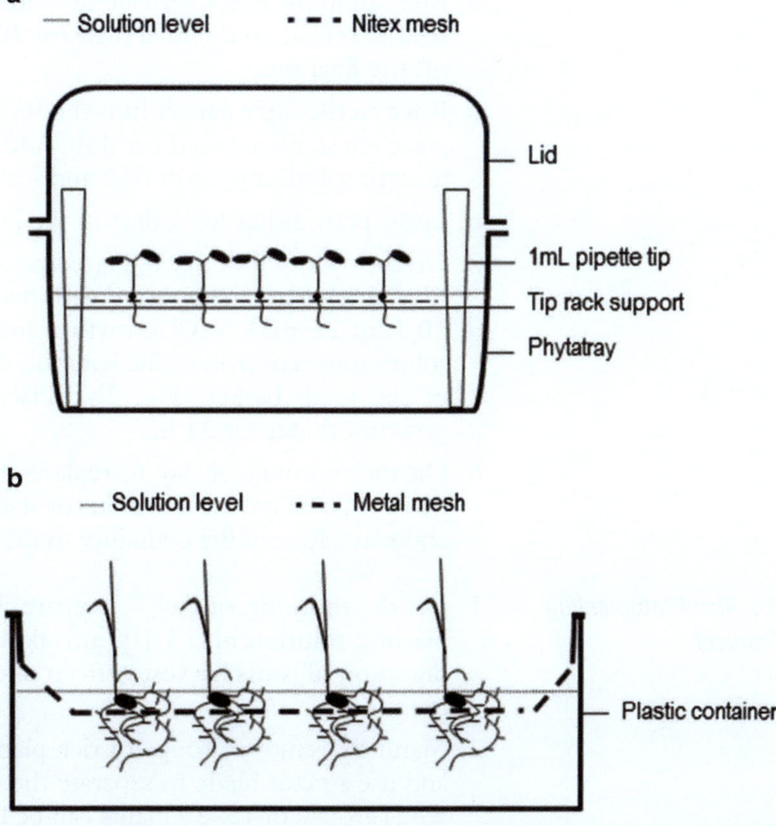

Fig. 2. Assembly of the *Arabidopsis* and rice supported hydroponic systems. (**a**) *Arabidopsis* seeds are pipetted directly onto the Nitex mesh (250 μm) that rests on top of the tip rack support. The plant roots grow through the mesh and are easily harvested by scrapping a surgical blade across the lower side of the mesh. (**b**) Germinated rice seedlings are placed directly into the mesh baskets; 5 mm holes within the mesh allow the rice roots to grow down into the solution. Plastic containers with an inner ridge should be used to hold the mesh basket approximately 5 cm above floor of the container.

5. Centrifuge the filtrate for 5 min at $500\times g$ in a clinical centrifuge. Pour off most of the solution and resuspend the protoplast pellet in a remaining 1–2 mL of protoplasting solution (see Note 5).
6. Proceed with FACS as soon as possible (see Note 6).

3.3. Rice Germination, Growth, and Application of Salinity Stress

All work until day 5 should be carried out under sterile conditions. Perform each procedure as early as possible in the morning.

1. Place approximately 75 dehusked rice seed into a 50 mL centrifuge tube (75 seedlings are sufficient to fill one mesh basket, scale up as required).
2. Add 40 mL of 70% ethanol and mix for 1 min. Pour off the ethanol.
3. Add 40 mL 30% bleach solution and mix on a benchtop orbital shaker set at 100 rpm for 30 min. Pour off the bleach.

4. Rinse multiple times with sterile water until clean, allowing the seed to remain in the final rinse for 20 min on the shaker. Pour off the final rinse.
5. Place sterile filter papers into the 95 × 95-mm petri dishes and place ten sterilized seed per dish. Add 20 mL of autoclaved rice growth solution to each dish and seal with parafilm.
6. Leave petri dishes for 4 days in the growth cabinet.
7. On the morning of day 5, place up to 75 rice seedlings into an Elfa® mesh basket supported within a plastic container holding 10 L of 25 mM NaCl growth solution (or standard growth solution for controls). The solution should just cover the base of the mesh basket (Fig. 2b). Place the container into the growth cabinet for 24 h.
8. On the morning of day 6, replace the solution with 10 L of 50 mM NaCl growth solution (or standard growth solution for controls). Return the container to the growth cabinet for 24 h.

3.4. Rice Protoplasting Protocol

1. On the morning of day 7, prepare 30 mL of the rice protoplasting solution into a 100 mL flask (30 mL is sufficient for digestion of roots harvested from one mesh basket, scale up as required).
2. Manually remove groups of rice plants from the mesh basket and use a razor blade to separate the roots from the shoot and seed (groups of 15–20 plants can be handled easily). Chop the roots on the lid of petri dish and scrape the root tissue into a few milliliter of rice protoplasting solution to keep the root tissue moist.
3. Once all of the roots have been harvested, use the razor blade to gently mince the root tissue for 30 s and then transfer the tissue to 30 mL of rice protoplasting solution.
4. Apply a vacuum for 7 min at 25 Hg to facilitate infiltration of the enzymes. Stop the pump and let the chamber come back to normal pressure.
5. Incubate for approximately 2 h at 26°C on a benchtop orbital shaker set at 100 rpm. Protoplasting should be monitored under the light microscope every 30 min until protoplasts are seen to be releasing from the root tissue (Fig. 3; also see Note 7).
6. Filter the solution two times through a 40 μm cell strainer fitted on top of a 50 mL centrifuge tube to remove debris.
7. Centrifuge the filtrate for 5 min at 500 × *g* in a clinical centrifuge. Pour off most of the solution and resuspend the protoplast pellet in a remaining 1–2 mL of protoplasting solution (see Note 5).
8. Proceed with FACS as soon as possible (see Note 6).

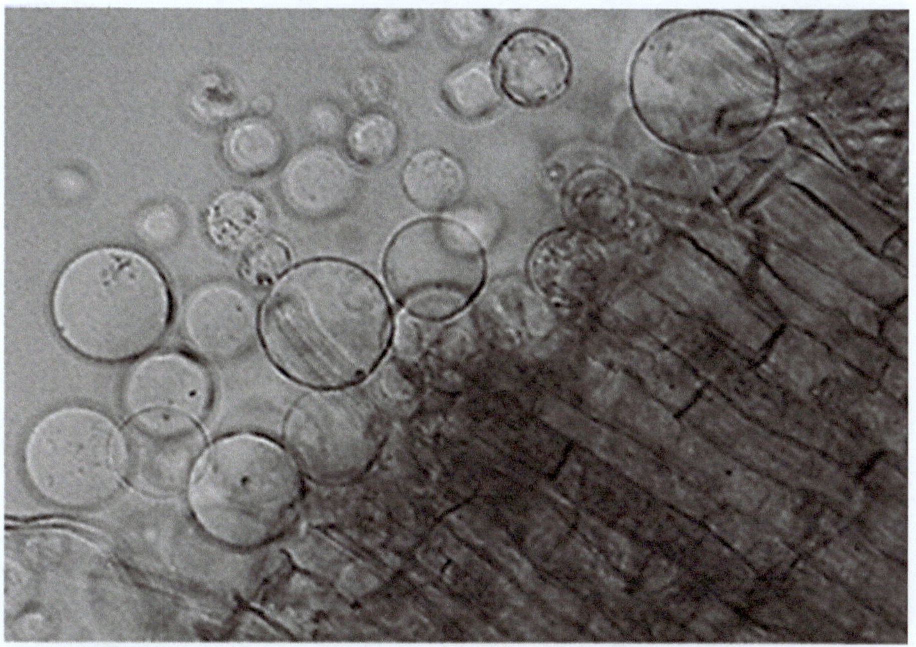

Fig. 3. Rice protoplasts being released from the cut edge of a root, indicating that the digestion reaction has proceeded sufficiently to begin filtration and centrifugation to collect the protoplasts for FACS.

3.5. FACS Collection of GFP-Expressing Protoplasts

The sorting ability of the FACS instrument is based on its capacity to excite and detect specific fluorescence wavelengths emitted by cells in suspension. Cell populations emitting relatively more fluorescence at a specific wavelength (i.e., GFP fluorescence; 488 nm excitation and 530/30 nm emission) can be isolated and collected. However, the autofluorescent properties of protoplast suspension interfere with a distinction of GFP-positive cells based on emission at 530/30 nm alone. Autofluorescent emission at 610/20 nm and a 488 nm excitation is therefore taken as a compensational measure of autofluorescence allowing for the identification of sorting events with GFP-specific fluorescence.

1. Load the protoplast solution into the FACS. The loading tube should be under constant agitation to maintain a homogenous solution.
2. Run protoplasts through the FACS using a 70-μm (or 100 μm) nozzle (PBS buffer for sheath fluid) at event rates up to 20,000 events/s and a system pressure up to 70 psi (events/s and system pressure will depend on the nozzle utilized). Laser excite the cells at 488 nm and set a green channel to measure GFP emission at 530/30 nm and a red channel to measure autofluorescence at 610/20 nm. Use a scatter plot with red emission intensity (autofluorescence) vs. green emission intensity (GFP fluorescence) (Fig. 4a, c).

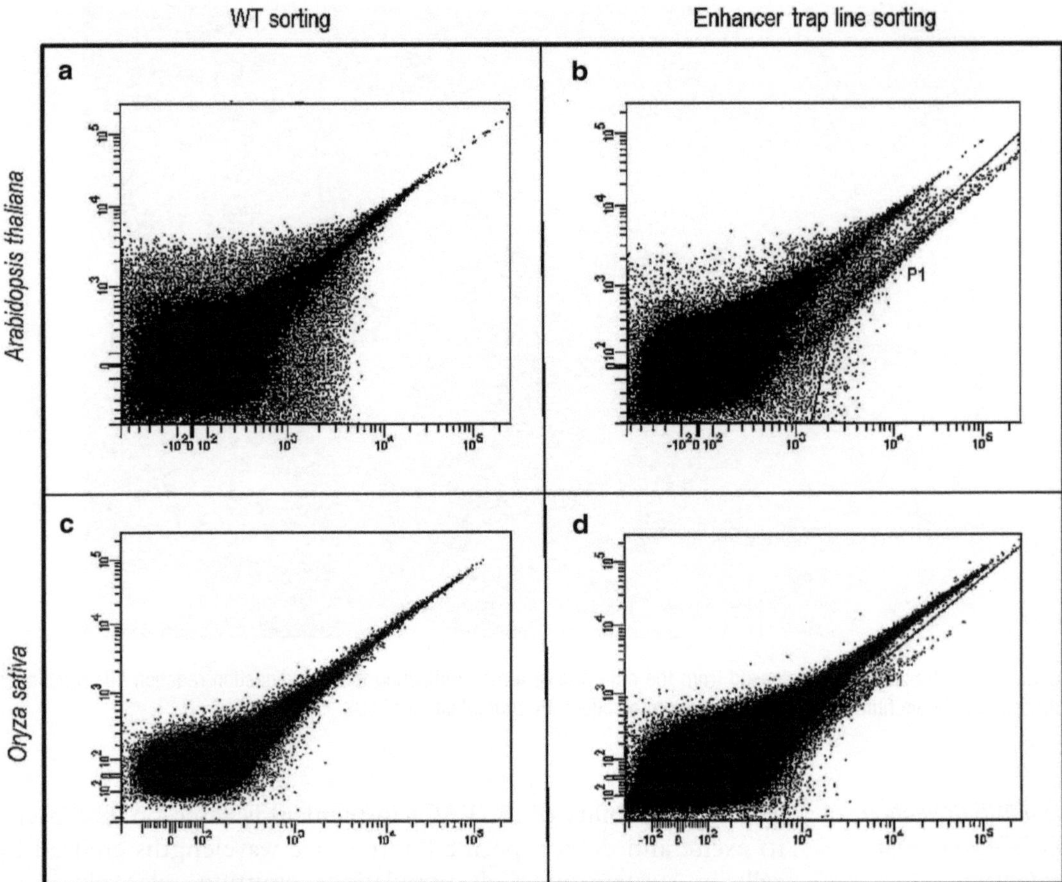

Fig. 4. Scatter plots of green fluorescence intensity on the *x*-axis (arbitrary units; excitation 488 nm/emission 530/30 nm) and red fluorescence intensity on the *y*-axis (arbitrary units; excitation 488 nm/emission 576/26 nm) from representative *Arabidopsis* and rice FACS experiments. (**a**) Scatter plot of protoplasts from wild-type *Arabidopsis* plants not expressing GFP with most cells having a typical autofluorescence pattern of equal green and red fluorescence. (**b**) Sort of protoplasts from an *Arabidopsis* GAL4-GFP enhancer trap line with a gate (P1) set to collect the GFP tail where cells display relatively more green than red fluorescence. (**c**) Sort of protoplasts from wild-type rice plants not expressing GFP with most cells having a typical autofluorescence pattern of equal green and red fluorescence. (**d**) Sort of protoplasts from a rice GAL4-GFP enhancer trap line, with a gate (P1) set to collect the GFP tail where cells display relatively more green than red fluorescence.

3. A non-fluorescent control protoplast solution (protoplasts prepared from wild-type plants not expressing GFP) should initially be run to determine gate settings for capture of GFP fluorescent protoplasts (Fig. 4a, c).
4. When sorting protoplasts from the enhancer trap lines, a clear tail of protoplasts exhibiting proportionally high green fluorescence should be visible (Fig. 4b, d; also see Note 8). Make sure the gate is well positioned to collect GFP positive protoplasts while avoiding as many false positives as possible. (In this study, BDFACSDiva software version 6.1.3 was used with the BD FACSAria cell sorter set to sort precision mode

"Purity" to ensure a stringent sorting of GFP protoplasts.) Sort into a desired collection buffer (for our purposes, we used RNA extraction buffer).

5. The number of cells to collect by FACS is dependent on the type of study and RNA extraction kit used. For rice transcriptomic studies using microarrays, we sorted between 3,000 and 17,000 GFP positive cells into RLT buffer from the RNeasy Plant Micro Kit (see Note 9). Maintain a ratio of 100 μL cell suspension to 350 μL RLT buffer to ensure efficient RNA extraction (refer to manufacturer's instructions). For RT-PCR studies, more cells may be needed.

4. Notes

1. Use ultrapure water for all rinses and to prepare solutions (2 MΩ at room temperature). Store solutions at room temperature unless specified otherwise.
2. The addition of $CaCl_2$ to the 25 and 50 mM NaCl growth solutions ensures that Ca activity remains constant upon addition of salt, so that any measured effects of salinity are due to the addition of NaCl and not NaCl-induced decreases in Ca activity of the external solution (14).
3. If more *Arabidopsis* protoplasts are required, it is recommended to increase the number of Phytatrays. Overloading Phytatrays with more than 1,000 seed leads to poor plant growth and greater contamination rates.
4. Work in sterile conditions when changing hydroponics *Arabidopsis* solutions to avoid problems with bacterial contamination. The rice-supported hydroponics setup does not need to be kept sterile.
5. A cell pellet may not always be visible after centrifugation and resuspension. It is recommended to observe a drop of the resuspended pellet under the microscope to check protoplast density and condition before proceeding with FACS.
6. Minimize holding times of the protoplasts to avoid transcriptional changes related to the protoplasting procedure itself. If necessary, protoplasts can be kept for approximately 60 min at 15–20°C prior to beginning of FACS.
7. We suggest monitoring digesting rice roots under the microscope to better assess how enzymatic digestion is proceeding. Digestion time can be reduced if protoplasts are visible.
8. The absence of GFP positive protoplasts as a GFP tail on the sorting scatter plot may be due to poor efficiency of protoplasting.

Check pH of the protoplasting solution and temperature for any errors and monitor digestion via microscopic examination.

9. Quickly vortex or vigorously shake the sorted protoplasts in RLT buffer to ensure cell lysis. Either proceed with RNA extraction directly or freeze the lysed cells in liquid nitrogen and store at −80°C.

Acknowledgments

This work was supported by grants from the Australian Research Council (DP0770966 to M.T. and A.A.T.J.) and the U.S. National Institutes of Health (NIH grant R01 GM078279 to K.D.B.)

References

1. Birnbaum K et al (2003) A gene expression map of the *Arabidopsis* root. Science 302:1956–1960
2. Brady SM et al (2007) A high-resolution root spatiotemporal map reveals dominant expression patterns. Science 318:801–806
3. Dinneny JR et al (2008) Cell identity mediates the response of *Arabidopsis* roots to abiotic stress. Science 320:942–945
4. Gifford ML et al (2008) Cell-specific nitrogen responses mediate developmental plasticity. Proc Natl Acad Sci USA 105:803–808
5. Nawy T et al (2005) Transcriptional profile of the *Arabidopsis* root quiescent center. Plant Cell 17:1908–1925
6. Borges F et al (2008) Comparative transcriptomics of *Arabidopsis* sperm cells. Plant Physiol 148:1168–1181
7. Zhang C et al (2008) Global characterization of cell-specific gene expression through fluorescence-activated sorting of nuclei. Plant Physiol 147:30–40
8. Birnbaum K et al (2005) Cell type-specific expression profiling in plants via cell sorting of protoplasts from fluorescent reporter lines. Nat Methods 2:615–619
9. Bargmann BOR, Birnbaum KD (2010) Fluorescence activated cell sorting of plant protoplasts. J Vis Exp (36):e1673
10. Haseloff J (1999) GFP variants for multispectral imaging of living cells. Methods Cell Biol 58:139–151
11. Johnson AAT et al (2005) Spatial control of transgene expression in rice (*Oryza sativa* L.) using the GAL4 enhancer trapping system. Plant J 41:779–789
12. Møller IS et al (2009) Shoot Na^+ exclusion and increased salinity tolerance engineered by cell type-specific alteration of Na^+ transport in *Arabidopsis*. Plant Cell 21:2163–2178
13. Plett D et al (2010) Improved salinity tolerance of rice through cell type-specific expression of AtHKT1;1. PLoS One 5:e12571
14. Tester M, Davenport R (2003) Na^+ tolerance and Na^+ transport in higher plants. Ann Bot 91:503–527

Chapter 19

Transformation Using Controlled cDNA Overexpression System

Gábor Rigó, Csaba Papdi, and László Szabados

Abstract

The controlled cDNA overexpression system (COS) was developed to identify novel regulatory genes in model plants as well as in other species that might have a particular valuable trait. The COS system (Papdi et al. Plant Physiol 147:528–542, 2008) is composed of a random cDNA library prepared in a T-DNA plant expression vector, under the control of the estradiol-inducible XVE promoter. Large-scale genetic transformation of *Arabidopsis thaliana* generates a transgenic plant population with randomly inserted cDNA clones. Overexpression of the inserted cDNA can create selectable phenotypes, allowing the facile identification and cloning of the responsible genes. Here we describe protocols to create and use the COS system for diverse purposes in plant biology.

Key words: cDNA library, Transformation, Overexpression, COS, Large-scale screen, *Arabidopsis*

1. Introduction

Genetic transformation of plants has become a standard tool in plant research and is essential for analysis of gene function. In most plant species, genetic transformation is performed on somatic cells. This requires subsequent in vitro selection and plant regeneration. Transformation itself can be mediated by *Agrobacterium* or via direct DNA transfer using protoplasts or ballistic gene transfer (1). Tissue culture-based genetic transformation is usually a time-consuming and labor-intensive process, and only a limited number of transgenic plants can be produced. Facile transformation is an important technical criterion when high frequency DNA transfer is required for generating large numbers of transgenic plants (2). In planta genetic transformation methods have been established for

Sergey Shabala and Tracey Ann Cuin (eds.), *Plant Salt Tolerance: Methods and Protocols*, Methods in Molecular Biology, vol. 913, DOI 10.1007/978-1-61779-986-0_19, © Springer Science+Business Media, LLC 2012

Arabidopsis thaliana, the most important model for plant research, and they offer an efficient and facile method for gene transfer (3, 4). As the tissue culture step is omitted in the in planta methods, scaling up the transformation procedure becomes feasible. *Agrobacterium*-mediated in planta floral transformation is routinely used for genetic transformation of *Arabidopsis*, especially when large numbers of transgenic plants need to be generated. Procedures of *Agrobacterium*-mediated tissue culture transformation have also been optimized for rice and are used successfully to generate T-DNA insertion lines for genetic screens (5–7).

T-DNA insertion lines offer numerous possibilities for identifying genes and analyzing their function. Insertion in exons, 5′ or 3′ UTR regions, or in important promoter regions can produce loss of function mutations (8, 9). Promoters or enhancer sequences positioned close to the T-DNA left or right border regions can enhance the expression of neighboring genes and offer tools for activation tagging (6, 7, 10). Use of promoterless reporter genes offer a convenient tool for promoter or gene trapping when in situ gene fusions are generated and used to study gene function and expression with the same mutant line (5, 11–14). Large-scale transformations have been used to generate T-DNA insertion lines, which are subsequently employed for genetic screens or to establish mutant collections for reverse genetic approaches (9, 11, 15–19).

Transfer and expression of random or customized cDNA libraries in plants offer an alternative possibility to generate plants with gain or loss of function phenotypes. cDNA libraries under the control of the constitutive CaMV35S promoter have been used to generate *Arabidopsis* and rice transgenic lines with dominant phenotypes (20–22). The controlled cDNA overexpressing system (COS) employs a chemically inducible expression system, where transcription of the inserted cDNA clones is controlled by an externally added inducer (23). cDNA library transformation can generate dominant phenotypes due to the overexpression of full-length cDNA or suppress endogenous gene activity by transcription of truncated cDNA (24). For example, screening for ABA insensitive germination of cDNA overexpressing lines leads to the identification of the novel *AtPP2AC* and the HSP17.6A genes that encode a protein phosphatase and small heat shock protein, respectively (23, 25). In addition, large-scale transformation of *Arabidopsis* root cultures with an *Arabidopsis* cDNA library led to the identification of the *ESR1* gene that encodes a putative AP2/EREBP transcription factor. This confers cytokinin-independent plant regeneration to cultured roots (26).

cDNA library transformation can be used to express cDNAs in plant species different from the RNA source. This includes natural variants carrying valuable genetic traits, such as tolerance to biotic or abiotic stresses. Ectopic expression of full-length rice cDNA library in *Arabidopsis* represents a novel genetic tool to identify and study rice genes in a heterologous system (21, 27). cDNA expression library from the halophyte plant, *Thellungiella salsuginea,* was recently transformed into *Arabidopsis*, allowing the identification of novel genes that could confer salt tolerance to *Arabidopsis* (28, 29). These examples show that random cDNA library transformation with subsequent genetic screens offers a convenient tool to identify plant genes in model plants as well as in other species (24, 30).

2. Materials

2.1. Preparation of COS cDNA Library

1. For RNA isolation, use TRI Reagent from Sigma Cat. No.: T9424 (31).
2. Materials for mRNA isolation: FastTrack™ MAG Maxi mRNA Isolation Kit (Invitrogen Cat. No.: K1580-02).
3. Materials for full-length cDNA library construction: use Superscipt® Full Length cDNA Library Construction Kit (Invitrogen Cat. No.: A11181 or A13268).
4. LB Agar plates: 1 L medium contains 10 g tryptone, 5 g yeast extract, 10 g NaCl, pH 7.2. For solid media add 20 g/L Agar. Supplement medium with appropriate antibiotics (for *Escherichia coli* use 25 mg/L Kanamycin or 50 mg/L Spectinomycin).
5. Equipment for electroporation. Several electroporation apparatuses are available, e.g., Bio-Rad Gene Pulser or Micropulser™ 165-2100. For electroporesis, use electroporator cuvettes with 0.1 cm gap (e.g., Bio-Rad Cat. No.: 165-2089).
6. Large-scale plasmid isolation: Plasmid Plus Giga Kit (QIAGEN Cat. No.: 12991) or equivalent.
7. Enzymes for Gateway cloning. To generate cDNA library in pDONR222, use Gateway® BP Clonase® II enzyme mix (Invitrogen, Cat. No.: 11789-020). To clone the library into plant binary vector, use Gateway® LR Clonase® II enzyme mix (Invitrogen, Cat. No.: 11791-020).
8. YEB medium: 1 L medium contains 5 g peptone, 5 g beef extract, 1 g yeast extract, 5 g sucrose, pH 7.2. For solid media add 20 g/L Agar. After autoclaving, add $MgSO_4$ to a final concentration of 2 mM (32).
9. Materials to prepare electrocompetent *Agrobacterium*: 3 L Erlenmeyer flask, sterile MQ grade water, sterile 10% glycerol in MQ water, liquid nitrogen (33). The *Agrobacterium* strain GV3101/pMP90 gives reproducible results in our laboratory (32).

10. For the selection of GV3101/pMP90 strain, use 100 mg/L Rifampicin and 25 mg/L Gentamycin together with the T-DNA vector selective antibiotics such as 100 mg/L Spectinomycin, or 100 mg/L Kanamycin, or 100 mg/L Carbenicillin.
11. To store *Agrobacterium* library in −80°C until use, you need sterile 87% glycerol.

2.2. Large-Scale Genetic Transformation of Arabidopis (Floral Dip)

1. Seeds of wild-type (Col, Ws, Ler, etc.) *Arabidopsis* or any genotypes of interest.
2. *Agrobacterium* strain with the binary vector of cDNA library. For *Arabidopsis* transformation, we generally use the GV3101/pMP90 strain (RifR, GmR) (32).
3. Antibiotics. For the selection of GV3101/pMP90 strain, use 100 mg/L Rifampicin and 25 mg/L Gentamycin together with the T-DNA vector selective antibiotics (in case of pER8-based vectors use 100 mg/L Spectinomycin).
4. Gardening pots (~12 cm diameter) filled with commercially available gardening soil mixed with perlite or vermiculite (10:1 ratio).
5. Greenhouse or plant growth chambers.
6. Infiltration solution: distilled water, 5% sucrose, and 0.01% Silwet L77 (Lehle Seeds, Round Rock, TX, USA)
7. Plant culture medium: half-strength Murashige and Skoog medium (½ MS), pH 5.7, containing 0.5% sucrose and 0.7% agar (34).
8. Antibiotics for plant selection: Kanamycin, Hygromycin, Phosphinothricin, depending on the marker used.

2.3. Selection of Transformed Plants

1. For in vitro Hygromycin selection, use germination medium (½MS) (34), supplemented with 200 mg/L Cefotaxime, 15 mg/L Hygromycin (or another antibiotic, depending on the marker used).
2. For herbicide selection, use gardening tray (e.g., 40 × 60 cm), filled with commercially available gardening soil:vermiculite mix (10:1).
3. 100 mM Gibberellic acid (GA_3) solution dissolved in 96% ethanol. Store the solution at −20°C.
4. For Basta selection, use herbicide solution of 300 mg/L Basta, prepared from commercially available herbicide (e.g., Finale 14 SL containing 150 g/L glufosinate-ammonium from Bayer), diluted in distilled water. For spraying, use a 2 L Plant Spray Bottle. Keep the diluted solution at 4°C, for up to 1 month.

3. Methods

The following protocols describe the construction of a plant cDNA library in a plant expression vector and large-scale transformation into wild-type *Arabidopsis* plants. The procedure has a number of standard steps that have been published and it can be easily completed using commercially available molecular kits. We refer to these steps and the type of kits we have used successfully in our experiments. Due to space constraints, we do not describe the detailed protocols in all the cloning steps, but cite the adequate method published in standard manuals or commercial kits. The outline of the experimental strategy is depicted in Fig. 1 (see Note 1).

3.1. Construction of cDNA Library in Plant Expression Vector

1. RNA isolation (see Note 2). High quality total plant RNA can be isolated using several methods. The TRI Reagent Kit from Sigma (Cat. No.: T9424) is routinely used in our laboratory and gives satisfactory results. Using this method, the amount and the quality of the purified RNA is suitable for most subsequent applications. Check the RNA quality and quantity, using a conventional spectrophotometer or NanoDrop (Thermo Fisher Scientific). It should have an A260/280 ratio ≥1.7. For the mRNA purification step, you will need at least 500 μL RNA solution, at concentration of 1 μg/μL.
2. mRNA isolation is performed with Invitrogen FastTrack™ MAG Maxi mRNA Isolation kit. The protocol is optimized for isolation of mRNA from 500 μg of total RNA and is dissolved in 500 μL DEPC-treated water. Quality and quantity of the RNA preparations should be tested by gel electrophoresis and spectrophotometry (Fig. 2).
3. Reverse transcription. The most convenient way to produce cDNA is by using one of the numerous available commercial kits. We have good experience with the Superscipt® Full Length cDNA Library Construction Kit from Invitrogen. Using this kit, 5–10 μg high quality mRNA or 0.5–1 mg total RNA is needed for the construction of one cDNA library. The kit uses oligo(dT) based 3′ primer, combined with high quality Superscript® III Reverse Transcriptase and cap antibody 5′ region selection to enrich the full-length clones in the cDNA library.
4. Cloning the cDNA library into pDONR vector using the Gateway® Technology. Using the BP Clonase® enzyme reaction, the attB-flanked cDNA library can be cloned into a donor vector (pDONR222) to produce Gateway® entry library. For the BP reaction mix, add 7 μL attB-flanked cDNA, 2 μL pDONR222 (150 ng/μL), and 3 μL TE buffer. Remove the BP Clonase®II enzyme mix from −20°C and place on ice

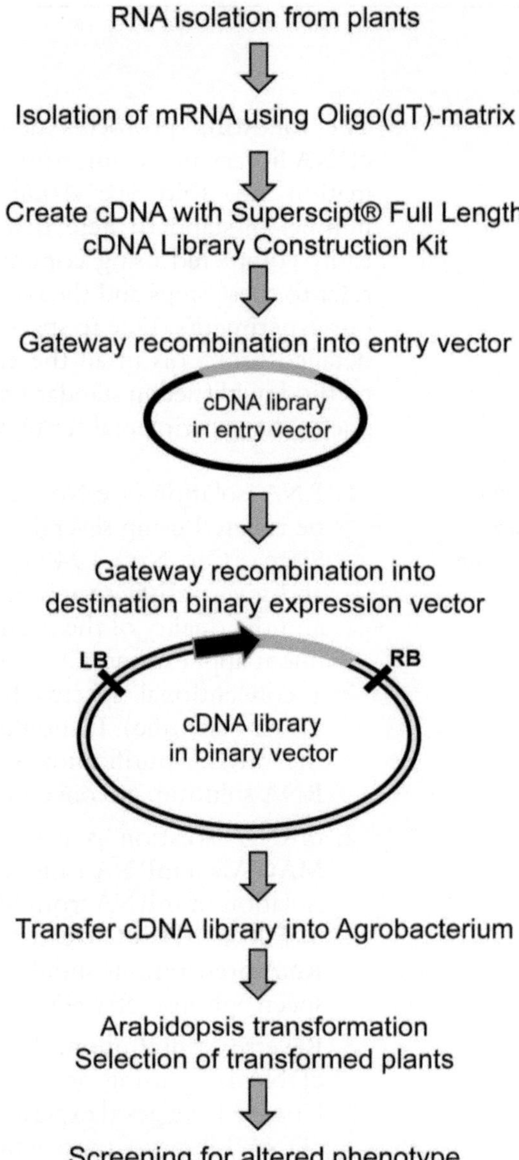

Fig. 1. Schematic representation of the COS library construction.

for 2 min. Gently vortex the BP Clonase®II enzyme and add 3 μL to the reaction. Mix the tube by pipetting up and down. Incubate the reaction mixture overnight at 25°C.

5. Before electroporation into electrocompetent *E. coli* cells, remove salt and other impurities by ethanol precipitation of cDNA library. Add 1 μL glycogen (20 μg/μL), 50 μL 7.5 M NH_4OAc, H_2O to 100 μL, and 375 μL 100% ethanol to the reaction mixture. Keep the tube at –20°C overnight or on dry ice –80°C for at least 1 h. Centrifuge the sample at 16,000 × *g*

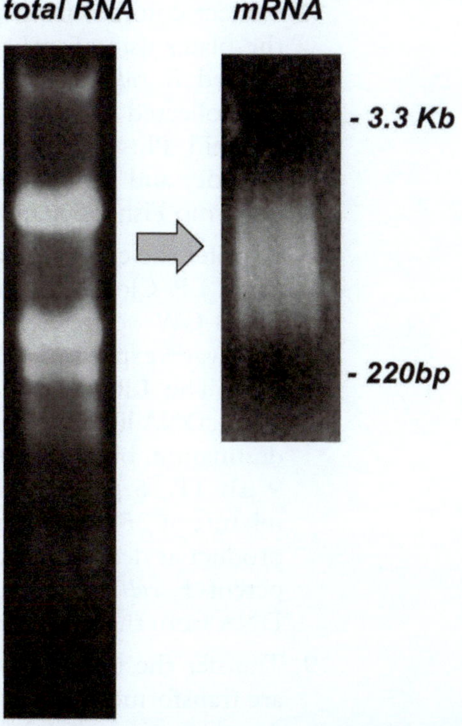

Fig. 2. Total and mRNA samples used for preparation of the COS library.

at +4°C for 30 min. Carefully remove the supernatant, and then wash the pellet twice with 150 μL 70% ethanol. Gently dry the pellet, avoiding the over-drying. Resuspend the pellet in 20 μL of TE buffer.

6. Electroporation of the cDNA library into ElectroMAX™ DH10B™ T1 Phage Resistant Cells. The electrocompetent cells (provided by the Superscipt® Full Length cDNA Library Construction Kit, or obtained from Invitrogen, Cat. No.: 12033-015) are thawed on ice and mixed with plasmid DNA. Cells are transferred to a chilled 0.2 cm electroporation cuvette and placed in an electroporator (e.g., Bio-Rad Gene Pulser or MicroPulser), set to 2.5 kV (voltage), 25 μF (capacitance), and 200 Ω (resistance). After the pulse, add 750 μL of LB (pre-heated at 37°C), transfer cells into a centrifuge tube and shake for 45 min at 37°C for recovery. Plate the transformed *E. coli* cells onto 15 cm diameter Petri plates containing LB medium supplemented with 25 mg/L Kanamycin. Adjust the titer of the cDNA library to get 10–15,000 colonies in one plate. Calculate the amount of the LB plates to get about 10^6 independent colonies in the whole experiment. We routinely use 100 plates in one experiment.

7. Collect colonies from plates by washing the *E. coli* cells off the plates using 5 mL of liquid LB media, and pipetting the washed *E. coli* suspension into a centrifuge tube. Centrifuge the collected cells and purify plasmid DNA using QIAGEN Plasmid Plus Giga Kit or equivalent. Check the plasmid quantity and quality, using spectrophotometer or NanoDrop (Thermo Fisher Scientific) and agarose gel eletrophoresis.
8. Transfer the cDNA library into the Gateway expression vector using LR Clonase reaction. We use the chemically inducible pER8-GW vector for library construction (23), but other Gateway® expression vectors are suitable for this purpose as well. The LR reaction contains the following components: 2 μL cDNA library in pDONR™222 vector (25 ng/μL), 3 μL destination binary vector (150 ng/μL, pER8-GW vector), 9 μL TE, 6 μL LR Clonase® II enzyme mix. Incubate the mixture at 25°C, overnight. Precipitate and wash the reaction product as described in step 5 and transform into electrocompetent *E. coli* cells as described in step 6. Purify the plasmid DNA from the recovered colonies according to step 7.
9. Transfer the library into *Agrobacterium*. *Agrobacterium* cells are transformed in the same way as described in step 6, except Agrobacteria are grown in YEB medium at 28°C (35). Transform the electrocompetent *Agrobacterium* cells with the purified cDNA library to get at least 10^6 independent colonies on selective YEB plates, supplemented with appropriate antibiotics. To select the cDNA library cloned in the pER8-GW vector, use 100 mg/L Spectinomycin. Ensure that there are 10–15,000 transformed *Agrobacterium* colonies on one plate. In a typical experiment, you will need around 100 selective YEB plates. Incubate the plates on 28°C for 2 days, and collect the colonies by washing them off with liquid YEB medium. Mix the *Agrobacterium* slurry and add an equal amount of sterile glycerol (87%). Dispense 1.0 mL aliquots of the *Agrobacterium* suspension into 2 mL microcentrifuge tubes, freeze them in liquid nitrogen and store at –80°C (see Note 3).

3.2. Large-Scale Genetic Transformation of Arabidopsis (Floral Dip Method)

In planta *Arabidopsis* transformation is the method of choice when large-scale experiments are planned. Several methods with slight variation have been published and can be scaled up for large-scale cDNA library transformation (4, 36–38). Here we describe a simple method, that gives reliable and reproducible transformation efficiencies.

1. Grow 7–10 wild-type *Arabidopsis* plants in pots containing soil and perlite or vermiculite mixture under short-day illumination (8 h light/day). For large-scale transformation, use 200–500 pots in one experiment. After developing 6–8 cm rosette leaves, transfer the plants to long days (14–16 light hours/day)

for bolting. If you do not have adequate facilities for this, plants can be grown in long-day illumination from the beginning. In this case, grow 15–20 plants in each pot.

2. When plants begin to bolt, cut the first few inflorescence shoots. Plant transformation can be performed 5–7 days later, when a higher number of secondary bolts have emerge.
3. Inoculate 1 L YEB medium (supplemented by appropriate antibiotics) in 3 L Erlenmeyer flask with 3 mL starter culture of *Agrobacterium* strain carrying the cDNA library (see step 9 in Subheading 3.1). Incubate overnight with 250 rpm horizontal shaking at 28°C. This amount of *Agrobacterium* is enough for the transformation of at least 50 pots of *Arabidopsis.*
4. Spin down the *Agrobacterium* cells (3,000 × *g*/15 min/28°C), resuspend the pellet in infiltration solution, and set the OD_{600} to 0.8–0.9.
5. Dip *Arabidopsis* shoots into the *Agrobacterium*-infiltration solution for 2 min. Remove plants from the solution and lay them horizontally on a tray covering them with aluminum foil to maintain the inoculated plants in dark and with high levels of humidity. Use the same solution for the infiltration of more pots. The next day, transfer the plants back to long-day illumination. Repeat the Agrobacteria infiltration procedure 7–10 days later.
6. Let the plants flower and set seed. After seed ripening harvest seeds and collect them in bulk. Store seeds in dry, cool place.

3.3. In Vitro Selection of Transformed Arabidopsis Plants

1. For surface sterilization, transfer 5 mL seeds of infiltrated *Arabidopsis* plants to a 50 mL conical tube. Fill up the tube with 70% ethanol and incubate for 1 min, by inverting the tube a couple of times then let the seeds settle. Change the ethanol to 50% household chlorine bleach solution containing 0.01% Triton X-100 and mix the tube continuously for 10 min. Allow the seeds to settle for 1 min, remove the bleach solution, and wash the seeds at least five times with sterile distilled water.
2. Prepare 15 cm diameter Petri plates containing 70 mL ½ MS agar culture media supplemented with 200 mg/L Cefotaxime and the selection antibiotics, depending on the marker used. For Hygromycin selection, we routinely use 15 mg/L Hygromycin B. In one selection experiment, 100 selective plates can be safely handled.
3. In a 500 mL Erlenmeyer flask, mix 200 mL ½ MS media with 200 mL sterile distilled water containing the sterilized *Arabidopsis* seeds and add 200 mg/L Cefotaxime and 15 mg/L Hygromycin B. Pour 10 mL from this to each previously prepared agar plate.

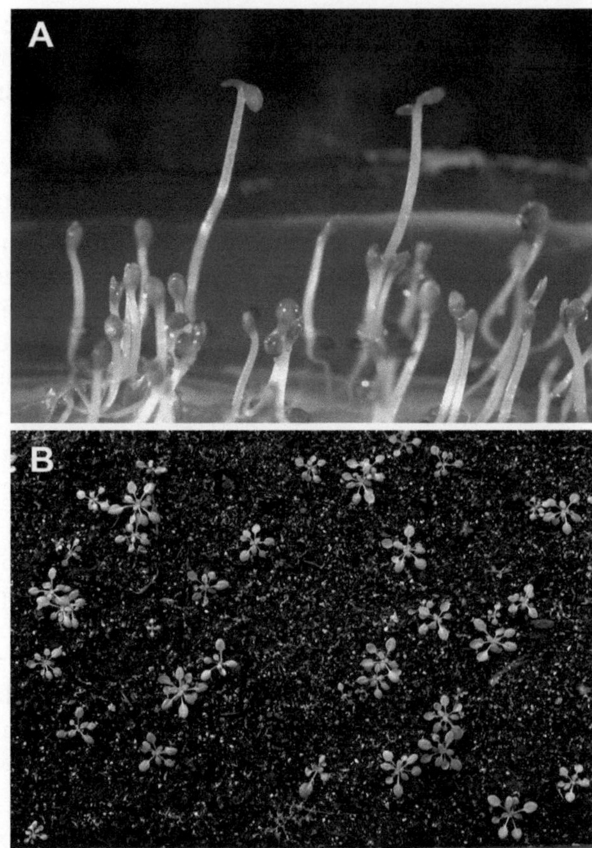

Fig. 3. Selection of transformed Arabidopsis plants. (**A**) Selection on hygromycin plates under dim light. Hygromycin resistant seedlings have elongated hypocotyls. (**B**) Selection in soil by spraying with Basta. Only resistant plants grow.

4. Stratify the seeds by keeping the solidified agar plates in the dark at 4°C for 2 days. Then transfer plates to the growth chamber or culture room at 20–22°C, but keep them under dim light (15–20 μmol/m^2/s^1) during germination.
5. Select transformed plants. When Hygromycin selection is used, resistant plants can be identified by hypocotyl elongation in dim light (Fig. 3). Five days later, collect transformant plants with the elongated hypocotyls and transfer them to fresh plates and standard growth conditions (temperature: 20–22°C, light: 100 μmol/m^2/s^1, 8/16 h light cycle). Rooted plants can be transferred to soil and greenhouse conditions to flower and set seeds.

3.4. Selection of Transgenic Plants with Herbicide Resistance Marker

1. Transfer 0.5–1 mL of transformed seeds into 50 mL Falcon tube, and resuspend in 40 mL distilled water, containing 100 μM GA_3. Stratify the seed by keeping them at 4°C for 2 days. This step can synchronize and enhance the germination.
2. Prepare soil mix soil: vermiculite (10:1) fill the gardening tray and irrigate well.

3. Pour seeds onto soil, using a hole-punched (with needle) 50 mL Falcon tube cap and cover with plastic foil until seedlings with green cotyledons emerge (about 5 days). Keep the trays in dim light. Remove plastic foil or cover after 1 week.
4. Prepare Basta solution in a Plant Spray Bottle, at a concentration of 300 mg/L of glufosinate-ammonium. Spray the 10 days old seedlings with BASTA.
5. Repeat spray two times a week, until the Basta resistant plants can be distinguished (Fig. 3). Transfer plants to pots, and allow them to flower and set seed under long-day light cycle (15/9 h illumination).

3.5. Genetic Screens with the COS Library

Genetic screens with COS system are similar to other mutant screens performed with EMS or T-DNA mutagenized stocks (24, 39–41). Overexpression of the inserted cDNA is expected to cause dominant phenotype. Therefore, heterozygous plants are suitable for screening. When an inducible promoter is used to control the transcription of the inserted cDNA, induction should be applied during the screening procedures (23).

1. Seeds of selected plants can be collected individually or in bulk. Depending on the type of selection, bulks of seeds from 25 to 100 lines can be prepared. Fresh seed stocks can be used for screening, or stored in a cool, dry place for up to 1 year. When seed stocks are stored for longer periods, keep them at –20°C.
2. Screening conditions should be carefully established and optimized before large-scale experiments are initiated. T2 generation seed stocks can be used efficiently for genetic screens, when conditions allow the visual identification of mutant phenotypes. Alternatively, T1 generation seeds can be screened, when antibiotic selection of germinating seedlings can be followed by screening for the desired phenotype.
3. Selected plants should be transferred to a greenhouse to set seed. Constitutive overexpression of essential genes might influence fertility, therefore flowering and seed set should be carefully monitored.
4. Identification and cloning of the inserted cDNA is straightforward and easy task with the COS system. Genomic DNA is prepared from a small amount of plant material (e.g., one leaf) and the inserted cDNA is amplified by PCR reaction using primers annealing to the promoter and terminator sequences, flanking the insert. The nucleotide sequence of the PCR fragment is determined and the identity of the cDNA is determined by sequence homology search (see Note 4).
5. Cloning the recovered PCR fragment can be done by Gateway technology. As the Gateway recombination sites are conserved in the cDNA library, the cDNA of the PCR fragment can be

cloned into the pDONR222 vector, using the gateway BP reaction. Nucleotide sequence of the cloned inserts should be verified by sequencing. Subsequent cloning of the recovered insert into different destination vectors can be done by employing the Gateway LR reaction (see Note 5).

6. Verification of the selected phenotype can be done by outcrossing the selected plants with wild-type *Arabidopsis.* Cosegregation of the observed phenotype and the inserted cDNA is a proof for causative relationship.
7. Alternatively, the cloned cDNA insert can be cloned into a destination expression vector, and the new construct can be introduced into wild-type *Arabidopsis* plants. The transformation vector can be the same, as that used for the construction of the COS library, or distinct, carrying a different expression cassette. If morphology or development of independently transformed plants or transgenic lines is similar, then the inserted gene is considered to be responsible for the observed phenotype.

4. Notes

1. Gene identification with the controlled cDNA overexpressing system is a complex program, which has a number of consecutive phases. All phases of the program are composed of several individual steps, which requires attention, precision, and/or is labor-intensive: preparation of the cDNA library, transfer of the library into *Agrobacterium*, large-scale genetic transformation, screening of the desired phenotype, gene identification and cloning. In order to improve efficiency and reduce the possibilities of errors, use of molecular kits is recommended.
2. A cDNA library can be prepared from any plant species, different type of tissues, or plants subjected to all types of treatments. Therefore, the range of RNA source is very large. This can be explored to identify novel genes and gene variants in specific traits.
3. Preparation of the cDNA library is a key step in research programs that use the COS system for gene identification. Therefore, care must be taken to generate as good library as possible. To enhance the frequency of certain cDNA fractions in the library, source RNA can be prepared from specific organs or tissues, from plants subjected to a number of treatments. Moreover, cDNA libraries can derive from any plant species, offering a rich source of genes for basic and applied plant science.

4. Once a transgenic line is identified, in which the inserted cDNA is responsible for the observed phenotype, gene identification is a fast and straightforward procedure. cDNA insert can be amplified from the transgenic plants by standard PCR reactions, using vector-specific primers annealing to flanking regions of the cloning sites. We routinely use PCR primers, which anneal to promoter and 3′ termination sequences of the pER8-GW expression vector (23). Gene identification is done by determination of the nucleotide sequence of the purified PCR fragment.
5. Due to the conserved gateway recombination sites in the COS system, cloning of the inserted gene can be performed in several days. The PCR fragment, amplified from the selected plant, can be cloned into pDONR222 vector using the gateway BP reaction. The cloned insert then can be subsequently transferred to any destination vector. Gene identification and cloning of cDNA inserts can therefore be performed in a few days.

Acknowledgments

Authors are indebted for Mary Prathiba Joseph for critical reading and correcting the manuscript. Research was supported by OTKA Grants no. K-68226, K-81765, HuRo Cross border Cooperation Programme HURO/0801/167 and COST Action FA0605.

References

1. Hansen G, Chilton MD (1999) Lessons in gene transfer to plants by a gifted microbe. Curr Top Microbiol Immunol 240:21–57
2. Koncz C, Martini N, Mayerhofer R et al (1989) High-frequency T-DNA-mediated gene tagging in plants. Proc Natl Acad Sci USA 86: 8467–8471
3. Bechtold N, Jaudeau B, Jolivet S et al (2000) The maternal chromosome set is the target of the T-DNA in the in planta transformation of *Arabidopsis thaliana*. Genetics 155: 1875–1887
4. Tague BW, Mantis J (2006) In planta *Agrobacterium*-mediated transformation by vacuum infiltration. Methods Mol Biol 323: 215–223
5. Hsing YI, Chern CG, Fan MJ et al (2007) A rice gene activation/knockout mutant resource for high throughput functional genomics. Plant Mol Biol 63:351–364
6. Jeong DH, An S, Kang HG et al (2002) T-DNA insertional mutagenesis for activation tagging in rice. Plant Physiol 130:1636–1644
7. Wan S, Wu J, Zhang Z et al (2009) Activation tagging, an efficient tool for functional analysis of the rice genome. Plant Mol Biol 69:69–80
8. Bouche N, Bouchez D (2001) *Arabidopsis* gene knockout: phenotypes wanted. Curr Opin Plant Biol 4:111–117
9. Koncz C, Nemeth K, Redei GP et al (1992) T-DNA insertional mutagenesis in *Arabidopsis.* Plant Mol Biol 20:963–976
10. Weigel D, Ahn JH, Blazquez MA et al (2000) Activation tagging in *Arabidopsis.* Plant Physiol 122:1003–1013
11. Alvarado MC, Zsigmond LM, Kovacs I et al (2004) Gene trapping with firefly luciferase in *Arabidopsis.* Tagging of stress-responsive genes. Plant Physiol 134:18–27

12. Koo J, Kim Y, Kim J et al (2007) A GUS/luciferase fusion reporter for plant gene trapping and for assay of promoter activity with luciferin-dependent control of the reporter protein stability. Plant Cell Physiol 48:1121–1131
13. Meissner R, Chague V, Zhu Q et al (2000) Technical advance: a high throughput system for transposon tagging and promoter trapping in tomato. Plant J 22:265–274
14. Yamamoto YY, Tsuhara Y, Gohda K et al (2003) Gene trapping of the *Arabidopsis* genome with a firefly luciferase reporter. Plant J 35:273–283
15. Alonso JM, Stepanova AN, Leisse TJ et al (2003) Genome-wide insertional mutagenesis of *Arabidopsis thaliana*. Science 301:653–657
16. Rios G, Lossow A, Hertel B et al (2002) Rapid identification of *Arabidopsis* insertion mutants by non-radioactive detection of T-DNA tagged genes. Plant J 32:243–253
17. Rosso MG, Li Y, Strizhov N, Reiss B, Dekker K, Weisshaar B (2003). An Arabidopsis thaliana T-DNA mutagenized population (GABI-Kat) for flanking sequence tag-based reverse genetics. Plant Mol Biol 53:247–259
18. Samson F, Brunaud V, Balzergue S et al (2002) FLAGdb/FST: a database of mapped flanking insertion sites (FSTs) of *Arabidopsis thaliana* T-DNA transformants. Nucleic Acids Res 30:94–97
19. Szabados L, Kovacs I, Oberschall A et al (2002) Distribution of 1000 sequenced T-DNA tags in the *Arabidopsis* genome. Plant J 32:233–242
20. Ichikawa T, Nakazawa M, Kawashima M et al (2006) The FOX hunting system: an alternative gain-of-function gene hunting technique. Plant J 48:974–985
21. Kondou Y, Higuchi M, Takahashi S et al (2009) Systematic approaches to using the FOX hunting system to identify useful rice genes. Plant J 57:883–894
22. LeClere S, Bartel B (2001) A library of *Arabidopsis* 35S-cDNA lines for identifying novel mutants. Plant Mol Biol 46:695–703
23. Papdi C, Abraham E, Joseph MP et al (2008) Functional identification of *Arabidopsis* stress regulatory genes using the controlled cDNA overexpression system. Plant Physiol 147:528–542
24. Papdi C, Joseph MP, Pérez-Salamó I, et al (2009). Genetic technologies for the identification of *Arabidopsis* genes controlling environmental stress responses. Funct Plant Biol 36:696–720
25. Kuhn JM, Boisson-Dernier A, Dizon MB et al (2006) The protein phosphatase AtPP2CA negatively regulates abscisic acid signal transduction in *Arabidopsis*, and effects of abh1 on AtPP2CA mRNA. Plant Physiol 140:127–139
26. Banno H, Ikeda Y, Niu QW, Chua NH (2001). Overexpression of Arabidopsis ESR1 induces initiation of shoot regeneration. Plant Cell 13: 2609–2618
27. Yokotani N, Ichikawa T, Kondou Y et al (2009) Tolerance to various environmental stresses conferred by the salt-responsive rice gene ONAC063 in transgenic *Arabidopsis*. Planta 229:1065–1075
28. Du J, Huang YP, Xi J et al (2008) Functional gene-mining for salt-tolerance genes with the power of *Arabidopsis*. Plant J 56:653–664
29. Szabados L, Salamó IP, Papdi C et al (2009) Functional gene mining in *Arabidopsis* and related species. Presented at PlantGem Congress, Lisbon, Portugal
30. Salamó I, Szabados L (2011) Identification of novel regulatory factors of plant stress responses using new genetic approaches. Presented at international conference on plant gene discovery technologies, Vienna
31. Sambrook J, MacCallum P, Russel D (2001) Molecular cloning: a laboratory manual. Cold Spring Harbor's Laboratory Press, New York
32. Koncz C, Martini N, Szabados L, et al (1994) Specialized vectors for gene tagging and expression studies. In: Gelvin SB (ed) Plant molecular biology manual. pp 1–22
33. Wise AA, Liu Z, Binns AN (2006) Three methods for the introduction of foreign DNA into *Agrobacterium*. Methods Mol Biol 343:43–53
34. Murashige T, Skoog F (1962) A revised medium for rapid growth and bio assays with tobacco tissue culture. Physiol Plant 15:473–497
35. Wang K (2006) *Agrobacterium* protocols. Humana, Totowa
36. Bechtold N, Pelletier G (1998) In planta *Agrobacterium*-mediated transformation of adult *Arabidopsis thaliana* plants by vacuum infiltration. Methods Mol Biol 82:259–266
37. Clough SJ, Bent AF (1998) Floral dip: a simplified method for *Agrobacterium*-mediated transformation of *Arabidopsis thaliana*. Plant J 16:735–743
38. Katavic V, Haughn GW, Reed D et al (1994) In planta transformation of *Arabidopsis thaliana*. Mol Gen Genet 245:363–370
39. Alonso JM, Ecker JR (2006) Moving forward in reverse: genetic technologies to enable genome-wide phenomic screens in *Arabidopsis*. Nat Rev Genet 7:524–536
40. Koiwa H, Bressan RA, Hasegawa PM (2006) Identification of plant stress-responsive determinants in *Arabidopsis* by large-scale forward genetic screens. J Exp Bot 57:1119–1128
41. Papdi C, Leung J, Joseph MP et al (2010) Genetic screens to identify plant stress genes. Methods Mol Biol 639:121–139

Chapter 20

Transcriptome Analysis of Membrane Transporters in Response to Salinity Stress

Prasad Senadheera and Frans J.M. Maathuis

Abstract

Exposure to high ambient levels of NaCl affects plant water relations and creates ionic stress. To a large extent, responses to such stress depend on the action of membrane transporters, particularly those that move cations such as Na^+ and K^+. A genomics approach can greatly help with the identification of important membrane transporter genes. This can be done by comparing transcriptomes of salinized and non-salinized plants, by comparing tolerant and non-tolerant species, or by using intraspecies variation. This chapter describes a protocol using oligo-microarrays to compare salinity treated (50 mM NaCl) and non-treated rice roots, presenting protocols for growth, RNA isolation, cDNA synthesis and labeling, and a summary of data collection, analysis, and interpretation. Although focused on rice root tissue, the described procedures can be applied to many different treatments, tissues, and plant species.

Key words: Microarray, Transcriptomics, Membrane transporter, Salinity

1. Introduction

Soil salinity is a global issue that challenges the food security of future generations and is likely to be aggravated by global warming (1). Virtually 40% of irrigated land worldwide is salt affected and about ten million ha of arable land is lost annually to salinization (2). Soil salinization is the result of an excessive presence of mineral nutrients such as Na^+, Mg^{2+}, Ca^{2+}, Cl^-, SO_4^{2-} and typically, saline soils show electrical conductivity >4 dS/m, with the most prevalent ions being Na^+, Cl^-, and SO_4^{2-} (3).

Most crop plants are glycophytes so are severely affected by salt stress (4). Yet crops differ in tolerance. For example, cereals include

Sergey Shabala and Tracey Ann Cuin (eds.), *Plant Salt Tolerance: Methods and Protocols*, Methods in Molecular Biology, vol. 913, DOI 10.1007/978-1-61779-986-0_20, © Springer Science+Business Media, LLC 2012

the salt sensitive rice (*Oryza sativa*) and the far more tolerant barley (*Hordeum vulgare*), whereas wheat (*Triticum aestivum*) is moderately tolerant (reviewed in ref. (4)). Initially, exposure to salinity causes osmotic stress, with other deleterious effects such as ion toxicity due to accumulation of Na^+ ions, occurring later. In combination, water, ion, and metabolic homeostasis are disturbed by salinity, leading to a significant yield loss.

Salinity tolerance in plants is largely dependent on membrane transporters. These control salt uptake at the root level, regulation of influx into cells, long distance transport, and compartmentation at the cellular and tissue level (5). Regulation of the movement of Na^+ and K^+ ions across the plasma membrane and tonoplast not only determines the cytosolic concentrations of sodium and chloride ions but also helps maintain a physiologically conducive cytosolic K^+/Na^+ ratio (6). Hence, manipulation of genes and proteins encoding the membrane transporter proteins that transport Na^+ and K^+ has enormous potential to improve salt tolerance in crop plants (reviewed in ref. (7)).

Hundreds of membrane transporters have been identified that have been shown or are believed to be important in relation to salt uptake and transport in plants. For example, Na^+ uptake at the soil-root boundary is thought to occur predominantly via non-selective cation channels such as CNGCs (cyclic nucleotide gated channels) and carriers from the HKT (high affinity K^+ transporter) family (8–10). Na^+ efflux into the vacuole and apoplast occurs via antiport systems such as NHX1 (Na:H exchanger) at the tonoplast and SOS1 at the plasma membrane (6, 11). SOS1 (Salt-Overly-Sensitive) may also mediate xylem loading of Na^+ along with other antiporters such as CHXs (cation:H exchangers) (12, 13). K^+ uptake mainly occurs through inward rectifying K^+ selective channels such as AKT1 and carriers from the HAK/KUP (high affinity K^+ uptake) family (14, 15).

To critically evaluate the role of this multitude of proteins in salt tolerance is a mammoth task. However, this could be greatly facilitated by comparative genomics such as between salt-treated and non-treated plants, between species that differ in tolerance or within species using cultivars that vary in their tolerance. DNA microarrays provide a high throughput technique to analyze and compare changes in gene expression of membrane transporters and we will describe protocols used for rice, based on oligo-arrays. Genome-wide transcript profiling in rice shoot (16) and root (17, 18) at the seedling stage as well as in the panicle initiation stages (19), using rice varieties with contrasting tolerance to salt revealed several differentially regulated transporter genes. These may provide targets for genetic engineering and molecular breeding to improve rice salt tolerance. Nonetheless, the described methodology is also applicable to other species.

2. Materials

2.1. Plant Growth (See Note 1)

1. Germination plates with perlite or acid washed sand for germination.
2. Plastic tubs of 5–8 L capacity with a lid with perforations of 2 cm diameter at 5 × 5 cm spacing for hydroponic cultivation of rice seedlings.
3. Sponges to fix rice plants in the hydroponic tub.
4. Deionized water.
5. Plant growth medium (prepared as stocks in deionized water at the given concentration).
   - Potassium nitrate; 1.25 M KNO_3.
   - Calcium nitrate; 0.5 M $Ca(NO_3)_2{\cdot}4H_2O$.
   - Magnesium sulphate; 0.5 M $MgSO_4{\cdot}7H_2O$.
   - Ion salt of ethylenediaminetetra acetic acid; 42.5 mM FeNaEDTA.
   - Potassium di-hydrogen phosphate 0.625 M KH_2PO_4.
   - Micronutrient solution containing 0.16 mM $CuSO_4$, 0.38 mM $ZnSO_4$, 1.8 mM $MnSO_4$, 45 mM H_3BO_3, 0.015 mM $(NH_4)6Mo_7{\cdot}O_{24}$, and 0.01 mM $CoCl_2$.
6. $Na_2O_7SiO_3$ solution (Sigma, USA).
7. Sodium chloride (Promega, USA).
8. Concentrated H_2SO_4.
9. pH meter.
10. Aerator.
11. Conductivity meter.

2.2. Tissue Collection (See Note 2)

1. Liquid N.
2. Deionized water.
3. Aluminum foil.
4. Stable permanent marker pen.

2.3. RNA Isolation

1. Trizol® reagent (Invitrogen).
2. 3 M Na acetate (pH 5.3), RNAase free: prepare the 3 M Na acetate in diethylpyrocarbonate (DEPC) treated H_2O in an RNAase-free container; adjust the pH with acetic acid and autoclave before using.
3. Isopropanol.
4. Chloroform.

5. RNase-free water—autoclaved.
6. Ethyl alcohol.
7. Pipettes and filtered pipette tips (1,000, 200, 20, and 2 μL).
8. Mortar and pestles—autoclaved.
9. Microtubes (6/sample)—autoclaved.
10. Gloves.
11. Refrigerated microcentrifuge.
12. Water bath.

2.4. cDNA Production and Labeling (See Note 6)

1. RNA.
2. dNTP (10 mM dATP, dTTP, dGTP, and 2 mM dCTP).
3. Cy5 or Cy3 dUTP (1 mM; Amersham Biosciences, UK).
4. Random 15-mer primer (0.5 μg/μL; Operon).
5. 5× First strand buffer.
6. 0.1 M DTT (dithiothreitol).
7. 0.5 M EDTA.
8. 1 M NaOH.
9. 1 M Tris–HCl (pH 8).
10. TE (Tris EDTA; 10 mM Tris and 1 mM EDTA, pH 8) buffer.
11. Reverse transcriptase (Superscript II, Life Technologies, UK).
12. Water bath or heating block.

2.5. Microarray Hybridization

1. Microarray slide(s) (Fig. 1).
2. Microscope slide holders (Fig. 2; these can accommodate up to five microarray slides, otherwise ordinary 50 mL tubes can be used for each array slide).
3. 50 mL tubes.
4. Hybridization chamber (Fig. 3).
5. Microarray gene frame (Fig. 4; ABgene, Surrey, UK) (see Note 8).
6. Coverslips (e.g., Hybri-Slip, Sigma, UK).
7. Liquid Blocking Reagent (Amersham; Cat # RPN3601).
8. 2% SDS (sodium dodecyl sulfate).
9. 20× SSC buffer (saline sodium citrate).
10. DEPC treated H_2O.
11. 2× SSC + 0.5% SDS.
12. 0.5× SSC.

Fig. 1. Microarray. A 45,000 element microarray slide is shown (same dimensions as an ordinary microscope slide). *Inset* shows magnification of separate grids, each containing 200 × 200 probes.

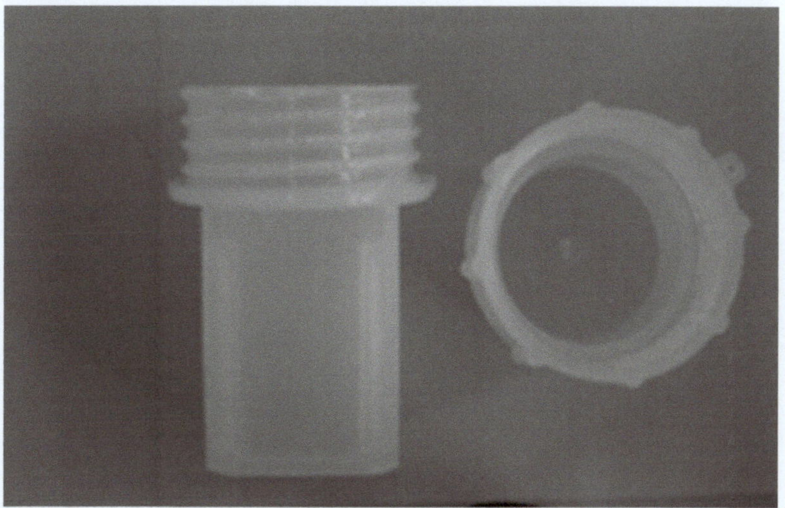

Fig. 2. Microscope slide container. Up to five individual slides slot into these containers making them ideal for wash steps of multiple slides.

13. 0.05× SSC.
14. Water bath.
15. Heating block.

2.6. Microarray Data Retrieval

1. Microarrays.
2. Scanner equipment.
3. Software.

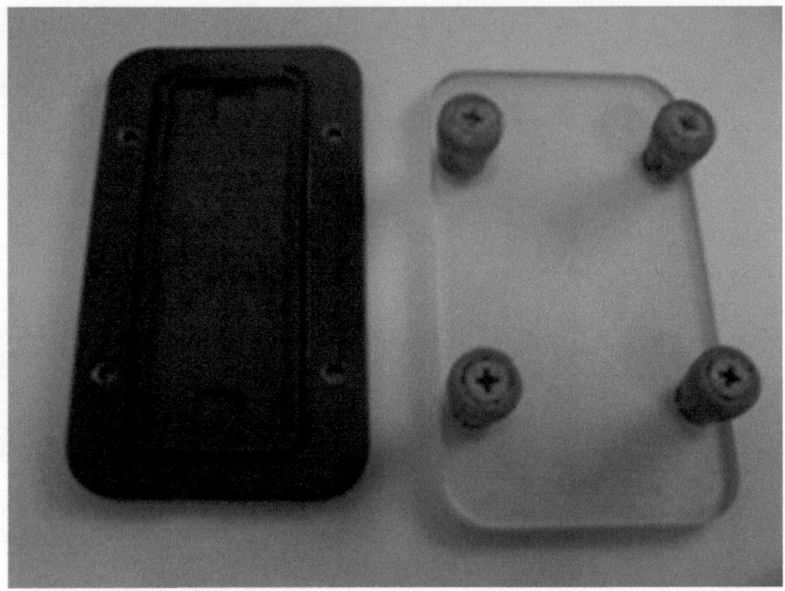

Fig. 3. Hybridization chamber. Aluminum hybridization chamber consisting of a base that holds the array slide, a seal and a perspex lid that is clamped on top and fastened using the four thumb screws.

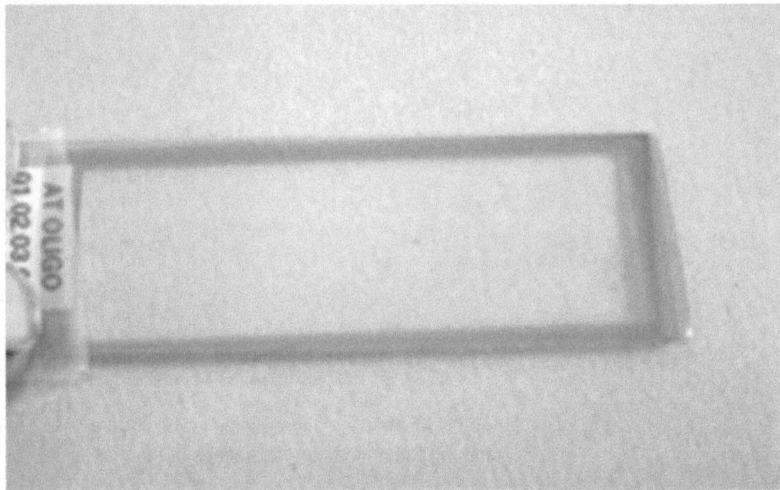

Fig. 4. Gene frame. Adhesive raised edges can be applied to increase the volume of the hybridization mix that is added. In the shown example, the volume would be 200 μL.

3. Methods

3.1. Plant Growth (See Note 1)

1. Germinate rice seeds, previously soaked in deionized water over night in moist sand at 30°C.
2. Prepare the hydroponic medium by adding 1 mL/L of the plant nutrient solutions listed under 5, and 0.1 mL/L of $Na_2O_7SiO_3$.

3. Adjust pH to 5.5–6.5 using concentrated H_2SO_4.
4. Transplant the seedlings 10 days after sowing (DAS) into the hydroponic medium by fixing one seedling in each hole using a sponge band wrapped around the base of the culm. Growth conditions are set at: mean air temperature between 25 and 32°C, relative humidity around 80% and light averaged 500–1,000 μmol/m^2/s at noon.
5. Seedlings are cultivated in hydroponic solution for 5 days before exposing them to salt stress by adding NaCl (e.g., 50 mM final concentration). Hydroponic solution is continuously aerated and renewed every second day for 12 days.
6. Electrical conductivity of the nutrient solutions is to be maintained approximately at ($[Na^+]/10$) dS/m (e.g., 5 dS/m for 50 mM NaCl) in the treatment and <1 dS/m in the control.

3.2. Tissue Collection (See Note 2)

1. Harvest plant tissues at the appropriate growth stage (e.g., 25 DAS) for RNA extraction. Dissect the necessary tissue samples (using scissors or razor blades) and wash in cool de-ionized water for 1 min. After excess water is removed by gentle blotting, freeze the tissues in liquid N. Roots of about five plants of each variety per replication are pooled for RNA extraction.

3.3. RNA Isolation

1. Isolate tissue RNA from control and salt treated plants using Trizol® reagent (see Note 3). All reagents must be of molecular biology grade. RNA is very easily contaminated and degraded by ambient RNAses, so all reagents and equipment should be devoid of RNase. Maintaining cleanliness is of the utmost importance. Use filtered pipette tips and gloves throughout the procedure.
   (a) Cool mortar, pestles, and microtubes in liquid N.
   (b) Grind the root sample (100–200 mg) to fine powder in liquid N and transfer into a microtube.
   (c) Add 1 mL Trizol immediately and vortex (if used from frozen, allow to thaw slightly before this step).
   (d) Incubate the macerated tissue for 10 min at room temperature (20°C).
   (e) Centrifuge for 10 min at 12,000 × *g* at 4°C and decant the liquid into a fresh tube.
   (f) Add 200 μL chloroform and shake vigorously for 15 s, then incubate for 3 min at room temperature.
   (g) Centrifuge for 15 min at 12,000 × *g* in 4°C and transfer 500 μL of the aqueous supernatant to a fresh tube.
   (h) Add 250 μL of both 3 M Na acetate solution and isopropanol. Mix by inverting and incubate for 10 min at room temperature (20°C).

(i) Centrifuge for 10 min at 15,000 × *g* at 4°C. A small pellet should be visible.

(j) Discard supernatant carefully and add 1 mL 75% ethanol prepared using RNase-free water.

(k) Centrifuge for 5 min at 7,500 × *g* at 40°C and discard the supernatant.

(l) Repeat the above step and remove the ethanol carefully.

(m) Leave the tubes open in a clean place to allow the ethanol to evaporate. Alternatively, drying can be done by inverting the opened tubes on tissue paper. Prolonged drying can make re-dissolving the RNA pellet difficult.

(n) Add 20–100 μL RNase-free water to dissolve the RNA. Resuspend the pellet by pipetting and incubating on ice for 15 min and then store the RNA at −80°C.

2. RNA is pooled from three to four independent sets of five plants for each experiment. Repeat this procedure three times or more, according to the requirement for each cultivar and treatment.
3. Check concentration and quality of RNA using spectrophotometry and/or gel electrophoresis. If necessary, RNA can be purified using RNAeasy spin columns (Qiagen, London, UK) according to the manufacturers' protocol or precipitation-based procedures (see Notes 4 and 5). An absolute minimum of 1 μg of RNA per sample is required for the subsequent steps described below.

3.4. cDNA Production and Labeling (See Note 6)

1. Transfer 50–100 μg total RNA to a microtube.
2. Add 3 μL dNTP mixture.
3. Add 1.5 μL of Cy5 or Cy3 dUTP to each sample (see Note 7).
4. Add 2 μL random 15-mer primer.
5. Heat the mixture for 5 min at 65°C.
6. Add 17 μL 5× first strand buffer.
7. Add 8 μL 0.1 M DTT.
8. Add 1 μL RNAse inhibitor.
9. Add 1 μL reverse transcriptase.
10. Incubate at 42°C for 2 h.
11. Add μL 0.5 M EDTA and 5 μL 1 M NaOH.
12. Heat the mixture for 10 min at 65°C.
13. Add 25 μL 1 M Tris–HCl and 0.1 mL TE.

Labeled target cDNA is cleaned, for example on QIAquick spin columns (Qiagen, UK), to remove non-incorporated nucleotides. At the end of the cleanup procedure, the presence of faint reddish and blue coloration on the cellulose filter usually indicates successful probe labeling. If necessary, label incorporation can be quantified using a spectrophotometer to measure the dye incorporation. Incorporation of Cy5 is measured at 650 nm and incorporation of Cy3 at 550 nm. The absorption values can be used to calculate the amount of incorporated dye in your sample at many websites including http://www.pangloss.com/seidel/Protocols/percent_inc.html and http://www.prontosystems.com/technical_support/calculator/#ResultsView. A "percent incorporation" value of 1 or more is sufficient to obtain a good signal.

3.5. Microarray Hybridization

Slide preparation: Some oligo-based slides require probe rehydration and/or immobilization. This should be carried out according to the manufacturers' instructions and will not be discussed here.

Hybridization setup: After elution of the cDNA from the cleanup column, it is vacuum dried and re-dissolved in hybridization buffer, the composition of which will depend on array type and manufacturer but typically contains SSC and SDS. Make sure all solutions and materials are dust free.

1. Add 20 μL 20× SSC.
2. Add 12 μL Liquid Block.
3. Add 8 μL 2% SDS.
4. Add both labeled targets.
5. Add dH_2O to reach a final volume of 200 μL.
6. Mix ingredients in tube.
7. Denature labeled target by incubating tube at 95°C for 2 min.
8. Transfer tube immediately to ice.
9. Clean hybridization cassette with distilled water and dry.
10. Place microarray slide in cassette (DNA side up).
11. Carefully apply the labeled target without touching the array.
12. Cover the hybridization chamber with cover slip (22 × 60 mm), avoiding trapped air.
13. Apply downward pressure on the hybridization cassette lid and tighten.
14. Incubate slide at 55°C for 8–12 h.
15. After hybridization, remove cassette lid.
16. Warm to 55°C enough 2× SSC + 0.5% SDS solution to fill microscope slide holder (~50 mL).
17. Remove the microarray slide from the cassette and place it in the microscope slide holder filled with 2× SSC + 0.5% SDS.

18. Wash slides in 2× SSC + 0.5% SDS at 55°C for 5 min, and subsequently at room temp in 0.5× SSC and 0.05× SSC for 3–5 min (make sure the slide is completely immersed in buffer).
19. Spin dry the slides by placing them in a holder (e.g., 50 mL tube) and spinning ~1,000 rpm for 2 min in a centrifuge.

3.6. Microarray Data Retrieval

1. After hybridization and washing of array slides, the fluorescence signals from Cy3 and Cy5 need to be read and quantified. This is best done immediately due to signal deterioration. Alternatively, slides may be stored temporarily in zero light conditions. Many companies manufacture appropriate equipment for array scanning (e.g., Axon Instruments, Braintree, UK, or Packard BioChip Technologies, Pangbourne, UK). It is good practice to scan arrays using different gain settings in order to increase the dynamic range of the detectable signals. Most scanners will be associated with software packages that allow global signal normalization (e.g., ratio between Cy3 and Cy5 signals) and image output in various formats.
2. The images produced by array scanning can be further analyzed with a variety of software packages. We have successfully used Scanalyze software (freely available from B Eisen, Stanford, USA) and GSI Lumonics QuantArray 2.1 (Packard, see above). These packages will ensure that fluorescence intensities are measured and averaged over a defined spot area. Background signals are determined and averaged over a defined area in the immediate spot environment and subtracted from the spot signals. Spots can then be flagged for P ("present") or A ("absent"), according to signal/background ratios, for inclusion in downstream analyses. The signal/background ratio cut-off for P/A classification is determined separately for each array. In addition to data exclusion based on signal/background ratios, all images/spots can be visually inspected and flagged B ("bad") in the event of aberrations in fluorescence or background signals.
3. Data can also be analyzed using online facilities. For example, "SNOMAD" software (available at http://pevsnerlab.kennedykrieger.org/snomadinput.html) is freely accessible for signal correction (to correct for differences in fluorescence properties of the labels) background subtraction, global mean normalization, calculation of average signals and standard deviations, etc. as previously described (e.g., (20)). Transcripts that are included for further analysis can then be ranked according to changes between control and treatment ("fold change ranking").

3.7. Data Verification

Although microarray data are now well established and form a fantastic mining source, many journals still require array data to be compared to other methods. Thus, RT-PCR analyses are often carried out on a subset of array data that typically covers the entire dynamic range. Preferably, this would be quantitative (Q) PCR on 5–10 transcripts starting with the RNA of various samples and treatments. Typically, 1 μg of total RNA, isolated from the treated and control samples, is used to synthesize first strand cDNA as described above, and then subjected to Q-PCR. A good comparison between fold changes derived with arrays and Q-PCR gives confidence that the array data represent genuine, treatment-induced changes in transcript level.

3.8. Further Data Analyses

1. When the focus is on membrane transporter genes, it may often be necessary to check whether transcripts of "hypothetical" and "unknown" genes encode integral membrane proteins. Prediction of transmembrane domains can be carried out using bulk sequence data retrieval from TIGR at: http://www.tigr.org/tdb/e2k1/osa1/batch_download.shtml and subsequent sequence analysis according to TMHMM at http://www.cbs.dtu.dk/services/TMHMM/.
2. Cluster analysis can potentially identify groups of transcripts that show similarity in expression profile. Again, many options are open to the experimenter, using commercial software, free downloadable packages, or online analysis. GeneCluster and TreeView software (B Eisen, Stanford, USA) is very suitable for both K-means and hierarchical clustering as are online platforms such as SMD (http://smd.stanford.edu/resources/restech.shtml) and Goulphar (http://transcriptome.ens.fr/sgdb/tools/).
3. Putative *cis*-acting elements can help identify overrepresented DNA binding domains and co-regulation of sets of transcripts. Libraries of "strictly upstream regions" (SURs) are now available for all Arabidopsis and rice genes (e.g., Osiris: http://www.bioinformatics2.wsu.edu/cgi-bin/Osiris/cgi/home.pl) that show common *cis* motifs. Alternatively, the promoter sequences can be searched for previously unidentified motifs to discover new putative regulatory motifs. Such analyses can be carried out for instance using TAIR tools (http://arabidopsis.org/tools/bulk/motiffinder/index.jsp) or Bioprospector (http://ai.stanford.edu/~xsliu/BioProspector/).

4. Notes

1. The protocol described here is for growth of rice. Naturally, growth and treatment protocols will depend on plant species and the desired experimental conditions.
2. Because transcript levels can change within minutes, it is imperative that tissue collection is carried out as quickly as possible while avoiding stress. If tissues are collected from more than

one set of plants, other factors should be kept constant, especially the time of sampling.

3. The Trizol method is reproduced here only as an example, there are many appropriate protocols for RNA isolation that are suitable for microarray studies.
4. RNA quality is critical for good labeling and hybridization, so gel-based control of RNA integrity may be advisable.
5. If tissues contain large quantities of starch or other polysaccharides, the protocol requires modification to remove polysaccharides that would interfere with target production. Polysaccharides can be removed by RNA precipitation: dissolve RNA in ~0.5 mL DEPC treated water or TE buffer and at −20°C for 2 h. After defrosting, remove precipitate by centrifugation. If necessary this step can be repeated.
6. The subsequent steps (described under "D" and "E") can also be outsourced to various microarray facilities that carry out cDNA synthesis, labeling, and hybridization. Examples of these facilities can be found at the Arizona, USA, microarray facility (http://ag.arizona.edu/microarray/), the NASC Affymetrix facility in the UK (http://affymetrix.arabidopsis.info/), or RMOS (for rice) in Japan (http://cdna01.dna.affrc.go.jp/RMOS/).
7. It is good practice to alternate the labeling between treatment and control samples.
8. "Gene frame" or similar accessories can be used to apply a larger volume of hybridization mix. However, hybridizations can also be carried out without this system by simply applying a small volume (~20 μL) of mix to the array and covering with a covering slip.

References

1. Brinkman R, Sombroek G (1996) The effects of global change on soil conditions in relation to plant growth and food production. FAO corporate document repository. http://www.fao.org/docrep/w5183e/w5183e05.htm
2. Pessarakli M, Szabolcs I (1999) Soil salinity and sodicity as particular plant/crop stress factor. In: Pessarakli M (ed) Handbook of plant and crop stress. Marcel Dekker, New York, pp 1–16
3. Flowers TJ (2004) Improving crop salt tolerance. J Exp Bot 55:307–319
4. Munns R, Tester M (2008) Mechanisms of salt tolerance. Annu Rev Plant Biol 59:651–681
5. Blumwald E, Aharon GS, Apse MP (2000) Sodium transport in plant cells. Biochim Biophys Acta 1465:140–151
6. Zhu JK (2002) Salt and drought stress signal transduction in plants. Annu Rev Plant Biol 53:247–273
7. Maathuis FJM (2007) Monovalent cation transporters; establishing a link between bioinformatics and physiology. Plant Soil 301:1–15
8. Gobert A, Park G, Amtmann A, Sanders D, Maathuis FJM (2006) *Arabidopsis thaliana* cyclic nucleotide gated channel 3 forms a non-selective ion transporter involved in germination and cation transport. J Exp Bot 57:791–800
9. Laurie S, Feeney KA, Maathuis FJM, Heard PJ, Brown SJ, Leigh RA (2002) A role for HKT1 in sodium uptake by wheat roots. Plant J 32:139–149
10. Golldack D, Quigley F, Michalowski CB, Kamasani UR, Bohnert HJ (2003) Salinity

stress-tolerant and -sensitive rice (*Oryza sativa* L.) regulate AKT1-type potassium channel transcripts differently. Plant Mol Biol 51:71–81

11. Apse MP, Sottosanto JB, Blumwald E (2003) Vacuolar cation/H^+ exchange, ion homeostasis, and leaf development are altered in a T-DNA insertional mutant of AtNHX1, the *Arabidopsis* vacuolar Na^+/H^+antiporter. Plant J 36(2):229–235
12. Chinnusamy V, Schumaker K, Zhu JK (2004) Molecular genetic perspectives on cross-talk and specificity in abiotic stress signaling in plants. J Exp Bot 55:395
13. Morsomme P, Boutry M (2000) The plant plasma membrane H^+-ATPase: structure, function and regulation. Biochim Biophys Acta 1465:1–16
14. Pilot G, Pratelli R, Gaymard F, Meyer Y, Sentenac H (2003) Five-group distribution of the shaker-like K^+ channel family in higher plants. J Mol Evol 56:418–434
15. Szyroki A, Ivashikina N, Dietrich P, Roelfsema MRG, Ache P, Reintanz B, Deeken R, Godde M, Felle H, Steinmeyer R, Palme K, Hedrich R (2001) KAT1 is not essential for stomatal opening. Proc Natl Acad Sci USA 98: 2917–2921
16. Walia H, Wilson C, Condamine P, Liu X, Ismail AM, Zeng L, Wanamaker SI, Mandal J, Xu J, Cui X, Close TJ (2005) Comparative transcriptional profiling of two contrasting rice genotypes under salinity stress during the vegetative growth stage. Plant Physiol 139: 822–835
17. Senadheera P, Singh RK, Maathuis FJM (2009) Differentially expressed membrane transporters in rice roots may contribute to cultivar dependent salt tolerance. J Exp Bot 60(9):2553–2563
18. Cotsaftis O, Plett D, Johnson AA, Walia H, Wilson C, Ismail AM, Close TJ, Tester M, Baumann U (2011) Root-specific transcript profiling of contrasting rice genotypes in response to salinity stress. Mol Plant 4(1):25–41
19. Walia H, Wilson C, Zeng L, Ismail AM, Condamine P, Close TJ (2007) Genome-wide transcriptional analysis of salinity stressed japonica and indica rice genotypes during panicle initiation stage. Plant Mol Biol 63: 609–623
20. Maathuis FJM (2006) cGMP modulates gene transcription and cation transport in *Arabidopsis* roots. Plant J 45:700–711

Chapter 21

Marker-Assisted Selection in Plant Breeding for Salinity Tolerance

M. Ashraf, N.A. Akram, Mehboob-ur-Rahman, and M.R. Foolad

Abstract

Marker-assisted selection (MAS) is the process of using morphological, biochemical, or DNA markers as indirect selection criteria for selecting agriculturally important traits in crop breeding. This process is used to improve the effectiveness or efficiency of selection for the traits of interest in breeding programs. The significance of MAS as a tool for crop improvement has been extensively investigated in different crop species and for different traits. The use of MAS for manipulating simple/qualitative traits is straightforward and has been well reported. However, MAS for the improvement of complex/polygenic traits, including plant tolerance/resistance to abiotic stresses, is more complicated, although its usefulness has been recognized. With the recent advances in marker technology, including high-throughput genotyping of plants, together with the development of nested association mapping populations, it is expected that the utility of MAS for breeding for stress tolerance traits will increase. In this chapter, we describe the basic procedure for using MAS in crop breeding for salt tolerance.

1. Introduction

Of the various strategies to minimize the deleterious effects of high soil or water salinity on plant growth and crop production, the genetic approaches of developing plants with improved salt tolerance are being exploited world wide. These approaches entail development of new crop varieties that are capable of tolerating high levels of salts in the soil or in the irrigation water and produce economically acceptable yields. Compared to the technological approach of implementing large engineering schemes for irrigation, drainage, and reclamation with high quality water, production of crop plants with improved salt tolerance (ST) is considered a more convincing approach to utilizing marginal lands afflicted with high salinity. However, development of plants with enhanced ST is a challenging

Sergey Shabala and Tracey Ann Cuin (eds.), *Plant Salt Tolerance: Methods and Protocols*, Methods in Molecular Biology, vol. 913, DOI 10.1007/978-1-61779-986-0_21, © Springer Science+Business Media, LLC 2012

endeavor, for multiple reasons. Plant response to salt stress is multifaceted and appears to be controlled by more than one gene and is highly influenced by environmental variation (1, 2). The determination of salinity tolerance is a complex process. Direct selection in natural field conditions is complicated because uncontrollable climatic agents negatively influence the repeatability and precision of such trials (3). To the extent of our knowledge, there is not a single authentic field or greenhouse screening procedure that could be used repeatedly generation after generation. Furthermore, for phenotypic selection for ST to be effective, specialized personnel and facilities are required. Selection and breeding for ST is also complicated, because the trait is developmentally regulated and varies with plant ontogeny. Tolerance at each stage of plant development is inadequately associated with tolerance at other developmental stages (4–7). Specific stages throughout the ontogeny of plants such as emergence, seedling survival, germination, and vegetative and reproductive growth should be appraised individually for the estimation of the overall ST and identification of genetic variants. Each developmental stage may require a different screening procedure. Therefore, sequential or simultaneous screening may become impossible or impractical.

A promising approach for crop improvement for ST is the identification of marker(s) associated with genes or QTLs (quantitative trait loci) contributing to ST and to use them as indirect selection criteria (3, 5) for selecting for ST in a process known as marker-assisted selection (MAS). During the last two decades, molecular markers technology has allowed the identification and genetic characterization of QTLs with significant effects on plant ST during different developmental stages (5, 8–12). Theoretically, the use of DNA markers associated with genes or QTLs associated with ST at the various developmental stages can facilitate development of plants with improved ST throughout the ontogeny of the plant. Selection based on DNA markers genetically associated with ST can potentially overcome at least some of the limitations associated with phenotypic selection (PS) for ST. A major advantage of DNA markers is that they are "neutral" in phenotypic reactions, that is, they do not have any pleiotropic effect, nor are they influenced in their segregation and inheritance by the growing conditions of the plant. Furthermore, DNA markers can be detected at all growth stages, offering the possibility of selecting plants on the basis of convenience to the breeder, in contrast to the season-bound nature of PS.

However, MAS is not a silver-bullet approach to selecting for agriculturally important traits, including ST. When it comes to breeding for ST, MAS can be more effective than PS under certain conditions, including when (1) PS for ST is not feasible (e.g., lack of selection environment), (2) ST is not expressed clearly in the population, (3) PS for ST is ineffective due to low heritability of the

ST-related traits, (4) ST is developmentally regulated or phenotypically not obvious, (5) ST is governed by recessive or incompletely dominant genes, (6) there are too many genotype x environment interactions, and (7) transferring ST from wild to cultivated genetic backgrounds. Furthermore, with MAS it may be possible to select multiple traits that may be directly or indirectly related to ST.

Over the last two decades, numerous research programs worldwide have identified and reported molecular markers associated with QTLs/genes for ST in various plant species and during different developmental stages (4, 10, 13–18). However, there has been a limited use of markers and MAS for breeding for ST in any crop species. This has been due to numerous factors, including (1) markers being population-specific, (2) markers developed based on ST evaluation under controlled greenhouse or growth chamber conditions that may not reflect ST under field conditions, (3) markers identified based on populations derived from wide crosses (e.g., crosses with wild species) that may not be directly useful in breeding populations (i.e., lack of correspondence between QTLs identified in interspecific populations and those existing in breeding populations), (4) lack of marker validation by repeating experiments, and (5) lack of marker polymorphism in breeding populations. Thus, further research is needed to alleviate these issues and facilitate the use of markers for ST breeding in different crop species. Nonetheless, in this chapter we describe the basic procedure for identifying markers associated with ST and approaches for using them in improving plant ST using MAS.

2. Materials

2.1. Instruments

1. Isoelectric focusing apparatus: (BIO-RAD Protean® IEF Cell, Italy).
2. Polyacrylamide gel electrophoresis (BIO-RAD, Model: Mini-PROTEIN, Italy).
3. Agarose gel electrophoresis (Wealtec Corp., Model: GES, Kennesaw, GA).
4. Ultra sonicator (Hielscher, Model: UP200S, Teltow, Germany).
5. pH meter (InoLab pH/Cond 720, WTW 82362 Weilheim, Germany).
6. Vortex machine (SeouLin Bioscience 452-2 Seoulin Bldg. SongNae-Dong, KangDong-Ku, Seoul, Korea, Model: LV6).
7. Shaking-incubator (GeneMachines, San Carlos, CA).
8. Centrifuge machine (Sigma Model 3K30, Germany).

9. Spectrophotometer (Model Hitachi-U 2001, Tokyo, Japan).
10. Gel documentation apparatus (UVP GelDoc-It TS 310 Imaging System with built-in computer, Model: 97-0258-02, Englewood, the United States).
11. Vacuum pump (Beijing Beiyi Woosung Vacuum Technology Co., Ltd. Vacuum Pump, Vane manufacturers, Model: MVP144, China).
12. PCR machine (BIO-RAD DNA Engine Peltier thermal cycler, Model PTC0200, CA, USA).
13. Magnetic stand (Stratagene, San Diego, CA).
14. Kits (Invitrogen, Life Technologies, USA).

2.2. Reagents or Chemicals

All the reagents or chemicals used for MAS can be purchased from the following companies:

1. Sigma-Aldrich Chemie GmbH, Steinheim, Germany.
2. BDH Chemicals Ltd. Poole, England.
3. Merck, Darmstadt, Germany.
4. MP Biomedicals, de Kayserberg Illkirch, France.
5. BDH, AnalaR, England.
6. QIAGEN GmbH, Hilden, USA.
7. Bio Basic Inc. Kearnay, United States.
8. BIO-RAD California, USA.
9. MBI Fermentas GmbH, Opelstrasse, Germany.

2.2.1. Isolation of Genomic DNA (Doyle and Doyle, 1990 with slight modification)

1. Prepare extraction buffer A using: cetyl trimethylammonium bromide (2%), Tris–HCl (100 mM; pH 8.0), EDTA (20 mM), NaCl (1.4 M), PVP (1% (w/v)), ascorbic acid (0.1%w/v), and mercaptoethanol (10 mM).
2. Prepare extraction buffer B using: Tris–HCl (100 mM; pH 8.0), EDTA (50 mM), NaCl (100 mM), and mercaptoethanol (10 mM).
3. Prepare Tris-EDTA buffer by dissolving: Tris–HCl (10 mM; pH 8.0), EDTA (1 mM).
4. Other reagents: Sodium dodecyl sulfate (SDS) (20%; w/v).
5. Potassium acetate (5 M; store at –20°C).
6. Sodium acetate (3 M; pH 5.2).
7. Ethanol (70%).
8. Absolute isopropanol (store at –20°C).

2.2.2. For RAPDs, PCR Reaction Mixture (Williams et al. (19) with Slight Modifications) Contains The Following

Template DNA* (15 ng)	3 μL
dNTPs (2.5 mM)	4 μL
Buffer** (10×)	2.5 μL
$MgCl_2$ (25 mM)	3 μL
Primer (15 ng/μL)	2 μL
Taq polymerase (5 U/μL)	0.2 μL
Double distilled deionized H_2O	10.3 μL
Total volume	25 μL

*Concentration of genomic DNA varies from species to species

**T [(750 mM Tris–HCl (pH 8.8), 200 mM $(NH_4)_2SO_4$, 0.1% tween 20)]

2.2.3. For AFLPs, Prepare the Following Reaction Mixtures with Slight Modifications from the Procedure described by Vost et al. 1995

1 Restriction reaction mixture in a 0.2 mL tube 5X reaction buffer 5 μl Sample DNA (80ng/μl)* 1μl EcoRI/MseI combination 1μl Distilled water 18 μl Total volume 25 μl

5X reaction buffer	5 μL
Sample DNA (80ng/μl)*	1 μL
EcoRI/MseI combination	1 μL
Distilled water	18 μL
Total volume	25 μL

2. Prepare the preamplification reaction mixture (Pre-Mix) in a 0.2 mL tube. Preamplification reaction mix = 20 μl 10X PCR Buffer + Mg 1.75 μl Taq Polymerase (5U/ul) +0.25 μl DNA template (diluted 10 times)* +3 μl Total volume = 25 μl * can be varied, depending upon the concentration of the restricted genomic DNA
3. Prepare two mixtures

   Mix-I

   EcoR1 primer (E-ACC) = 5 μL

   MseI primer (M-CCA) = 45 μL

   Total volume = 50 μL

   Use Mix-I as 5 μL per reaction

   Mix-II

   Distilled water = 79 μL

   10× buffer + Mg = 20 μL

   Taq polymerase = 1 μL

   Total volume = 100 μL

   Mix-II is used at the rate of 10 μL per reaction.

4. Prepare the amplification mixture by adding the followings reagents in 0.2 mL micro-centrifuge tube:

Components	Quantity
Mix-I	5 μL
Mix-II	10 μL
Template DNA	2 μL
Distilled water	3 μL
Total volume	20 μL

5. Then centrifuge the reaction mixture briefly

2.2.4. For SSRs, the Reaction Mixture (Saiki et al. (20); Saiki (21) with Slight Modifications) Comprises The Following

Template DNA* (15 ng)	2 μL
dNTPs (2.5 mM)	6.4 μL
Buffer (10×)	2 μL
$MgCl_2$** (25 mM)	1.6 μL
Primer-F (30 ng/μL)	1 μL
Primer-R (30 ng/μL)	1 μL
Taq polymerase (5 U/μL)	0.2 μL
Double distilled deionized H_2O	5.8 μL
Total volume	20 μL

*, ** can be varied, depending upon the species

3. Methods

Major steps for marker-assisted selection for salt tolerance:

3.1. Identification of Parental Lines Contrasting in Salt Tolerance

1. Primary step in conducting any genetic/genomic or breeding research is the identification of genetic variability for the trait(s) of interest (see Note 1).
2. Then parental lines with contrasting traits are used for making cross hybridization and development of mapping populations (see Note 2).

3.2. Development of Mapping Population

1. Carry out cross hybridization between parental lines.
2. Use segregating population for genetic mapping, including early filial (e.g., F_2, F_3) or backcross populations (e.g., BC_1, BC_2), inbred backcross populations (e.g., BC_1S_1, BC_1S_2, BC_2S_1), recombinant inbred line (RIL) populations, or double haploids (DHs).

3. First common step in the development of all these populations is the generation of F_1 progeny from crosses between two contrasting parents, i.e., different in ST as well as molecular markers (see Note 3).

3.2.1. Early Filial or Backcross Population

Derive early filial population by self-crossing F_1 progeny to produce F_2 or backcross population by crossing F_1 progeny with one of the parents to produce BC_1.

3.2.2. Inbred Backcross Population (e.g., BC_1S_1, BC_1S_2, BC_2S_1)

Derive it by self-fertilizing BC_1 or BC_2 population (see Note 4).

3.2.3. Recombinant Inbred Line Populations or F_2 Derived Lines

A RIL population is developed by self-breeding an F_2 population for several generations, using a single-seed descent approach (see Note 5).

3.2.4. Double Haploids

1. The DH approach is used to develop a series of inbred lines from crosses between two parents (similar to those in a RIL population) in a rather short period of time.
2. For making DHs, a number of approaches have been pursued such as chromosome elimination, in situ pollination with irradiated pollen, pollen from a triploid, and in vitro culture of gametophytes. However, the most common method is to apply colchicine to haploid plants derived from pollen culture of an F_1 progeny (see Note 6).

3.2.5. Nested Association Mapping Populations

Cross one parent genotype (usually a reference genotype) with a set of approximately 25 genotypes representing much of the diversity available in the gene pool (e.g., see Fig. 4; Note 7).

3.3. Identification of Genetic Markers Polymorphic Between the Parental Lines

3.3.1. Protein (Isozyme) Markers (See Note 8)

1. Extract a specific enzyme from the plant tissue.
2. Different forms are separated on the basis of electrical charge, shape, and molecular size, using gel electrophoresis.

3.3.2. DNA Markers

A variety of DNA markers (hybridization-based DNA markers/ PCR-based markers) are being used these days in QTL mapping as well as MAS.

1. Restriction fragment length polymorphic DNA (RFLPs) hybridization-based DNA marker (see Notes 9 and 10)

   The main steps (Fig. 1) in running RFLP markers are as follow:

   (a) Freeze the plant sample in liquid nitrogen.

   (b) Weigh 0.3 g of plant tissue.

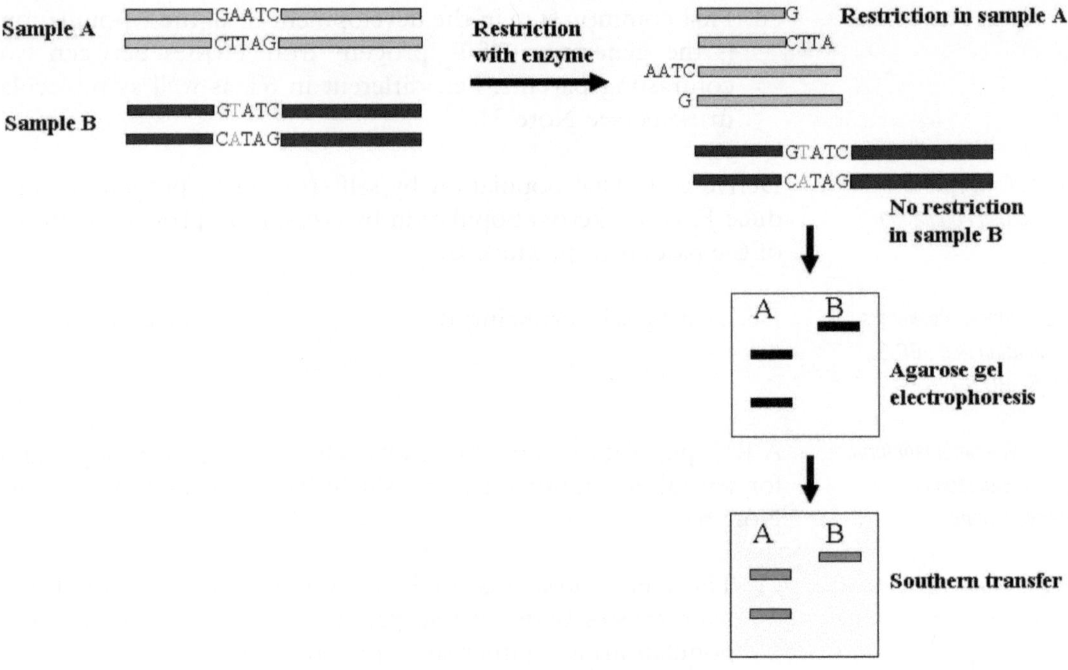

Fig. 1. Schematic illustration of restriction fragment length polymorphism (RFLP)—a first type of DNA marker assay proposed after the discovery of restriction endonucleases.

(c) Homogenize the tissue into a paste either by sonicating or grinding in liquid nitrogen.

(d) Shift the homogenized plant sample into micro-centrifuge tube (size = 1.5–3 mL).

(e) Add 0.3 mL extraction buffer A, 0.9 mL extraction buffer B, and 0.1 mL SDS.

(f) Vortex the sample.

(g) Incubate it for 10 min at 65°C.

(h) Keep the tube on ice and add 0.41 mL cold potassium acetate (CH_3CO_2K) solution.

(i) Homogenize the mixture by shaking, and place the centrifuge tube back on ice for 3 min.

(j) Centrifuge it at 13,000 × g for 15 min.

(k) Shift the supernatant (100 μL) to another micro-centrifuge tube.

(l) Add 540 μL ice cold absolute isopropanol and incubate it in ice for 20 min.

(m) Again centrifuge at 10,000 × g for 10 min.

(n) Discard the supernatant.

(o) Wash the pellet once in 500 μL ethanol (70%) and dry it.

(p) Resuspend the dry pellet in 600 μL of TE, add 60 μL sodium acetate (3 M; pH 5.2) and 360 μL ice cold absolute isopropanol and incubate on ice for 20 min (repeat this step twice).

(q) Resuspend the pellet in 50 μL DNA purification buffer and carry out agarose gel electrophoresis.

(r) To confirm the presence of DNA, use diphenylamine (DPA) as an indicator.

(s) The concentration of DNA can be analyzed either spectrophotometrically or by agarose gel electrophoresis.

(t) Measure the intensity of absorbance of the DNA solution at wavelengths 260 and 280 nm which is an identification of DNA.

(u) Cleave DNA by specific restriction enzyme (endonucleases).

(v) Fractionate DNA fragments on a gel matrix by electrophoresis.

(w) Transfer DNA fragments to a nylon membrane by Southern blotting.

(x) DNA hybridization (using known or random probe DNA that is radiolabeled).

(y) Autoradiography

General polymerase chain reaction (PCR) (see Note 11). The three major steps in conducting PCR and amplification of specific DNA sequences include:

(i) DNA denaturation (92–94°C).

(ii) Primer annealing (50–60°C).

(iii) Primer extension (usually 72°C).

Primers are usually 18–22 bases in length.

2. Randomly amplified polymorphic DNAs (RAPDs) PCR-based marker (see Note 12) (Fig. 2).

(a) Isolate genomic DNA (as in Section 3.3.2 (1)).

(b) Initiate DNA thermocycler for heating at 94°C for 5 min, 40 cycles of 1 min at 94°C, 1 min at 36°C, and 2 min at 72°C.

(c) Keep the reaction mixture at 72°C for 10 min.

(d) Cool at 4°C till the reaction tubes are removed from the machine.

(e) Separate PCR amplified DNA fragments on an agarose gel [(1.5% (w/v))].

(f) Stain with ethidium bromide (20 μg per 100 mL) at 3 V/cm for 4–5 h or silver staining.

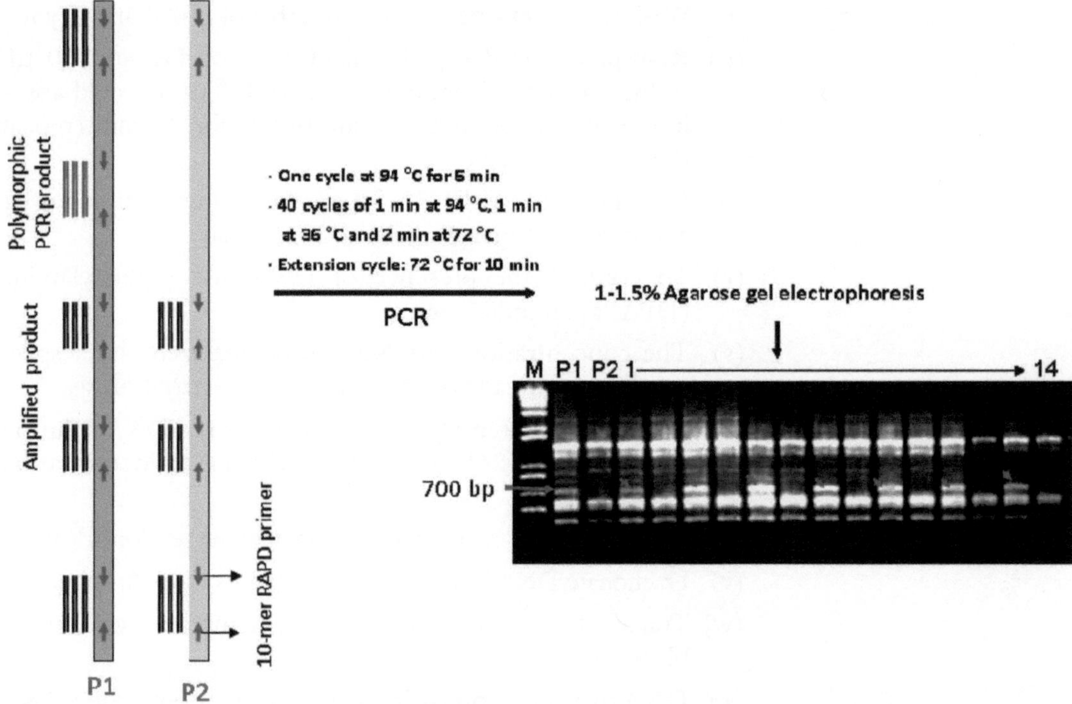

Fig. 2. RAPD fragment of 700 bp is present in P1 while it is absent in P2. From 1 to 14 are F_2 individuals showing the amplification of DNA marker in each alternate lane (like P1). *Conversion into SCAR marker*: This fragment can be eluted from the gel followed by cloning. These clones can be sequenced. Design reverse and forward primers of ~20 bases long. The resultant PCR product will be called as SCAR marker.

(g) A sample without template DNA is used as a control to detect any contamination.

(h) Use gel documentation apparatus for electrophoretic profile photograph under UV radiation.

(i) By comparing a standard ladder, DNA fragments can be quantified.

3. Amplified fragment length polymorphisms (AFLPs) PCR-based marker (see Notes 13 and 14)

   The basic requirements for undertaking AFLP analysis for genotyping complex genomes are (Vos et al. (22) with slight modifications Fig. 3):

   1. Mix the restriction reaction mixture gently and centrifuge it, followed by incubating at 37°C for 3 hours.
   2. Stop the reaction by incubating at 70°C for 15 min, and place the tubes on ice.
   3. Add 24 μl of adapter ligation solution and 1 μl of T4 DNA ligase. Mix the reaction mixture in the tube and centrifuge it briefly.

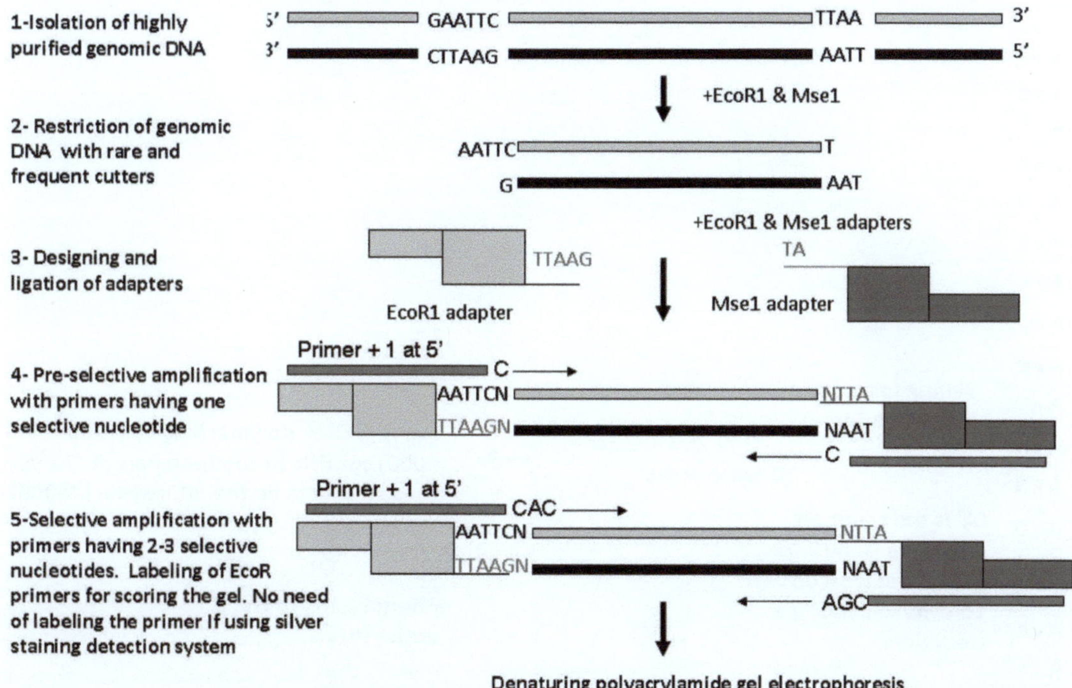

Fig. 3. Schematic illustration of generation of amplified fragment length polymorphism (AFLP); depicting that this genotyping system enjoys the advantages of both RAPD and RFLP.

4. Keep the tube(s) at 20°C for 2 hrs.
5. Dilute the mixture (add 1µl to 0.5 ml tube and 9µl TE buffer).
6. Prepare the Pre-Mix in 0.2mL tube.
7. Perform 30 cycles of PCR (each cycle comprising of 94oC for 30 s, 56°C for 30 s and 72oC for 30 s) followed by one cycle at 72oC for 5 min (final extension). Keep the tubes at 4oC until removed.
8. Make dilution (1:35) of the product by transferring 2µl of the amplification reaction mixture in a new 0.5-ml tube. Then add 70µl TE buffer. This is sufficient for 30 selective amplification reactions.
9. Prepare two mixtures (Mix-I and Mix-II).
10. Prepare the selective amplification reaction mixture in a 0.2 mL.
    (a) Perform one cycle at 94°C for 30 s; 65°C for 30 s; and 72°C for 60 s.
    (b) Perform 12 cycles by lowering annealing temperature 0.7°C in each cycle.
    (c) After this, perform 23 cycles at 94°C for 30 s, 56°C for 30 s, and 72°C for 60 s.

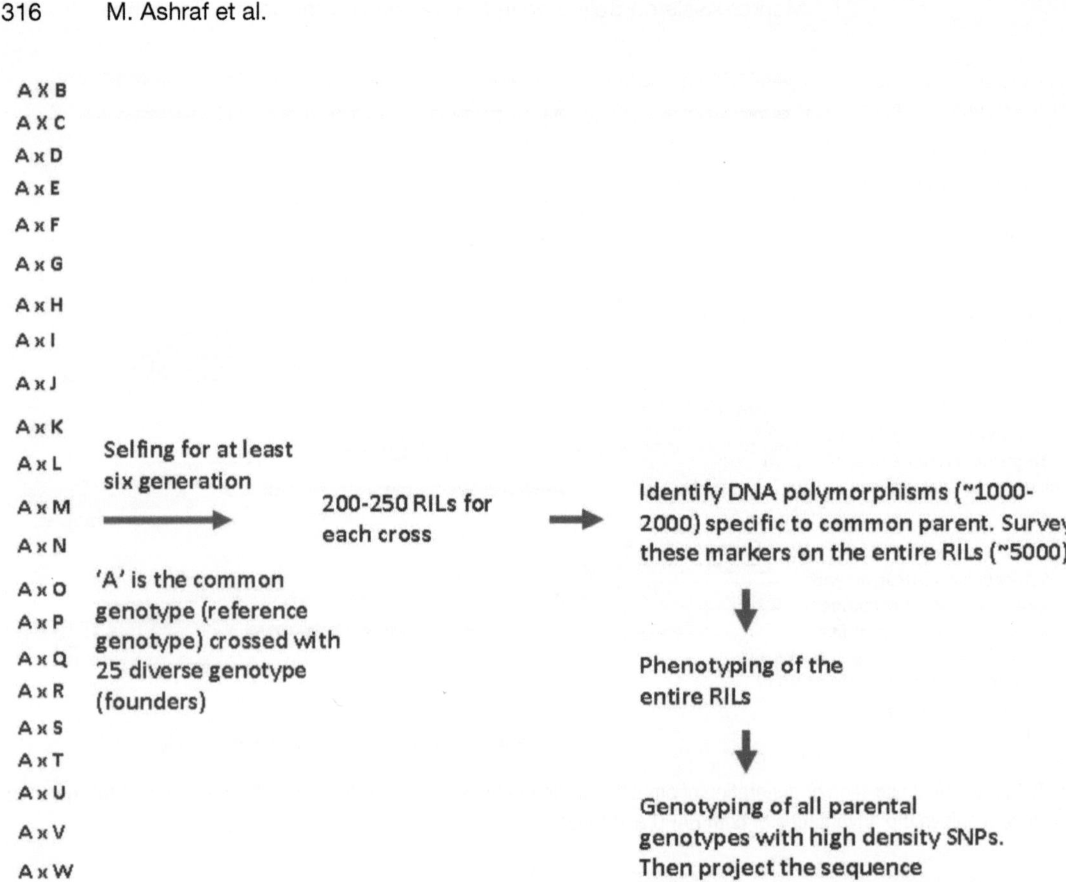

Fig. 4. Schematic illustration of development of NAM populations.

(d) This can be followed by one cycle at 72°C for 10 min.

(e) Keep it at 4°C before the tubes are removed from the PCR machine.

(f) Perform electrophoresis using polyacrylamide sequencing gels (7 M urea and 5% acrylamide).

(g) Stain gels with silver nitrate.

4. Simple Sequence Repeats (SSRs) PCR-based marker (see Notes 15 and 16)

Isolation of SSRs

Microsatellite or simple sequence repeats (SSRs) are robust type of DNA markers. Among the conventional DNA-based marker systems, SSRs are very popular because of their co-dominant nature, abundance, and hyper-variability in various genomes. A variety of methods available for SSR identification/isolation are: Screening of small insert genomic libraries followed by screening of SSRs with PCR or colony hybridization (slight modifications in protocol proposed by Zhang et al. (23)).

PCR screening method

(a) Isolation of genomic DNA (as described earlier).

(b) Digest 2–10 μg of high quality genomic DNA with restriction enzymes, EcoR I and Hind III (20 U each) for 2–3 h in single reaction mixture, as well as 50 ng of RNAs. It is noteworthy that choice and number of restriction enzymes for chopping the genomic DNA depends upon the kind of organism under study.

(c) The digestion mixture must contain diverse population of fragments of ~300–1,000 bp. Elute small size fragments (300–1,000 bp) from low melting 1.0% agarose gel after doing electrophoresis.

(d) Clone the purified fragments into a plasmid pUC19.

(e) Transfer the plasmids containing small inserts into competent *Escherichia coli* DH5 α cells.

(f) For screening purpose, drop every ten recombined clones in a PCR tube followed by amplification of these inserts with combination of (AC)8 and an M13 universal primer. PCR cycling conditions include one cycle of 95°C for 5 min (initial denaturation) followed by 35 cycles of 95°C for 30 s, 55°C for 45 s, 72°C for 45 s, and final extension with 72°C for 5 min.

(g) Separate the PCR products in 1.2% agarose gels.

(h) Perform PCR again to pools bands (amplification) with the same primer combination.

Colony hybridization

Transfer the clones (as described above) to nitro-cellulose membranes followed by screening with (GA)20 and (CA)20 oligonucleotide probes (labeled with Gene Images 3′-oligolabeling and ECL detection system according to the manufacturer's protocols (Amersham Biosciences)).

Development of targeted enrichment of DNA for microsatellites

(a) Digest ~2-μg genomic DNA with restriction enzymes HaeIII, RsaI, and DraI for 3 h in a single reaction mixture. Also add 50 ng of RNaseA. This reaction will yield a diverse population of blunt-ended fragments of ~550 bp.

(b) Purify digested DNA using a purification kit (Easy to use for a starter); many kits of various companies are available (QIA-quick purification column, etc.), eluted with 50 μL of 5 mM Tris-pH 8.0 followed by complete drying under vacuum.

(c) Prepare a double-stranded adaptor molecules AP11/12 by mixing equal molar quantity of oligonucleotides AP11 (5′CTCTTGCTTAGATCTGGACTA3′) and AP12 (5′pTAGTCCAGATCTAAGCA-AGAGCACA3′; p = 5′ phosphate). Then heat the reaction to 94°C followed by cooling to 25°C over a period of 5 h.

(d) Resuspend the digested genomic DNA in a 30-μL ligation reaction including 100 ng of AP11/12 double-stranded adaptor and 30 Weiss units of T4 DNA ligase, and then ligate at 14°C for 16 h.

(e) Perform ten cycles of PCR (10–26; depending upon the amplification) using 2 μL of the ligation reaction as a template at 55°C annealing temperature in a 50-μL reaction volume using the single primer AP11.

(f) Add the pre-amplified product (~100 ng) to a single reaction mixture containing 6× SSC (0.9 M NaCl, 90 mM sodium citrate, pH 7), 0.1% SDS. Also add 200 ng each of biotinylated oligos b(TA)30, b(CA)20, b(GA)20, b(AGA)15, b(TGA)15, and b(ACA)15 (b = 5′ biotinylation) in this reaction mixture.

(g) Denature at 95°C for 5 min followed by annealing of the biotinylated oligonucleotides at 60°C for 1 h.

(h) Add 200 μg of fresh streptavidin-coated paramagnetic beads (Promega, Madison, WI) equilibrated with 6× SSC, and incubate at 60°C with gentle agitation for 15 min. Remove the liquid by separation using a magnetic stand. Wash the beads twice in 300 μL of 6× SSC, 0.1% SDS for 15 min at room temperature with a slight shaking. Again wash the beads twice in 300 μL 6× SSC, 0.1% SDS for 15 min at 60°C with a slight shaking. Then briefly wash the beads twice with 6× SSC at room temperature.

(i) Elute the captured DNA by adding 100 μL of 60°C 0.1 M NaOH followed by neutralization with 100 μL of 1 M Tris-pH 7.5.

(j) Desalt and equilibrate the captured DNAs with 10 mM Tris-pH 8.0, 1 mM EDTA-pH 8.0 to a final volume of ~50 μL using a 100-kDa MW cutoff size filtration column (Millipore, Bedford, MA).

(k) Use 5 μL of the desalted DNA as a template for 30 cycles of PCR in a 50-μL reaction volume using primer AP11.

(l) Clone 6–8 μL of the resulting PCR products (~60 ng) into a TA-cloning vector pCR4-TOPO through topoisomerase-mediated ligation (Invitrogen, San Diego, CA) and transform into chemically competent *E. coli* TOP10.

(m) Identify the recombinant colonies by positive selection through insertional inactivation of the ccdB (control of cell death) open reading frame. Transfer the colonies into 96-well microtiter plates for archival storage.

(n) Inoculate the recombinant bacterial colonies into 300 μL of 2× YT broth in a 96-well 0.6-mL-deep plate (Marsh Bioproducts, Rochester, NY).

(o) Agitate the cultures at 500 rpm in a HiGro high-density shaking-incubator for overnight (16–18 h) at 37°C.

(p) Centrifuge the culture tube at 14,000 rpm for 5 min and discard the supernatant. Dry the pellet for 2 min and then resuspend the pellet 100 μL solution (Tris 50 mM, EDTA 1 mM and RNAs 100 μg/mL) by vortexing for 5 min.

(q) Perform lysis of the cells with 150 μL solution (NaOH 0.2 N and SDS 1%) and mix well by inverting gently.

(r) Add 200 μL of solution (potassium acetate) to eppendorf tube and mix well followed by centrifugation at 14,000 rpm for 5 min.

(s) Collect the supernatant in a fresh eppendorf tube, then add two volumes of 100% ethanol.

(t) Keep the eppendorf tubes at −40°C for 20 min and centrifuge at 14,000 rpm for 10 min.

(u) Discard the supernatant and wash the pellet with 70% ethanol.

(v) After centrifugation, discard the supernatant.

(w) Dry the pellet.

(x) Add 20 μL of sterile distilled water to dissolve the pellet and store at 4°C.

(y) Check the plasmid concentration and size on 1% agarose gel using standard DNA markers.

(z) Perform restriction analysis for confirming the size of the cloned fragments.

(a1) Sequence the cloned fragments, and identify the SSR regions using various software.

(b1) It is noteworthy that the aforementioned cloning steps can be avoided if facility for direct sequencing is available.

(c1) Design primers for flanking regions followed by conducting PCR as mentioned in Subheading 2.2.4.

Public database mining (from whole genome sequence information or from ESTs).

Software like Repeat Reporter 1.5 is very helpful in searching various types of repetitions in the sequences, and SSRs containing sequences can be saved as FASTA format for grouping clusters using Vector NTI Suite 8.0 software package (24).

5. Cleaved amplified polymorphic sequence (CAPS) markers (PCR-based marker)

   CAPS markers are developed by converting hybridization-based RFLP markers to PCR-based markers (see Note 17).

   The process involves:

   1. Partially or completely sequence the RFLP clone
   2. Design PCR primers
   3. Amplify genomic DNA
   4. Digest the amplified products by one or more restriction enzyme

6. Sequence characterized amplified regions (SCARs) PCR-based marker

   SCAR markers are basically derived from RAPD markers (see Note 18), and the process involves:

   1. Elute and clone the RAPD fragment followed by sequencing of the fragment
   2. Synthesize primers (usually 20-mer)
   3. PCR amplification
   4. Detect polymorphism using gel electrophoresis

7. Expressed sequence tags (ESTs) (see Note 19) PCR-based marker

   1. Construct a cDNA library
   2. Sequence the clones (partially or completely)
   3. Design primers
   4. PCR amplification of genomic DNA

8. Single nucleotide polymorphism (SNPs) PCR-based marker

   SNPs result from a single nucleotide alteration between species or paired chromosomes within an individual (see Note 20).

3.4. Development of Genetic Maps

1. Before conducting meaningful QTL mapping in a given population, development of a genetic map is required.
2. Through different mapping techniques, one can use the relationship of markers with the desired phenotype, from which the relative positions of the QTLs can be estimated.
3. The primer sequences for different crops are available online and these can be used for synthesizing/designing the primers that are provided by different companies.
4. To put markers in different linkage groups and to arrange the molecular markers within each linkage group, a number of different software packages are available online such as basic local alignment search tool (BLAST), QGene, QTL Cartographer, PLABQTL, Mapmaker/QTL, Map Manager, R/QTL, and twopartqtl (25, 26).

5. Calculate the position of each marker in centimorgans (cM) relative to the position of other markers on the chromosome. This converts the recombination percentage to genetic distance.
6. Examine polymorphic bands in the populations/genotypes depending on the absence or presence of the specific molecular marker used.
7. Determine the frequency of such genes using different software packages.
8. The marker map is required only for interval mapping.
9. In contrast, individual marker analysis could be carried out without the marker map.
10. If this is the case, association between marker genotype and phenotype can then be investigated using the composite interval mapping (CIM) (27, 28), interval mapping (IM), and single marker analysis (SMA).

3.5. Screening Population for ST (See Note 21)

It can be carried out on the basis of:

1. Morphological traits, e.g., growth, yield, or height.
2. Physiological traits, e.g., nutrient uptake, osmotic adjustment, hormonal regulation, photosynthetic processes, or antioxidant activities.
3. Molecular markers.

3.5.1. Greenhouse or Growth Chamber Experiments

A salt tolerant (donor) and salt sensitive (recipient) cultivars/varieties/lines of a crop can be selected under greenhouse or growth chamber experiments.

3.5.2. Field Experiments

Plants that show significant tolerance under greenhouse/controlled conditions should be also assessed under variable, more realistic, and ultimately, relevant field conditions (see Note 22).

3.5.3. Screening at Different Developmental Stages

Selection of plant growth stage for salinity tolerance is one of the crucial prerequisites for initiating a successful marker-assisted selection (MAS) programs because the degree of salt tolerance observed at one stage may or may not reflect that at other growth stages in most plant species.

3.6. QTL Analysis and Identification of Markers Associated with ST

Various types of statistical analyses to identify QTL-marker association.

Statistical analyses for QTL mapping are:

1. Linkage disequilibrium-based mapping, analysis of variance (ANOVA), principal component analysis-based mapping, and partial least squares regression.
2. Simple as well as multiple interval mapping.
3. QTL analyses could be performed using the software GridQTL.
4. ANOVA is the simplest method for QTL mapping.

3.7. Conducting MAS

1. QTL mapping can be achieved using bi-parental cross populations between two parents that have contrasting phenotypes for the trait of interest.
2. Crossing may be up to a number of generations up until the required trait is achieved. Dissect the quantitative traits and detect the chromosomal regions associated with such traits.
3. Identification of salinity tolerance trait linked to chromosomal regions is useful to enhance the selection efficiency in breeding programs.
4. By appraising the usefulness of these genomic regions, the breeder can precisely generate new combinations of genes of different origins.
5. Determine QTLs/genes in the identified region with plant species whose genome/DNA sequence is known (see Note 23).
6. If the genome sequencing information is not available, then sequencing of the identified region, as well as the determination of the functions of genes by their similarity to genes with known function, can be compared with known genomes using online tools.
7. Users enter a primary sequence and search for similar sequences within the database of genes from various organisms.
8. QTL mapping uses the linear model to find out the correlation between the putative QTL and phenotypic value.
9. QTL mapping depends on two types of data sets such as marker genotype data and phenotype data of the QTL.

3.8. Always Better to Combine MAS with PS

1. Identify both phenotype and genotype of the desired traits involved in salinity tolerance using molecular markers that map the possible location of QTLs/genes of interest.
2. The location identifies markers and their favorable genes.
3. Use the identified QTLs/genes with the desired effect in MAS.
4. For MAS, a minimum of five backcross generations are required to transfer a gene/QTL of interest from a donor to a recipient cultivar.
5. MAS can be carried out to identify individual plants carrying the donor alleles in the backcross generations/populations.
6. Use highly polymorphic markers for distinguishing a specific allele between a donor and a recipient parent.
7. Identify the polymorphisms between alleles, based on the number of molecular markers.
8. After QTL mapping or gene identification, select the salt tolerance genes for a particular trait using molecular markers so as

to produce salt tolerant cultivars/lines using molecular breeding techniques.

9. Select inbred lines and cross the recipient with the donor in the field/greenhouse.
10. Backcross the resulting F_1 plants with the donor to produce BC_1F_1 seeds.
11. Then backcross these plants to the donor again to obtain the BC_2F_1 generation.
12. Self-pollinate the selected BC_2F_1 individuals to produce BC_2F_2 seeds.
13. Use the BC_2F_2 seeds for MAS validation under saline conditions.
14. Confirm the presence of desired locus using different molecular markers particularly SSR markers as described in the earlier section.
15. For confirmation of the desired QTLs/genes, the DNA sequences of SSR primers can be retrieved online and PCR primers can be synthesized from international companies.

3.9. Use of Large Populations to Identify Desirable Combinations Is Necessary

1. Selection of a population size is a great link between what is feasible in practice and what is theoretically desirable (see Note 24).
2. QTLs that are universal in two populations, the approximation of population size could be decreased in a larger population, supporting the notion that the degree of QTL effects is overestimated in small populations.

3.10. High-Throughput Marker Analysis Is Needed

In most plant species, preferred DNA markers are those that are phenotypically neutral and frequent in number and could be used for the determination of the whole genome (29).

For choosing a suitable molecular marker, the important characteristics must be undertaken including:

1. Polymorphic nature
2. Co-dominant inheritance
3. Quantity and quality of DNA
4. Distribution in the genome
5. Random distribution in the entire genome
6. Easy and economical to detect
7. Reproducible
8. DNA markers should be less than 5 cM genetic distance, linked to target loci

4. Notes

1. In most crop plants, genetic variation for ST exists either within the cultivated species or most likely, among the related wild species. For most crop species, germplasm collections already exist within national and international gene banks. These can be easily obtained for screening for ST. However, in case of the absence of such collections, collecting new germplasm from appropriate climatic conditions would be necessary. For the identification of contrasting parental lines for ST, one must have a prior appropriate selection criteria that could be either morphological, physiological, biochemical, or molecular. Reproducible differential manifestation in plants with respect to their morpho-physiological or biochemical parameters is considered for the overall differential phenotypic expression of selected contrasting parental lines to be used for MAS (30–32).
2. A population employed for gene mapping is usually known as a mapping population. Generally, evaluation for selection of parents, mapping population, and mating design, as well as the type of markers used for genetic mapping, depends entirely on the aims of the experiments, the molecular map as well as the availability of molecular markers. The parents of mapping populations must have significant variability in both the DNA sequence (i.e., genetic markers) and the trait(s) of interest.
3. The major disadvantage is that phenotypic evaluation (e.g., evaluation for ST) of these populations is often based on single plant performance for ST may not be very reliable. Further, these populations are not genetically fixed, so they may not be maintained for repeated evaluation (i.e., they are not eternal). However, to improve the reliability of the phenotypic data, it is often possible to conduct the genetic mapping in one population (e.g., F_2 or BC_1), but phenotypic screening in both the parental (e.g., F_2 or BC_1) and progeny populations (e.g., F_3 or BC_1S_1). In the latter situation, samples of the progeny populations could be evaluated in multiple environments or years. This modification can significantly improve the quality of the phenotypic data, in particular, for low-heritability traits such as ST.
4. Inbred backcross increases homozygosity, but maintains more of the genome of the recurrent parent and less of the donor parent. Such populations are useful, particularly if the original parental crosses are between a cultivated and a wild species. The main advantage of such populations over early filial and backcross populations is that they can be used for both genetic mapping as well as for breeding purposes. Such a population contains most of the genome of a donor parent, collectively existing among different lines in the population. The majority

of the genome of each line is from the recurrent parent, with a small portion from the donor parent (33). The first stage of generating a BC_1S_1 population is similar to generating a backcross breeding population. The main difference is that many BC_1 individuals are self-fertilized to develop a BC_1S_1 population. Alternatively, a large number of BC_1 individuals are backcrossed to the recurrent parent to generate the BC_2 generation that are then self-fertilized to produce BC_2S_1. In such a population, each BC_2S_1 progeny family has on an average only 12.5% of its genome from the donor parent.

5. Generally, after several generations of self-breeding, a group of RI lines are developed, each having on average 50% of its genome from each of the two original parental lines. The RIL populations are potentially the most useful populations for genetic mapping for several reasons, including the eternal nature of the populations. Such populations can repeatedly grown and evaluated for multiple traits in different environments, providing quality phenotypic data for genetic mapping. The main disadvantage of these populations is the long time it takes to develop one RIL population.
6. The chromosome doubling of the haploid cells results in a doubled haploid (DH) genotype. The DH populations usually accelerate the breeding process, because the traditional approach requires several generations to obtain a population of homozygous lines. However, the DH populations have tremendous use for accelerating breeding activities, including using MAS. MAS, when combined with double haploidy, is a timesaving method of performing backcross conversion to select an elite line with a particular trait. By combining molecular markers and DHs, it is possible to pyramid resistance genes more quickly. Similar to RIL populations, DH populations can be used as permanent mapping populations, because they are stable and immortal. DHs offer a unique opportunity to improve selection efficiency, because selection is based on gametophytic instead of sporophytic cells. DHs provide a link between conventional breeding and genomics. DHs are also a key feature in genomic programs for integrating genetic and physical maps. They have widespread application in quantitative genetics and SNP discovery and in establishing chromosome maps, resulting in reliable information on the location of major genes and QTLs for important traits (34).
7. Most populations developed by crossing two parent genotypes offer a limited number of recombinants. It is the major reason for localizing QTLs at wider distances, often at 10–20 cM intervals. Thus, it limits the wider application of MAS in breeding programs. Secondly, most DNA markers identified on populations derived by involving two parent genotypes are mostly population-specific. This is another reason for limiting their

use in MAS. Also, the procedures of developing RILs are laborious and costly (35, 36). In order to overcome these limitations, the use of nested association mapping (NAM) populations can pave the way for identifying QTLs in crop species with low linkage disequilibrium (LD) (37). In NAM, one can combine the advantages of deploying robotic statistical tools for mapping QTLs coupled with a high-resolution power (smaller interval between linked QTLs) of association mapping.

8. Isozymes are enzymes with different molecular weight or electric charge that catalyze the same chemical reaction (38, 39). Isozyme markers are co-dominant; some are multi-allelic, and they also are faster and cheaper than RFLPs and RAPDs (40). The advantages of isozymes include: co-dominant markers, no known pleiotropic effects, and are fast, easy, and cheap. However, the major disadvantage is the limitation in the number of isozyme markers that have been characterized in any given plant species. Another disadvantage is the limited polymorphism among closely related genotypes. In addition, tissue variability is another drawback of using isozymes; some isozymes are better expressed in leaf tissues while others, in root tissues. Therefore, numerous samplings of a segregating population are necessary to score all the available isozymes (40). In most of the isozyme detection protocols, starch gels are employed. However, isoelectric focusing and polyacrylamide gels may also be used and this results in a higher resolution.

9. A restriction fragment length polymorphism (RFLPs) is generally characterized by the presence of alternative alleles linked with restriction fragments, and they vary in size from one another (41, 42). The polymorphism is the result of variation in restriction sites or variation in length of intervening sequences between the two restriction sites.

10. The main advantages of RFLPs include: highly reliable, co-dominant markers, no pleiotropic effects, and distributed throughout the genome. The main disadvantages are that they are laborious and costly, require extensive experience, often require radioactive labeling, limited polymorphism among closely related genotypes, and a large amount of good quality DNA is needed.

11. First described by Mullis et al. (43), PCR is a molecular operation for enzymatically amplifying/replicating small (up to 10 kb) segments of DNA in the absence of a living organism (44). It is used to amplify a well-defined short part of a DNA strand from a single gene or just a part of a gene.

12. The randomly amplified polymorphic DNA (RAPD), also known as arbitrarily primed PCR (AP-PCR), involves the use of arbitrary DNA sequences (usually 10-bases long) as primers to amplify random genomic DNA sequences by PCR. Usually,

during normal PCR, two primers are used while one primer is used in RAPD with an arbitrary sequence. In RAPD, PCR amplification of DNA fragments is done completely at random without sequencing of the DNA fragments to be amplified. It is most commonly used for DNA fingerprinting as well as for comparison among different organisms. However, in regular/ conventional PCR, there is a need of sequencing for amplification of specific fragments, so the correct primers are constructed within an order of sequence. In addition, in conventional PCR, the primer design should be specific to the flanking regions of the DNA of interest and it amplifies merely that region, while in RAPDs, small fragments of DNA called oligonucleotides are used as primers. Therefore, amplification in the RAPD process can occur in any part of the genome that has two complementary sequences to the primer within ~3 kb interval. PCR amplifications is carried out using RAPD primers that are available commercially. Generally, primers are tested for the occurrence of discrete and reliable bands (45). The main advantages of RAPDs include: no pleiotropic effects, distributed throughout the genome, no radioisotope use, faster and may be cheaper than RFLPs, often polymorphic among closely related genotypes (16, 46), requires little DNA, and generates large numbers of markers per reaction. The main disadvantages include: dominant markers, sensitive to contamination, problems with reproducibility (e.g., between runs, between PCR machines, and between labs), and it is population-specific.

13. In RAPDs, arbitrary and multiple primers are used, but the results obtained may not be very reliable. In AFLP (22), only two primers are used and it produces relatively more reliable results. Compared to RAPDs and RFLPs, the AFLP technique results in higher percentages of polymorphism (47) and it often can detect polymorphism between closely related genotypes. In complex genomes, a second round of PCR amplification (called selective PCR), with primers containing 2–3 nucleotides extra, is carried out to reduce the complexity of the DNA profile. The main advantages of AFLPs include no pleiotropic effects, distributed throughout the genome, many more markers than with RFLPs and RAPDs, no Southern blotting needed, and polymorphism among closely related genotypes. The main disadvantages of AFLP markers include dominant markers, it is expensive, may require automated data collation prior to analysis, may require radioisotope use, too many steps, requires extensive experience, not as reproducible as RFLPs.

14. For simple genomes like bacteria, fungi, etc., PCR products of first amplification are separated on PAGE. For genotyping complex genomes, a second round of PCR (selective amplification)

is required. For procedural advice, follow the procedure for using kits synthesized by Invitrogen, Life Technologies, USA. This company has two types of kits, i.e., for genotyping complex and simple genomes.

15. SSRs are also known as microsatellites, simple sequence length polymorphisms (SSLPs), and sequence-tagged microsatellites. They are tandem repeats of 2–4 bp, e.g., (AT)n, (TAT)n, and (AGAT)n, where polymorphism is based on the difference in the length of the repeat, with usually $n = 10$–60 repeats, and are flanked by conserved endonuclease restriction sites. Of all molecular markers, SSRs are more practical and useful because they can detect a high degree of allelic diversity, co-dominant inheritance, and their ability to automate the process. Furthermore, they can be very conveniently amplified by PCR (48). Primers for amplification of SSR loci are synthesized on the basis of sequences of the flanking regions of SSRs. The advantages of SSRs include: highly reproducible (due to the long PCR primers), frequent and randomly dispersed in the genome, PCR-based markers, required little DNA, high throughput once the markers are identified, polymorphic among closely related genotypes, multi-allelic loci, and highly polymorphic. The only disadvantage of the SSR markers is that their development may be costly (49).
16. Genomic sequence information, available publicly, is very economical and effective strategy for identification of microsatellite regions as it eliminates multiple methods and/or procedures which are involved in library construction, screening, etc. (24). Bioinformatics analysis of public databases has been used for SSR identification in species with abundant sequence information. In future genome sequence information of many crop species may be available which can set a firm stage for identification of SSRs at large scale. Similarly, sequence information of transcribed regions (ESTs, available in many databases) can be explored for identification of large number of EST-SSRs. SSR primer pairs (reverse and forward) are usually designed from the genome sequencing information available on various websites or these primers can also be designed from ESTs (EST-SSRs). New SSRs can also be identified by making shotgun genomic libraries, followed by screening of these libraries with probes containing SSRs. The positive clones can be sequenced for designing SSR primers.
17. CAPS markers, also known as PCR-RFLP markers, are basically an extension of RFLP technology. However, the detection of CAPS markers is usually faster than that of RFLP markers. The advantages of CAPS markers include: reproducible (due to the use of longer primers; usually 15–25 nucleotides), sequence-based (so no need for clone distribution), usually a

small amount of DNA is required, and they can be faster than RFLP or AFLP markers. A major disadvantage of CAPS markers is that they are less polymorphic than RFLP or AFLP markers.

18. Because of the use of longer primers (longer than those used for RAPD markers), SCARs are more reproducible than RAPDs. SCAR markers are basically derived from RAPD markers, and the process involves the following steps: elute and clone the RAPD fragment, sequence all or the ends, synthesize primers (usually 20-mer) followed by PCR amplification, and detect the polymorphism by gel electrophoresis. The main advantages of SCAR markers include: detecting single loci, amplification is less sensitive to reaction conditions, and loci can potentially be converted into co-dominant markers. It is very effective to identify genetic differences more quickly than the other PCR-based molecular markers (50).
19. ESTs are sequence-tagged sites that represent the functionally active part of the genes. Similar to CAPS, the main advantages of ESTs include: reproducibility (due to the use of longer primers; usually 18–22 nucleotides), sequence-based (so no need for clone distribution), usually a small amount of DNA is required, and they can be performed faster than RFLPs or AFLPs. ESTs have also been utilized for identifying genes conferring a vital role in plant stress tolerance (23), indicating their increasing utility in discovering new genes, developing linkage maps, estimating the genetic relationship among the species, comparative genomic studies for identifying syntenic/collinear regions of different species, etc. The greatest significance of EST sequencing is in studying the gene expression pattern in response to stressful environments (51).
20. By definition, a SNP is a position at which two alternative bases occur at appreciable frequency (>1%). SNPs can serve as genetic markers for identifying genes by linkage studies in families (e.g., in human populations), linkage disequilibrium, or association mapping. The main advantages of SNPs include: high abundance and distribution throughout the genome in various species including plants, and greater potential for automation compared to traditional markers (49, 52, 53). They are present throughout the genome in coding as well as non-coding regions, thus are abundantly found in entire genome (54). With today's sequencing technology, the SNPs assay is economical, fast, and efficient method to be used in MAS (55). Generally, SNP results rely on relatively costly reagents or equipment such as MALDI-TOF or TaqMan probe, making the application of SNPs in plant genetics and breeding somewhat narrow. So far SNPs are recommended

for genotyping due to their simplicity, stability, and abundance (56).

21. The efficiency of screening methods for enhancing ST in different crop plants needs re-evaluation under saline fields, where a number of abiotic stresses including salinity will, in combination, act on the plant (57). To attain maximum output in terms of crop yield production, the preliminary germplasm appraisal should be carried out either under a favorable environment using physiological or morphological criteria or under in vitro conditions (58). Furthermore, salt tolerance needs to be determined in the form of yield, the ultimate target of agronomists and breeders (59).
22. Usually, salinity tolerance of different crops is assessed at only one specific growth stage. However, the degree of salt tolerance observed at one stage may or may not reflect that at other growth stages in most plant species (60, 61).
23. With plant species whose genome/DNA sequence is known, we can determine QTLs/genes in the identified region. If the genome is not available, then sequencing of the identified region, as well as the determination of the functions of genes by their similarity to genes with known function, can be compared with known genomes using online tools.
24. During breeding, selection usually involves determination of one or more traits such as stress tolerance, agronomic traits, grain quality, and disease resistance in the greenhouse or field experiments (29). Thus, selection of a population size is a great link between what is feasible in practice and what is theoretically desirable. Various reports show that a small population size often results in the detection of a lower number of QTLs, but with large phenotypic effects (62). If a single trait is analyzed, an important strategy would be to minimize the workload using a large number of populations so as to find marker genotypes only for progeny depending upon the phenotypic traits (63). However, for QTLs that are universal in two populations, the approximation ofpopulation size could be decreased in a larger population, supporting the notion that the degree of QTL effects is over-estimated in small populations.

References

1. Flowers TJ (2004) Improving crop salt tolerance. J Exp Bot 55:307–319
2. Foolad MR (2004) Recent advances in genetics of salt tolerance in tomato. Plant Cell Tissue Organ Cult 76:101–119
3. Richards RA (1983) Should selection for yield in saline regions be made on saline or non-saline soils. Euphytica 32:431–438
4. Foolad MR (1999) Comparison of salt tolerance during seed germination and vegetative growth in tomato by QTL mapping. Genome 42:727–734
5. Foolad MR, Chen FQ, Lin GY (1998) RFLP mapping of QTLs conferring salt tolerance during germination in an interspecific cross of tomato. Theor Appl Genet 97: 1133–1144

6. Greenway H, Munns R (1980) Mechanism of salt tolerance in nonhalophytes. Annu Rev Plant Physiol 31:149–190
7. Maas EV (1986) Salt tolerance of plants. Appl Agric Res 1:12–26
8. Forster BP, Russell JR, Ellis RP et al (1997) Locating genotype and genes for abiotic stress tolerance in barley: a strategy using maps, markers and the wild species. New Phytol 137:141–147
9. Koyama H, Kawamura A, Kihara T et al (2000) Overexpression of mitochondrial citrate synthase in *Arabidopsis thaliana* improved growth on a phosphorus-limited soil. Plant Cell Physiol 41:1030–1037
10. Foolad MR, Zhang LP, Lin GY (2001) Identification and validation of QTLs for salt tolerance during vegetative growth in tomato by selective genotyping. Genome 44:444–454
11. Bonilla P, Dvorak J, Mackill D et al (2002) RLFP and SSLP mapping of salinity tolerance genes in chromosome 1 of rice (*Oryza sativa* L.) using recombinant inbred lines. Philipp Agric Sci 85:68–76
12. Takehisa H, Shimodate T, Fukuta Y et al (2004) Identification of quantitative trait loci for plant growth of rice in paddy field flooded with salt water. Field Crops Res 89:85–95
13. Monforte AJ, Asins MJ, Carbonell EA (1996) Salt tolerance in *Lycopersicon* species. IV. Efficiency of marker-assisted selection for salt tolerance improvement. Theor Appl Genet 93:765–772
14. Foolad MR (1997) Genetic basis of physiological traits related to salt tolerance in tomato, *Lycopersicon esculentum* Mill. Plant Breed 116:53–58
15. Munns R, Hare RA, James RA et al (2000) Genetic variation for improving the salt tolerance of durum wheat. Aust J Agric Res 51:69–74
16. Rahman M, Ullah I, Ashraf M et al (2008) Genotypic variation for drought tolerance in cotton. Agron Sustain Dev 28:439–447
17. Sabouri H, Rezai AM, Moumeni A et al (2009) QTLs mapping of physiological traits related to salt tolerance in young rice seedlings. Biol Plantarum 53:657–662
18. Siahsar BA, Narouei M (2010) Mapping QTLs of physiological traits associated with salt tolerance in 'Steptoe'×'Morex' doubled haploid lines of barley at seedling stage. Food Agric Environ 8:751–759
19. Doyle JJ, Doyle JL (1990) A rapid total DNA preparation procedure for fresh plant tissue. Focus Cited 12:13–15
20. Williams JG, Kubelik AR, Livak KJ et al (1990) DNA polymorphisms amplified by arbitrary primers are useful as genetic markers. Nucleic Acids Res 18:6531–6535
21. Saiki RK, Gelfand DH, Stoffel S et al (1988) Primer-directed enzymatic amplification of DNA with a thermostable DNA polymerase. Science 239:487–491
22. Saiki RK (1989) The design and optimization of the PCR. In: Erlich HA (ed) PCR technology—principles and applications for DNA amplification. Stockton, New York, pp 7–16
23. Vos P, Hogers R, Blecker M et al (1995) AFLP: a new technique for DNA fingerprinting. Nucl Acids Res 23:4407–4414
24. Zhang S, Chen QZ, Lu L et al (2006) Assessment of the variety resistance to *Pyricularia grisea* and *Rhizoctonia solani* induced under the natural condition in Hubei Province. J Huazhoung Agric Univ 25:236–240
25. Zhan AB, Bao ZM, Wang XL et al (2005) Microsatellite markers derived from bay scallop *Argopecten irradians* expressed sequence tags. Fish Sci 71:1341–1346
26. Lander G, Barlow A, Lincoln D et al (1987) MAPMAKER: an interactive computer package for constructing primary genetic linkage maps of experimental and natural populations. Genomics 1:174–181
27. Semagn K, Bjornstad A, Ndjiondjop MN (2006) An overview of molecular marker methods for plants. Afr J Biotechnol 5:2540–2568
28. Chakravarti A, Lasher LK, Reefer JE (1991) A maximum likelihood method for estimating genome length using generic linkage data. Genetics 128:183–193
29. Postlethwait JH, Johnson SL, Midson CN et al (1994) A genetic linkage map for the zebrafish. Science 264:699–703
30. Collard BCY, Mackill DJ (2008) Marker-assisted selection: an approach for precision plant breeding in the twenty-first century. Philos Trans R Soc Lond B 363:557–572
31. Singh RK, Mishra B (1997) Stable genotypes of rice for sodic soils. Indian J Genet 57: 431–438
32. Heggie L, Jansen MAK, Burbridge EM et al (2005) Transgenic tobacco (*Nicotiana tabacum* L. cv. Samsun-NN) plants over-expressing a synthetic HRP-C gene are altered in growth, development and susceptibility to abiotic stress. Plant Physiol Biochem 43:1067–1073
33. Riehle MM, Bennett AF, Long AD (2005) Differential patterns of gene expression and gene complement in laboratory-evolved lines of *E. coli*. Integr Comp Biol 45:532–538

34. Robbins M (2011) Introduction to inbred backcross (IBC) lines and populations. Plant Breeding and Genomics Home, http://www.extension.org/pages/32448/introduction-to-inbred-backcross-c-lines-and-populations
35. Khurana P, Chauhan H (2011) Doubled haploid bread wheat engineered for drought tolerance. ISB News Report, http://www.vt.edu
36. Doerge RW (2002) Multifactorial genetics: mapping and analysis of quantitative trait loci in experimental populations. Nat Rev Genet 3:43–52
37. Holland JB (2007) Genetic architecture of complex traits in plants. Curr Opin Plant Biol 10:156–161
38. Yu J, Holland JB, McMullen MD et al (2008) Genetic design and statistical power of nested association mapping. Genetics 178:539–551
39. Hunter RL, Merkert CL (1957) Histochemical demonstration of enzymes separated by zone electrophoresis in starch gels. Science 125:1294–1295
40. Tanksley SD (1983) Mapping polygenes. Annu Rev Genet 27:205–233
41. Patra N, Chawla HS (2010) Biochemical and RAPD molecular markers for establishing distinctiveness of basmati rice (*Oryza sativa* L.) varieties as additional descriptors for plant variety protection. Indian J Biotechnol 9: 371–377
42. Helentjaris T, King G, Slocum M et al (1985) Restriction fragment polymorphisms as probes for plant diversity and their development as tools for applied plant breeding. Plant Mol Biol 5:109–118
43. Paterson AH, Damon S, Hewitt JD et al (1991) Mendelian factors underlying quantitative traits in tomato: comparison across species, generations, and environments. Genetics 127:181–197
44. Mullis K, Faloona F, Scharf S et al (1986) Specific enzymatic amplification of DNA in vitro: the polymerase chain reaction. Cold Spring Harb Symp Quant Biol 51:263–273
45. Foolad MR, Arulsekar S, Rodriguez RL (1995) Applications of PCR to plant genome analysis. In: Gamborg OL, Phillips GC (eds) Plant cell, tissue, and organ culture. Springer, Berlin, pp 281–298
46. Shinde VM, Dhalwal K, Mahadik KR et al (2007) RAPD analysis for determination of components in herbal medicine. Evid Based Complement Alternat Med 4:21–23
47. Rahman M, Hussain D, Zafar Y (2002) Estimation of genetic divergence among elite cotton (*Gossypium hirsutum* L.) cultivars/genotypes by DNA fingerprinting technology. Crop Sci 42:2137–2144
48. Saunders JA, Mischke S, Hemeida AA (2000) The use of AFLP techniques for DNA fingerprinting in plants. Application Information Beckman Coulter
49. Oliveira EJ, Padua JG, Zucchi MI et al (2006) Origin, evolution and genome distribution of microsatellites. Genet Mol Biol 29:294–307
50. Rahman M, Zafar Y, Paterson AH (2009) Gossypium DNA markers types, number and uses. In: Paterson AH (ed) Genomics of cotton. Springer, Dordrecht
51. Ding X, Zhengtao W, Kaiya Z et al (2003) Allele-specific primers for diagnostic PCR authentication of *Dendrobium officinale*. Planta Med 69:587–588
52. Qutob D, Hraber PT, Sobral BW et al (2000) Comparative analysis of expressed sequences in *Phytopthora sojae*. Plant Physiol 123:243–253
53. Shaheen T, Rahman M, Zafar Y (2006) Chloroplast RPS8 gene of cotton reveals the conserved nature through out taxa. Pak J Bot 38:1467–1476
54. Shaheen T, Asif M, Zafar Y et al (2009) Single nucleotide polymorphism analysis of MT-SHSP gene of *Gossypium arboreum* and its relationship with other diploid cotton genomes *G. hirsutum* and *Arabidopsis thaliana*. Pak J Bot 41(1):117–183
55. Ganal MW, Altmann T, Röder MS (2009) SNP identification in crop plants. Curr Opin Plant Biol 12:211–217
56. Yong-Jun S, Yong L, Na-La-Hu W et al (2010) Mining and identification of SNPs from EST sequences in soybean and converting SNP markers into CAPS. Acta Agron Sin 36:574–579
57. de Souza GA, Softeland T, Koehler CJ et al (2009) Validating divergent ORF annotation of the *Mycobacterium leprae* genome through a full translation data set and peptide identification by tandem mass spectrometry. Proteomics 9:3233–3243
58. Jafari-Shabestari J, Corke H, Qualset CO (1995) Field evaluation of tolerance to salinity stress in Iranian hexaploid wheat landrace accessions. Genet Resour Crop Evol 42:147–156
59. Houshmand S, Arzani A (2005) Evaluation of grain yield and its component under drought stress during grain filling period in durum wheat. In: Proceedings of the 2nd International conference on integrated approaches to sustain and improve plant production under drought stress, Rome, Italy, vol 2, p. 45
60. Yamaguchi T, Blumwald E (2005) Developing salt-tolerant crop plants: challenges and opportunities. Trends Plant Sci 10:615–620
61. Singh RK, Flowers TJ (2010) The physiology and molecular biology of the effects of salinity

on rice. In: Pessarakli M (ed) Handbook of plant and crop stress, 3rd edn. Taylor and Francis, Florida, pp 901–942

62. Munns R (2010) Approaches to identifying genes for salinity tolerance and the importance of timescale. Methods Mol Biol 639:25–38
63. Schon CC, Friedrich UH, Susanne G et al (2004) Quantitative trait loci mapping based on resampling in a vast maize testcross experiment and its relevance to quantitative genetics for complex traits. Genetics 167:485–498
64. Paterson AH (1998) QTL mapping in DNA marker-assisted plant and animal improvement. In: Paterson AH (ed) Molecular dissection of complex traits. CRC Press, Boca Raton, pp 131–143.

Chapter 22

Transcriptomics on Small Samples

Stuart J. Roy, Simon J. Conn, Gwenda M. Mayo, Asmini Athman, and Matthew Gilliham

Abstract

Interrogating the cell-specific transcriptome forms an important component of understanding the role that specific cells play in assisting a plant to overcome abiotic stress. Among the challenges arising when extracting RNA from individual plant cells are: the isolation of pure cell populations; the small yield of material when isolating specific cell types, and ensuring an accurate representation of the transcriptome from each cell type after amplification of RNA. Here we describe two approaches for isolating RNA from specific cell types—single cell sampling and analysis (SiCSA) and laser capture microdissection. Isolated RNA can then be directly sampled qualitatively using reverse transcription PCR (RT-PCR) or amplified for profiling multiple specific genes using quantitative RT-PCR and genome-wide transcript analyses.

Key words: Single cell sampling and analysis (SiCSA), Laser capture microdissection, Transcriptome, RNA isolation

1. Introduction

Distinct plant cell types have specific roles in plant physiology and development, for example, in photosynthesis, nutrient uptake, and storage (1–4). It is therefore desirable to understand the basis of this contribution in order to understand how individual cell types define the function, performance, and development of specific plant tissues and organs. Under salt stress, cells within different tissues can perform contrasting roles, e.g., the control of ion transport to the shoot, the re-establishment of osmotic homeostasis, and the detoxification of reactive oxygen species (5). For example, in terms of ion transport, stelar cells may play a role in reducing salt transport to the shoot by retrieving Na^+ and Cl^- from the xylem, while root epidermal and cortical cells may enhance efflux of the

Sergey Shabala and Tracey Ann Cuin (eds.), *Plant Salt Tolerance: Methods and Protocols*, Methods in Molecular Biology, vol. 913, DOI 10.1007/978-1-61779-986-0_22, © Springer Science+Business Media, LLC 2012

same ions back to the soil (5–7). As such, it is unsurprising that under salt stress, individual cells within tissues will display considerable differences in their gene expression. However, to characterize the transcript abundance of genes in individual or small populations of cells, to gain a better understanding of the multiple salinity tolerance mechanisms available to the plant, specialized techniques and methodologies are required.

There are now numerous techniques and methodologies for extracting RNA from individual or small populations of cells for expression profiling. Single cell sampling and analysis (SiCSA) (8–10), laser capture microdissection (LCM) (10–13), and fluorescently activated cell sorting (FACS) (Johnson, this volume) are all techniques that have been used to successfully isolate RNA for transcriptomic analysis to address various biological questions. The cDNA reverse transcribed from RNA samples is amenable to RT-PCR. Alternatively, the RNA can be amplified prior to interrogation by quantitative real time RT-PCR, microarrays, or RNA-seq analysis (8–13).

We present methodologies for two techniques, SiCSA and LCM, for either qualitative and/or quantitative transcript profiling of specific cell types. There are advantages and disadvantages to both methodologies (Table 1) and these should be considered prior to using either technique. Whilst we have made every effort

Table 1
Pros and cons of single cell sampling and analysis (SiCSA) and laser capture microdissection (LCM)

Factor	SiCSA	LCM
Resolution	Commonly single cell (up to 100 cells)	>500 cells
Target cell types	Limited to easily accessible cell types	All cell types
Tissue preparation	None	Fixation, embedding, and sectioning (>1 week)
Artifacts	Stochasticity due to low sample size	Degradation of RNA
Cost of equipment & consumables per RNA sample[a]	<\$50k <\$10	>\$200k <\$20
Manual dexterity	Considerable	Considerable but simplified by computer interfaces
Destructiveness	Non-destructive, time courses, and additional measurements possible on same plant	Destructive

[a]RT-PCR, RNA amplification, quantitative PCR, microarray, and RNA-seq costs are additional and equal for SiCSA and LCM

to provide comprehensive methods for both techniques, we would still advise unfamiliar users to trial the methods with an experienced laboratory group prior to establishing them independently. Both techniques require a considerable initial outlay in infrastructure and contain manually dexterous steps that cannot be easily imparted on paper. However, this document will be a useful reference for more experienced users.

2. Materials

2.1. Equipment Required for Single Cell Sampling and Analysis (SiCSA)

1. Vibration free table (including air table or heavy metal plate resting on inner tube).
2. Stereo dissecting microscope with magnification of at least ×100 (e.g., SMZ645, Nikon, Tokyo, Japan).
3. Micropipette puller (e.g., PE-2 vertical microelectrode puller (Narishige, Tokyo, Japan)) or horizontal laser puller P-2000 (Sutter Instruments, Novato, CA, USA).
4. Microcapillary microforge (e.g., MF-900, Narishige).
5. Micromanipulator (e.g., M Series micromanipulator, Leica, Wetlaz, Germany).
6. 1 mm outer diameter, 0.58 mm inner diameter unfilamented borosilicate glass capillaries (Harvard Apparatus, Edenbridge, Kent, UK).
7. Tubing, foot-controlled pump (or mouth pipette).
8. Sample stage.

2.2. RNA Extraction from Samples Obtained by SiCSA

1. Chloroform.
2. 2% Tributylchlorosilane (Fluka, Sigma-Aldrich, St. Louis, MO, USA) in chloroform.
3. Salmon sperm DNA.
4. UltraPure™ DNase/RNase-free distilled water (Invitrogen Corporation, Carlsbad, CA, USA).
5. RNaseOUT recombinant ribonuclease inhibitor (Invitrogen).
6. RNase inactivation spray.

2.3. cDNA Synthesis of SiCSA Extracted RNA for Qualitative Transcript Analysis

1. T7-linked oligo dT_{18mer} with a mixed A, G, or C heel (AAACGACGGCCAGTGAATTGTAATACGACTCACTAT AGGCGCTTTTTTTTTTTTTTTTTTV, where V is either A, C, or G).
2. 10 mM dNTP Mix (Invitrogen). Contains 10 mM of dATP, dCTP, dGTP, and dTTP.
3. SuperScript® III RT (Invitrogen).

2.4. RNA Amplification of SiCSA Extracted RNA for Quantitative Transcript Analysis

1. RNA amplification kit using Eberwine method (Target AMP 2-round aRNA amplification kit (EpiCentre Biotechnologies)).
2. Bioanalyser and Total RNA pico chip (Agilent Technologies Inc., Santa Clara, CA, USA).
3. Superscript III (Invitrogen).
4. Random primers (hexamers, or nonamers).

2.5. Equipment Required for Laser Capture Microdissection (LCM)

1. Laser assisted microdissection microscope (e.g., Leica ASLMD laser assisted microdissection microscope).
2. Microtome (e.g., Leica RM2265 automated microtome).
3. Paraffin embedding station and cooling unit (e.g., Leica EG 1150H embedder and Leica EG 1150 cooling unit).
4. Water bath (e.g., Leica HI 1210 waterbath).
5. Slide warming tray (e.g., Ratek WT1 slide warming tray, Ratek, Boronia, VIC, Australia).
6. Laser dissection slides (e.g., PEN membrane slides, Leica).
7. Vacuum infiltrator.

2.6. Preparation of Tissue for LCM

1. Farmer's fixative (ethanol:glacial acetic acid 3:1).
2. 75, 85, and 100% ethanol (molecular sieve for dehydrating ethanol).
3. Isopropanol.
4. Paraffin (Paramat, BDH, VWR, Radnor, PA, USA).
5. Embedding moulds (Leica).
6. Xylene.
7. Oven.

2.7. RNA Extraction from Samples Obtained by LCM

1. TRIzol® (Invitrogen) or RNAqueous-Micro kit (Ambion, Austin, TX, USA).
2. UltraPure™ DNase/RNase-free distilled water (Invitrogen) or Turbo DNA-free (Ambion).
3. Chloroform.
4. Isopropanol.
5. RNase inactivation spray.

2.8. cDNA Synthesis of LCM Extracted RNA for Qualitative Transcript Analysis

1. Superscript III (Invitrogen).
2. T7-linked oligo dT_{18mer} with a mixed A, G, or C heel (AAACG ACGGCCAGTGAATTGTAATACGACTCACTATAGGC GCTTTTTTTTTTTTTTTTTTV, where V is either A, C, or G).
3. RNaseOUT recombinant ribonuclease inhibitor (Invitrogen).

2.9. RNA Amplification of LCM Extracted RNA for Quantitative Transcript Analysis

1. RNA amplification (MessageAMP II aRNA amplification kit, Ambion).
2. Bioanalyser and Total RNA pico chip (Agilent Technologies Inc., Santa Clara, CA, USA).
3. Superscript III (Invitrogen).
4. Random primers (hexamers, or nonamers).

2.10. PCR Components for Qualitative Transcript Analysis

1. Platinum® Taq DNA polymerase (Invitrogen).
2. 10 mM dNTP Mix (Invitrogen). Contains 10 mM of dATP, dCTP, dGTP, and dTTP.
3. T7 primer (GAATTGTAATACGACTCACTATAGGCG), gene-specific nested primers (see Note 1).

3. Methods

3.1. RNA Extraction from Samples Obtained by SiCSA: Capillary Preparation

1. Place a small 5 mL glass beaker upright in the middle of a glass Petri dish and place in an oven, preheated to 200°C (see Note 2).
2. Prepare sampling microcapillaries with an aperture of approximately 3 μm from 1 mm diameter borosilicate glass capillaries as detailed in Fricke, Chapter 5 (see Note 3).
3. Transfer the capillaries to the small beaker in the preheated oven, with their sampling tips up.
4. Place a larger covering beaker over the top of the capillaries and small beaker, being careful not to touch the sampling ends of the capillaries. Leave for 2 h at 200°C.
5. Carefully remove the hot large beaker, add 100 μL of 2% tributylchlorosilane (in chloroform) to the base of the small beaker (see Note 4), and immediately recover the capillaries with the large beaker.
6. Leave to silanize for 60–90 min at 200°C, removing large glass beaker for the last 30 min to allow the samples to dry.
7. Carefully replace the larger beaker, turn the oven off, and allow the sampling capillaries to cool to room temperature.
8. The capillaries can be stored for 2 weeks in the dark at room temperature until use. Leave the larger covering beaker in place to avoid RNase contamination of the tips.

3.2. RNA Extraction from Samples Obtained by SiCSA: Sampling of RNA for Qualitative Transcript Analysis

1. For each sample, prepare a double autoclaved 200 μL PCR compatible microfuge tube (see Note 5) containing 13 μL of an RNase inhibitor solution consisting of 11.25 μL of RNase-free water, 0.25 μL of 40 U/μL RNaseOUT recombinant ribonuclease inhibitor, 0.25 μL of 10 μM T7-linked Oligo dT_{18} primer (see Note 6), and 1 μL of 10 mM dNTPs.

2. Into a separate 200 μL microfuge tube, aliquot 20 μL of salmon sperm DNA. Prepare one tube per sample.
3. Mount the first silanized microcapillary onto the micromanipulator and carefully draw up salmon sperm solution into the tip. Leave the salmon sperm solution in the tip for 30 s before completely expelling the solution back into microfuge tube (see Note 7).
4. Place the plant on a sampling stage (see Fricke, Chapter 5), keeping the roots immersed in growth solution. If sampling from roots or from cross sections of seed (see Note 8), place the tissue into a small reservoir filled with water to keep it turgid and sample under water.
5. Guide the microcapillary to the cell to be sampled and gently pierce the cell, allowing cell sap to flow into the capillary tip. Apply gentle suction to remove cytosol, before withdrawing the capillary.
6. Once cell sap is collected, immediately immerse the tip into a microfuge tube containing the RNase inhibitor solution and expel contents. Keep on ice until all samples have been collected (see Notes 9–11).
7. Heat samples at 65°C for 5 min, then place back on ice.
8. Following denaturation and annealing of the RT primer to the sampled RNA, add 4 μL of Invitrogen 5× Superscript III first strand buffer, 1 μL of 0.1 M DTT, 1 μL of RNaseOUT, 0.25 μL of 200 U/μL Superscript III RT enzyme, and 0.75 μL water.
9. Incubate at 50°C for 1 h.
10. Terminate the reaction by heating the samples at 70°C for 15 min.
11. Store the cDNA at –20°C until required, proceed to Subheading 3.8 (see Note 12).

3.3. RNA Extraction from Samples Obtained by SiCSA: Sampling of RNA for Quantitative Transcript Analysis

1. Sample RNA as per Subheading 3.2; however, expel cell sap into 3 μL of nuclease free water containing (10 U) RNaseOUT™ RNase inhibitor, on ice. Note: do not use T7-linked Oligo dT_{18} primer or dNTPs in this solution (contrast with Subheading 3.2) (1).
2. Perform two rounds of RNA amplification using TargetAMP 2-round RNA amplification kit (Epicentre Biotechnologies) following the manufacturer's specifications (see Note 13). Take a small aliquot of RNA and use RT-PCR to confirm the absence of genomic DNA and contamination of transcripts from undesired cell types (10). Use random hexamer primers in the reverse transcript reaction (Superscript III) to produce cDNA (10).

3. Analyze amplified RNA using a Bioanalyzer (Agilent Technologies), using the picoRNA chip (see Note 14).
4. Samples are now ready for quantification through qPCR (see Note 1), microarray (see Note 15), or RNA-seq analysis.

3.4. LCM: Preparation of Sample

1. Using a razor blade, cut the plant material into 3 mm long piece. If roots are very thin, 6 mm long is acceptable. Material may be lost during decanting so dissect sufficient extra material to allow for this.
2. Place samples into either 5 or 10 mL plastic tubes (see Note 16) of ice-cold Farmer's fixative. To prevent damage, use a fine paintbrush to transfer tissue. Keep samples on ice.
3. Open tubes, place them again on ice, and vacuum infiltrate by increasing vacuum very slowly to 25 in. Hg for 5 min. Return very slowly to room pressure, cap and mix tubes gently, then repeat the process.
4. Carefully decant the solution and add 5 mL of fresh ice-cold Farmers fixative (see Note 17).
5. Wrap the tubes in aluminum foil and fix overnight at 4°C.
6. Decant Farmers fixative and dehydrate sample through an ethanol series of consecutive 3 h incubations in 75, 85, and 100% ethanol (see Note 18) at room temperature. Mix gently and thoroughly between changes (see Note 19). Leave the tissue overnight in 100% ethanol at 4°C.
7. The next day carry out two further 3 h incubations in fresh dehydrated 100% ethanol at room temperature, then leave overnight in fresh dehydrated 100% ethanol at 4°C.
8. Replace the ethanol with isopropanol by incubating in 50% ethanol: 50% isopropanol for 3 h, then replace with 100% isopropanol for a further 3 h. Carry out a further two 100% isopropanol changes during the day, mixing gently but thoroughly between changes to ensure all ethanol has been removed. Leave tissue overnight in fresh 100% isopropanol. All steps can be carried out at room temperature.
9. Decant solution and half fill tube with fresh 100% isopropanol. Add ¼ volume of paraffin pellets and incubate the samples at 60°C until melted, then mix samples gently but thoroughly.
10. Add a further ¼ volume of paraffin pellets and incubate the samples at 60°C until melted.
11. Add more paraffin pellets and repeat the process until the tube is full. Incubate samples for 5 h at 60°C, gently mixing every hour.
12. Gently decant isopropanol mix from the tube and replace with 100% molten paraffin wax (see Note 20). Leave overnight at 60°C

13. Replace with molten paraffin a minimum of six times, three changes over 2 days or two changes over 3 days. Mix samples gently and thoroughly between paraffin changes.
14. To embed the sample, fill an embedding mould with paraffin, keeping it on a hot plate to prevent it from solidifying.
15. Transfer a tissue sample to the embedding mould and carefully orientate it using a warm paint brush or needle.
16. Move mould to cold plate and quickly readjust sample orientation if necessary as the paraffin starts to set.
17. Allow the block to cool completely to room temperature until it is cool to the touch. Remove the block from the mould and store at 4°C until required.
18. To section the block on a paraffin microtome refer to the relevant equipment manual. Treat equipment with RNase inhibitor spray to remove RNases, and use RNase-free water.
19. Section blocks 7 μm thick to produce ribbons. Separate off several sections using needles, float them on RNase-free water in a waterbath (see Note 21) at 42°C, mount them on a plain glass slide, and check tissue quality under a microscope while the sections are still wet.
20. Separate off a ribbon of sufficient length to fit on a laser dissection membrane slide. Float the ribbon on the water bath and pick it up on a membrane slide as soon as possible to limit rehydration. Use an eyelash brush (see Note 22) to manipulate the ribbon, to prevent puncturing the membrane.
21. Drain the slide until just air-dry, by leaning it against a solid object with the lower end on clean tissue. Dry slide on a warming plate at 42°C overnight.
22. In a fume cupboard, de-wax the slide in a glass container filled with xylene for 10 min, then repeat in a fresh container of xylene.
23. Remove slide and allow it to air-dry in a fume cupboard. If xylene has seeped between the membrane and the glass due to puncturing of the membrane, dry the slide on a warming plate at 42°C in a fume cupboard, and prick sufficient small holes in the membrane to allow xylene to evaporate.

3.5. LCM: TRIzol® RNA Extraction (See Note 23)

1. Place a 200 μL microfuge tube in the tube holder and load it under the microscope stage of the laser dissection microscope (refer to relevant equipment manual), ready to capture dissected tissue into a dry tube cap (see Notes 24 and 25).
2. Using the laser, cut desired cells or groups of cells from the section (see Note 26); these should fall into and be visible in the microfuge cap. Depending on the plant tissue, it may be necessary to dissect cells from several sections to collect enough RNA. For example, when collecting RNA from either the inner

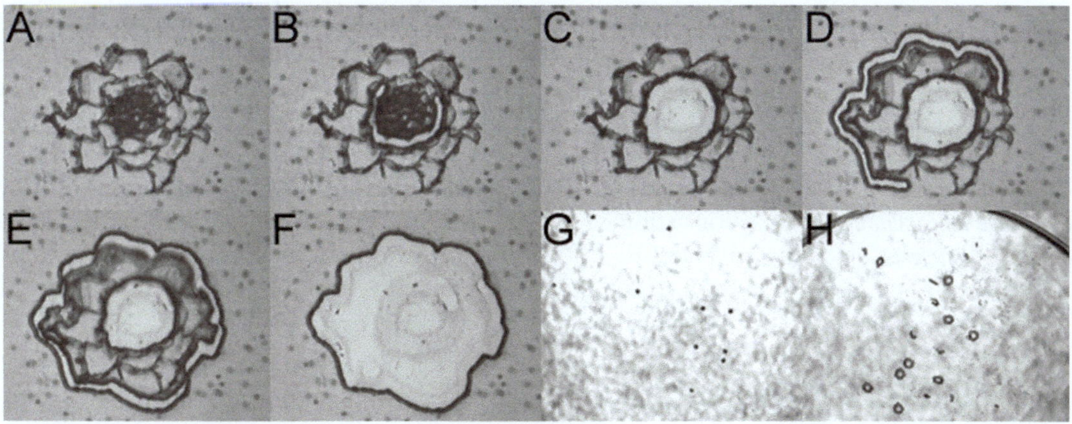

Fig. 1. Laser capture microdissection of *Arabidopsis* root sections, (**a–c**) showing the removal of the inner stelar cells by cutting through the endodermis, (**d–f**) and then the outer root cells. (**g–h**) Captured tissue is visible in the cap of the microfuge tube.

or outer root cells, 25–30 dissected tissue pieces were collected in the cap of one microfuge tube (Fig. 1).

3. Add 50 μL of TRIzol® to the cap of the microfuge tube, close the tube onto the cap gently, then vortex and centrifuge briefly to collect the contents at the base of the tube. Add a further 100 μL of TRIzol® to the main compartment of the tube, then vortex and centrifuge briefly.
4. Freeze samples in liquid nitrogen and store at −80°C until required.
5. When required, allow the samples to warm to room temperature, with periodic vortexing to mix the contents.
6. Add 30 μL of 100% chloroform and shake the microfuge tube for 30 s.
7. Leave samples at room temperature for 3 min, then centrifuge at 12,000 × *g* for 15 min at 4°C.
8. Remove the upper aqueous phase carefully by pipetting and transfer to a new tube (see Note 27).
9. Add 20 μg of glycogen and 150 μL of 100% isopropanol and incubate the sample at room temperature for 1 h.
10. Centrifuge at 12,000 × *g* for 10 min at 4°C.
11. Remove the supernatant carefully using a pipette. Note that you are unlikely to see the RNA pellet at the bottom of the tube (see Note 28).
12. Allow the pellet to air-dry for 5–10 min until the isopropanol has completely evaporated.
13. Re-suspend the RNA in 20 μL of nuclease free water (see Note 29).

14. Keep the sample at −80°C for long-term storage. Proceed straight to cDNA synthesis for qualitative transcript analysis (see Subheading 3.6) or to RNA amplification for quantitative transcript analysis (see Subheading 3.7) (see Note 30).

3.6. LCM: cDNA Synthesis for Qualitative Transcript Analysis

1. To 5 μL of RNA, add 8 μL of an RNase inhibitor solution consisting of 6.5 μL of RNase-free water, 0.25 μL of 40 U/μL RNaseOUT recombinant ribonuclease inhibitor (Invitrogen), 0.25 μL of 10 μM T7-linked Oligo dT_{18} primer (see Note 6), and 1 μL of 10 mM dNTPs.
2. Heat samples at 65°C for 5 min, then place back in ice.
3. Following denaturation and annealing of the RT primer to the sampled RNA, add 4 μL of Invitrogen 5× Superscript III first strand buffer, 1 μL of 0.1 M DTT, 1 μL of RNaseOUT, 0.25 μL of 200 U/μL Superscript III RT enzyme, and 0.75 μL water.
4. Incubate at 50°C for 1 h.
5. Terminate the reaction by heating the samples at 70°C for 15 min.
6. Store the cDNA at −20°C until required, proceed to Subheading 3.8 (see Note 12).

3.7. LCM: RNA Amplification for Quantitative Transcript Analysis

1. Sample RNA as per Subheading 3.3.
2. Proceed with two rounds of RNA amplification using MessageAmp™ II aRNA amplification kit (Ambion) according to manufacturer's instructions.
3. Following the first round of amplification, concentrate RNA from 100 to 10 μL by vacuum centrifugation and use this in the second round of amplification. Following the second round, concentrate 10 of 100 μL aRNA sample to 2 μL and analyze using an Agilent Bioanalyser (see Note 14). Concentrate the remaining 90 to 10 μL and store at −80°C prior to analysis by quantitative PCR (see Note 1), microarray (see Note 15), or RNA-seq.
4. Take a small aliquot of RNA and use RT-PCR to confirm the absence of genomic DNA and contamination of transcripts from undesired cell types (10). Use random hexamer primers in the reverse transcript reaction (Superscript III) to produce cDNA (10).

3.8. Nested Polymerase Chain Reaction of cDNA from Non-amplified RNA: SiCSA and LCM (See Note 31)

1. Thaw the necessary tubes of solutions and vortex to mix.
2. For each sample, prepare a 24 μL reaction mix in a 200 μL PCR compatible microfuge tube, with a final concentration of 1× PCR buffer, 1.5 mM $MgCl_2$ (see Note 32), 200 μM dNTPs, 0.2 μM Forward gene-specific PCR primer, 0.2 μM Reverse T7 PCR primer (see Notes 33 and 34), and 1 U of Platinum *Taq* DNA polymerase (Invitrogen) (see Note 35). Use water to bring volume to 24 μL.

3. Add 1 μL of cDNA from the RT reaction.
4. Using a thermal cycler, incubate the samples at 94°C for 2 min, followed by 30–35 cycles of: 30 s at 94°C to denature the DNA; 30 s at 50–70°C to allow primers to anneal (see Note 36); 1 min at 72°C for DNA extension (see Note 37). After all cycles are complete, incubate the samples at 72°C for 10 min for a final extension.
5. Samples can be stored safely at room temperature overnight and −20 or −80°C for longer storage.
6. PCR amplicons should be purified through a spin column to remove existing primers and eluted in water.
7. Add a 1 μL aliquot of the first purified PCR to a new reaction that contains the reagents outlined previously (step 2), with the exception of now using gene-specific forward and reverse primers which sit within the region amplified during the first experiment. The annealing temperature of this reaction is adjusted to suit the new primers and the extension time shortened to match the product length (usually 150–300 bases).
8. Amplicons are checked by gel electrophoresis.

4. Notes

1. Experience has taught us that it is often preferable to design PCR primers that amplify small regions of cDNA close to the 3′ end of the gene, with product sizes in the range of 200–500 bases. This is often desirable for quantitative PCR primers too (product size 100–300 bases). We have observed lower efficiency in product amplification when using primers that amplify regions greater than 1,000 bases or the primers anneal towards the 5′ end of the gene. This is perhaps due to RNA fragmentation during extraction, with the ends closest to the poly A tail more likely to be amplified during PCR after using an oligo dT primer in the RT reaction. Untranslated regions of DNA can also contain sequences that allow differentiation of close family members.
2. Due to the noxious fumes produced, the oven must be placed in a fume cupboard.
3. This requires trial and error. If the tip of the sampling capillary is too small, it will be easily blocked by cellular contaminants and if the tip is too wide the sap will not be retained within the capillary when removing the sampling tip from the cell.
4. Reserve an old pipette for this job as silane degrades plastics and metal.

5. RNases are highly active and stable enzymes. Great care should be taken to avoid introducing RNases into samples from plastic, hands, etc. This is particularly important when isolating RNA from small samples. Equipment and bench should be cleaned using 70% ethanol and RNase inactivation sprays (e.g., RNAZap). Always wear gloves when handling samples and solutions, and change gloves frequently. Keep tubes closed wherever possible. Use RNase-free disposable tubes and filter tips at every step. At all times, keep the reagents and RNase inhibitor solution on ice.
6. Having a mixture of A, G, and C as the first base before the run of 18 Ts allows the T7-linked Oligo dT primer to bind to the 5′ end of the mRNA's poly A tail.
7. The salmon sperm binds to any remaining DNA and RNA binding sites in the capillary, thereby inhibiting sampled DNA/RNA from sticking to the sides of the glass. This is essential to reduce contamination from other cells when guiding the capillary through plant material to reach the desired cell type (for example, when sampling through epidermal cells to reach mesophyll cells).
8. When extracting RNA from cells within seed tissue, use a razor blade to cut open the seed and place the seed under water, with the cells to be sampled facing up.
9. Due to the low cytosolic content of some cells, it is often necessary to sample from multiple cells per sample to collect sufficient mRNA (9, 10). We routinely amplified RNA from pooled samples of 30 epidermal or 3 mesophyll cells. Although samples were pooled, a new capillary was used for every cell extracted.
10. To sample from leaf mesophyll cells in cereals, it is possible to guide the sampling capillary carefully through an open stomata to sample from the mesophyll cell below. In other instances, such as sampling from *Arabidopsis* leaves, mRNA from mesophyll cells can be obtained by puncturing an overlying epidermal cell, expelling the epidermal sap from the tip, and then sampling from the mesophyll cell below (see Note 11).
11. To establish whether mesophyll mRNA has been contaminated with RNA from the overlying epidermis, look for the presence of transcripts of epidermal-specific genes. In addition to using known cell-specific transcripts, another means of addressing this problem is to sample transgenic plants that express the reporter gene *GFP* in epidermal cells (such as *Arabidopsis* lines available from Jim Haseloff, University of Cambridge). With these plants, it is possible to examine for contamination by *GFP* mRNA from epidermal cells, in sap collected from mesophyll cells (8–10). Fluorescent protein may also be detected if

sampling under fluorescent stereomicroscope to optimize the sampling technique.

12. We have successfully used three different reverse transcriptases: Qiagen's Sensiscript and Invitrogen's Superscript II and III. Superscript II and Sensiscript were utilized to reverse transcribe aRNA (following IVT) with random primers (hexamers or nonamers). Superscript III was utilized to primer mRNA (i.e., from SiCSA) with Oligo dT, T7-linked oligo dT, or gene-specific primers due to its higher optimum activity temperature
13. Two rounds of amplification using Target AMP-2 were sufficient to produce 1–2 μg of RNA. The input material needed for this could be as low as 3 mesophyll cells or 30 epidermal cells from *Arabidopsis thaliana*. Fresh sampling capillaries should be used for each mesophyll cell. We have tested several RNA amplification kits and found the Target AMP-2 kit to produce a linear amplification for numerous transcripts when the threshold exceeded ten copies per cell. This can be checked by quantitative PCR.
14. It is important to test that the product size distribution is normal with an average size >400 and the picoRNA chip on an Agilent Bioanalyser prior to proceeding for further analysis. Do not use a nanodrop to quantify RNA.
15. For microarray analysis, we tested the method of labeling the RNA with the fluorophores for 2-color microarrays. Firstly, the kit which incorporated aminoallyl-UTP into the RNA after the second round of amplification was used with NHS-ester containing Cy3 and Cy5. This produced a degree of labeling (DoL) of 1.8% (equivalent to 1.8 Cy-UTPs per 100 nucleotides), with an average product length of 449 nt. We subsequently tested the kit excluding the aminoallyl-dUTP and used a direct labeling kit incorporating the Universal Labeling System (ULS) to covalently attach Cy dyes to the N7 residue of guanines for both Cy3 and Cy5 (Kreatech Biotechnology, Product# EA-021). Following RNA purification using the *Kreapure* columns (Kreatech Biotechnology), it was found that this technique improved both the labeling efficiency (average DoL = 2.8%) and the length (525 nt) of the amplified RNA (aRNA), the latter a probable result of the absence of modified nucleotides in the second in vitro transcription reaction.
16. Paraffin will solidify in the neck of small centrifuge tubes.
17. Decant (rather than pipette) solutions to minimize damage. When decanting, retain a small amount of solution at the bottom of the tube to prevent tissue desiccation. Tissue will generally stay at the bottom of the tube during decanting.
18. Use a molecular sieve to remove the water from 99.9% ethanol to make 100% ethanol.

19. Between all solution changes, mix gently then decant solution until tissue moves to the top of tube; prevent any carry-over of previous solution.

20. It is important to work quickly but carefully to prevent paraffin wax solidifying in the tube. Paraffin is only partly miscible with isopropanol and mixtures are opaque. Tissue will be difficult to see, but usually settles to the bottom of the tube and should stay there if the tube is tipped smoothly. If desired, keep decanted solution at 60°C to retrieve lost tissue.

21. Alternatively, float the ribbon directly onto a water-covered slide on a 42°C hot plate, then tip gently to drain with filter paper.

22. Use superglue to glue an eyelash or eyebrow hair to a bamboo skewer, allow to dry, then treat for RNase.

23. We also have successfully used RNAqueous-micro Kit (Ambion), following the manufacturer's instructions, performing the optional DNase I treatment and DNase inactivation.

24. We found that occasionally having a solution in the cap of the microfuge tube ensured that more of the tissue sections extracted were collected and less were lost due to sticking to the outsides of the cap by static charge.

25. Nuclease free water (20 μL) can be used in the cap.

26. For some tissues, it is necessary to traverse the laser over the same area several times due to the tough plant cell walls. This was particularly so when dissecting root endodermal tissue. When dissecting inner and outer root cells, we used the laser to first destroy the endodermal cells, and thus capture the inner stelar root cells.

27. When removing the upper aqueous phase from the TRIzol/chloroform, care should be taken not to remove any of the lower solvent phase. Compared with TRIzol/chloroform extractions on whole tissue, the interphase is more difficult to distinguish.

28. Due to the small amount of sampled tissue, it is unlikely that you will observe an RNA pellet. To avoid accidently removing the RNA with the isopropanol, add glycogen and orient the microfuge tube into the centrifuge with its hinge facing outwards. After spinning, the RNA pellet will be at the base of the tube on the same side of the wall as the hinge. Extract the isopropanol by placing the tip of the pipette at the base of the tube on the opposite side from the pellet.

29. Due to the possibility of residual RNase activity in DNases, we frequently avoid DNase treatment on these small RNA preparations. To establish whether we have significant DNA contamination in our RNA, we make sure that primers for PCR

are designed to flank an intron:exon junction in the target gene (in addition to using T7-linked oligo dT_{18} primers for reverse transcription). By examining the PCR product on a gel, we can easily establish from the observed banding pattern whether any genomic product has been amplified in addition to a product from cDNA.

30. In our experience, if proceeding to amplification best results will be achieved if this is done within 24 h.
31. It is possible to directly amplify cDNA products from RNA extracted from a small number of cells, using a large number of PCR cycles. Usually around 45 cycles will allow observation of amplicons by agarose gel electrophoresis. It is therefore critical to design PCR primers that do not easily form primer dimers, which will impede the reaction and produce gels with bright PCR dimer smears.
32. The final concentration of Mg^{2+} in the 1× PCR buffer is 1.5 mM, which should be adequate for most PCRs. It is possible to optimize the Mg^{2+} concentration in the reaction by adding different volumes of the $MgCl_2$ supplied with the kit.
33. We employ a T7 promoter sequence at the 5′ end of the oligo dT_{18} primer used for reverse transcription and use a T7 primer (GAATTGTAATACGACTCACTATAGGCG) and a gene-specific primer in the first round of PCR amplification. This makes certain only cDNA is amplified during subsequent PCR steps and ensures no genomic DNA amplicons are produced.
34. It is important to include a cell-specific control for each cell type sampled, to confirm sampling accuracy. Routinely, we incorporate additional reactions to detect transcripts that are present in one cell type but not the other.
35. We have also had success using Qiagen HotStar*Taq* DNA polymerase.
36. Use the annealing temperature predicted for the specific primers being used. During optimization of the reaction, it is advisable to set the annealing temperature 5°C below the predicted T_m of the primers.
37. For products longer than 1 kb, increase the extension time by 1 min for every 1 kb.

Acknowledgements

We would like to thank Grain Research Development Corporation, Australian Research Council, and University of Adelaide for funding.

References

1. Conn S, Gilliham M (2010) Comparative physiology of elemental distributions in plants. Ann Bot 105:1081–1102
2. Galbraith DW, Birnbaum K (2006) Global studies of cell type-specific gene expression in plants. Annu Rev Plant Biol 57:451–475
3. Tester M, Leigh RA (2001) Partitioning of nutrient transport processes in roots. J Exp Bot 52:445–457
4. Leegood RC (1985) Intracellular compartmentation of metabolites in leaves of *Zea mays* L. Planta 164:163–171
5. Munns R, Tester M (2008) Mechanisms of salinity tolerance. Annu Rev Plant Biol 59:651–681
6. Plett DC, Skrumsager Møller I (2010) Na^+ transport in glycophytic plants: what we know and would like to know. Plant Cell Environ 33:612–626
7. Teakle NL, Tyerman SD (2010) Mechanisms of Cl^- transport contributing to salt tolerance. Plant Cell Environ 33:566–589
8. Roy SJ, Gilliham M, Berger B et al (2008) Investigating glutamate receptor-like gene co-expression in *Arabidopsis thaliana*. Plant Cell Environ 31:861–871
9. Conn SJ, Conn V, Tyerman SD et al (2011) Magnesium transporters, *MGT2/MRS2-1* and *MGT3/MRS2-5*, are important for magnesium partitioning within *Arabidopsis thaliana* mesophyll vacuoles. New Phytol 190:583–594
10. Conn SJ, Gilliham M, Athman A et al (2011) Cell-specific vacuolar calcium storage mediated by *CAX1* regulates apoplastic calcium concentration, gas exchange, and plant productivity in *Arabidopsis*. Plant Cell 23:240–257
11. Li P, Ponnala L, Gandotra N et al (2010) The developmental dynamics of the maize leaf transcriptome. Nat Genet 42:1060–1067
12. Møller I, Gilliham M, Jha D et al (2009) Salinity tolerance engineered by cell type-specific over-expression of a Na^+ transporter in the *Arabidopsis* root. Plant Cell 21:2163–2178
13. Nelson T, Tausta SL, Gandotra N et al (2006) Laser microdissection of plant tissue: what you see is what you get. Annu Rev Plant Biol 57:181–201

Chapter 23

Plastid Transformation for Abiotic Stress Tolerance in Plants

K.C. Bansal, A.K. Singh, and S.H. Wani

Abstract

Abiotic stresses such as drought, salinity, and extreme temperatures are major limiting factors in plant growth and development and pose serious threat to global agricultural production. Here we describe a procedure, using a tobacco plastid transformation vector, to generate transplastomic plants with an enhanced ability to tolerate abiotic stresses such as salinity, drought, or cold stress. The procedure involves biolistic delivery of a plastid transformation vector into explants, antibiotic selection procedures, and identification of transplastomic lines. The plastid transformation vector contains an *aadA* gene that encodes resistance to spectinomycin as a selectable marker along with the gene of interest for developing transplastomic plants that are tolerant to abiotic stresses. Shoot buds appear over the surface of bombarded explants following spectinomycin selection. Transplastomic shoots are multiplied following several rounds of spectinomycin selection. Homoplasmic transplastomic lines are confirmed by spectinomycin and streptomycin double selection over a period of 4–5 weeks. The available reports suggest that transplastomic technology is a useful tool for expressing genes in plastids or chloroplasts for enhancing abiotic stress tolerance in plants.

Key words: Plastid transformation, Abiotic stress, Salt stress, Drought stress

1. Introduction

Increasing environmental stress factors such as salinity, drought, high temperature, and cold/freezing, pose serious threats to global agricultural production (1). Abiotic stresses adversely affect the physiological status of an organism by altering its metabolism, growth, and development. The situation is further aggravated with changing climate. Thus, it is of the utmost important to use modern tools of agricultural biotechnology for enhancing plant's capacity to tolerate a range of abiotic stresses, especially in the era of global climatic change (2, 3). Thus far, conventional breeding

Sergey Shabala and Tracey Ann Cuin (eds.), *Plant Salt Tolerance: Methods and Protocols*, Methods in Molecular Biology, vol. 913, DOI 10.1007/978-1-61779-986-0_23, © Springer Science+Business Media, LLC 2012

methods have met with limited success in improving crop productivity under adverse environmental stresses. The development of transgenic plants through nuclear transformation has shown promise in increasing crop production under water deficit or salt-affected conditions (4). However, introduction of transgene(s) into the nuclear genome has led to growing environmental risks, especially pollen-mediated transgene outflow to weedy or wild relatives of the transgenic plants. While the area under transgenic crops is steadily increasing at the global level, unintended genetic and epigenetic effects of transgenic crops produced via nuclear transformation need to be tackled for gaining public confidence in accepting the benefits of transgenic technology. Plastid genetic engineering is fast emerging as an alternative but effective tool for transferring genes to plastids for improving crops. Recently, Clarke and Daniell (3) have described the usefulness of plastid transformation technology for introducing agronomic traits into crop plants. Genetic engineering via plastid transformation has proven advantageous over nuclear transformation in several ways including transgene integration at a pre-determined site on a plastid genome via homologous recombination and high levels of foreign protein accumulation in plastids (5). Stacking of transgenes for engineering metabolic pathways in plants has also become feasible through transplastomic technology (6). Further, maternal inheritance of plastids has led to transgene containment—a significant advantage over nuclear transformation (7–9). Thus, numerous benefits leading to improved safety and the ease of selectable marker gene elimination have made this technology more attractive than the conventional nuclear transformation methods for expressing genes encoding useful agronomic traits. Plastid transformation has been reported in many plant species (10). Recently, we developed a plastid transformation system in eggplant (11) and work is underway to introduce the *otsB-A* operon for enhanced tolerance to salt stress. Tolerance to salinity has been achieved by Kumar et al. (12) in carrot cultured cells, roots and leaves by expressing the *betaine aldehyde dehydrogenase* gene. Lee et al. (13) reported that accumulation of trehalose in transplastomic tobacco confers drought tolerance. According to a recent report by Martret et al. (14), expression of dehydroascorbate reductase, glutathione-*S*-transferase, and glutathione reductase in the tobacco chloroplasts led to enhanced tolerance to abiotic stresses.

Here we describe a protocol for stable plastid transformation for enhanced tolerance to abiotic stresses such as salinity and drought in plants. Although the plastid transformation vector reported in this procedure was initially successfully used for tobacco, this vector can be used to develop transplastomics in other solanaceous plant species such as eggplant and tomato. In order to achieve plastid transformation in other plant species, plant-specific plastid transformation vectors should be constructed.

2. Materials

2.1. Explant for Bombardment

1. To produce explants for particle bombardment, grow plants under in vitro conditions at 16/8 h light/dark photoperiod with temperatures of 22–25°C.
2. Explants are prepared from uniform in vitro grown seedlings.

2.2. Culture Media Components

1. Composition of Murashige and Skoog (15) basal medium: Macro salts (20× stock): 33 g NH_4NO_3, 38 g KNO_3, 3.4 g KH_2PO_4, 7.4 g $MgSO_4 \cdot 7H_2O$, 8.8 g $CaCl_2 \cdot 2H_2O$. Make up to 1 L with distilled water and store at 4°C after autoclaving. Use 50 mL stock solution of macro salt for 1 L MS medium. Micro salts (1,000× stock): 10 g $MnSO_4$, 6.2 g H_3BO_3, 5.8 g $ZnSO_4 \cdot 7H_2O$, 0.8 g KI, 0.25 g $Na_2MoO_4 \cdot 2H_2O$, 0.025 g $CuSO_4 \cdot 5H_2O$, 0.025 $CoCl_2 \cdot 6H_2O$. Make up to 1 L with distilled water. Autoclave and store at 4°C. Use 1 mL stock solution of micro salts for 1 L MS medium. Iron solution (500× stock): 18.65 g Na_2EDTA, 13.90 g $FeSO_4 \cdot 7H_2O$. Make up to 1 L with distilled water. Stock solution should be protected from the light, autoclaved, and stored at 4°C. Use 2 mL iron stock solution for 1 L MS medium. Vitamins/inositol solution (250× stock): 25 g inositol, 0.025 g thiamine HCl, 0.125 g pyridoxine HCl, 0.125 g nicotinic acid, 0.5 g glycine. Make up to 1 L with distilled water and store solution at 4°C. Use 4 mL stock solution of vitamins/inositol for 1 L MS medium.
2. Add 30 g sucrose to make 1 L MS medium.
3. Zeatin riboside at 1 mg/mL: Dissolve 10 mg of zeatin riboside (Sigma, St. Louis, MO, USA) in 300 μL 1 M NaOH to dissolve the powder. Once completely dissolved, make up to a final volume of 10 mL with molecular biology grade water. Filter sterilized the solution with a 0.22 μm filter unit and store the stock solution at –20°C.
4. Spectinomycin at 20 mg/mL: Dissolve 200 mg of spectinomycin (Sigma, St. Louis, MO, USA) in 10 mL of molecular biology grade water and filter sterilize; aliquot in 1.0 mL volumes and store at –20°C.
5. Streptomycin at 20 mg/mL solution: Dissolve 1 g of streptomycin (Sigma, St. Louis, MO, USA) in 50 mL of molecular biology grade H_2O. Sterilize solution using a 0.22 μm filter unit and store the stock solution at –20°C.

2.3. Culture Media

1. MS0 seed germination medium: composition as basal MS medium above, with 0.8% (w/v) agar (Bacteriological grade, Hi-media, Mumbai, India) for solidification.

2. MS1 pre-culture medium: MS basal medium with 1.0 mg/L zeatin riboside and 0.8% agar.
3. MS2 selection medium: MS basal medium supplemented with 300 mg/L spectinomycin and 1.0 mg/L zeatin riboside. Add 0.8% (w/v) agar (Bacteriological grade, Hi-media, Mumbai, India) for solidification after adjusting pH of the medium.
4. MS3 double selection medium: MS basal medium supplemented with 300 mg/L each spectinomycin and streptomycin and 1.0 mg/L zeatin riboside. Add agar as mentioned above.
5. MS4 shoot elongation medium: MS basal medium supplemented with 300 mg/L spectinomycin, 1.0 mg/L zeatin riboside, and 0.2 mg/L NAA. Add 0.8% agar for solidification of medium.
6. ½-MS0 rooting medium: ½-strength MS basal medium with 20 g/L sucrose and add 0.8% agar to solidify the medium.

2.4. Biolistic Delivery System

The PDS-1000/He biolistic device (Bio-Rad), 1,100 psi rupture discs (Bio-Rad) and macro-carriers (Bio-Rad), and stopping screen (Bio-Rad).

2.5. Tungsten Particles Coated with Plasmid DNA (Plastid Transformation Vector Containing aadA Gene as Selectable Marker and Gene of Interest for Abiotic Stress Tolerance)

Sterile tungsten particles (50 mg/mL) suspended in ethanol. Purified plasmid DNA 1.0 μg/μL (see Note 1). 2.5 M $CaCl_2$ solution: Dissolve 3.68 g of $CaCl_2$ in 10 mL of water; filter sterilize. The solution should be stored at 4°C for short periods only. Do not freeze. Spermidine: Prepare a 0.1 M spermidine free base (Sigma, St. Louis, MO, USA) solution in sterile distilled water, filter sterilize. Store at −20°C (see Note 2). Ice-cold absolute ethanol.

2.6. Total DNA Extraction from Plant Tissues

CTAB (cetyltrimethyl ammonium bromide) method.

2.7. RNA Extraction

RNeasy Plant Mini Kit (Qiagen, Valencia, CA, USA). RNase-Free DNase set (Qiagen, Valencia, CA, USA).

3. Methods

3.1. Preparation of Explants for Bombardment

1. Grow plants aseptically on MS0 medium without any phytohormones.
2. Place explants (2–4 mm in length) on agar solidified MS medium supplemented with 1 mg/L zeatin riboside.

3.2. Preparation of Tungsten Suspension

Suspend 50 mg tungsten powder in 1 mL absolute ethanol and vortex until particles are clearly dispersed. Centrifuge briefly, remove supernatant, and resuspend in 1 mL of absolute ethanol. Repeat this step twice and finally resuspend tungsten particles in 1 mL sterile distilled water. Store tungsten suspension at –20°C (see Note 3).

3.3. Coating Tungsten Particles with Plasmid DNA

Take 50 µL of 50 mg/mL tungsten particle suspension. Add 10 µL plasmid DNA (1.0 µg/µL) and vortex immediately to assure good contact of the plasmid DNA with particles. Very quickly add 50 µL 2.5 M $CaCl_2$ and 20 µL 0.1 M spermidine to the particles and vortex for 1 min. Add in order the above-mentioned components (see Note 4). Precipitation of the plasmid DNA onto the particle occurs during the vortexing. Discard the supernatant after a short centrifugation and wash the particles with 200 µL ethanol. Redissolve in 30 µl ethanol and use 5 µL DNA-coated tungsten particles per bombardment.

3.4. Particle Bombardment of Explants

Sterilize the rupture disc, macro-carriers, and stopping screen with absolute ethanol for 5 min (see Note 5) and air-dry them in a laminar flow chamber. The gene gun should be placed in a laminar flow chamber. Sterilize the gene gun by wiping with 70% ethanol (see Note 6). Turn the gun on. Turn the pump on. Turn the gas cylinder on. Pipet 5 µL of the coated tungsten particle (micro-carriers) onto the center of macro-carrier and place it inside a sterile Petri dish to allow ethanol evaporation. Following the evaporation of the ethanol-coated particle will be attached to the macro-carrier. Place target cultures into the chamber. Place a rupture disk (1,100 psi) into the rupture disk-retaining cap. Operate the particle gun according to the instructions supplied with gene gun PDS 1000/He. After bombardment, release the shooting button and remove the vacuum by setting the vacuum button to vent. Take the petri plate out of the chamber.

3.5. Culture, Selection, and Plant Regeneration

1. After bombardment, incubate the bombarded explants for 7 days on MS medium supplemented with 1 mg/L zeatin riboside and then transfer onto selection medium (MS medium supplemented with 1 mg/L zeatin riboside and 300 mg/L spectinomycin) for induction and selection of transplastomic shoots. Spectinomycin resistant shoot buds appear from cut end of explants after 6–7 weeks. Several rounds of selection are required to get spectinomycin resistant shoots (see Note 7).
2. Subculture the regenerated shoots at intervals of 3 weeks and finally transfer to MS medium containing 0.5 mg/L zeatin riboside and 300 mg/L of both spectinomycin and streptomycin in order to get homoplasmy. Homoplasmic lines are obtained when spectinomycin resistant shoots are cultured on a double selection medium for about 6 weeks (see Note 8).

3. Transfer the transformed shoots to half-strength MS medium without any phytohormones for rooting. This normally occurs 2–3 weeks after culture.
4. Finally, the transplastomic plants are transferred to soil and grown to maturity for harvesting seeds (see Note 9).

3.6. Identification of the Transplastomic Lines

3.6.1. PCR and DNA Blot Analysis

The integration of the *aadA* gene or gene of interest into the chloroplast genome is verified by polymerase chain reaction (PCR) and DNA blot analysis. Extract total cellular DNA from leaves of transplastomic and wild-type plants using the CTAB method (16). Run the PCR products on a 1% agarose gel. Perform DNA blot hybridization using 10 μg of total cellular DNA. Run the digested DNA fragments on a 0.8% agarose gel. Transfer the agarose gel fractionated DNA to a nylon membrane (Hybond N^+, Amersham) for 16 h. Perform UV cross-linking of the transferred DNA. Label the probe with [α-^{32}P]-dCTP using the Random Primer Labeling Kit (Stratagene) and hybridize for 16 h at 65°C with rotation in a hybridization oven. Wash the membrane with 2× SSC + 0.1% SDS, 1× SSC + 0.1% SDS, and 0.1× SSC + 0.1% SDS at 60°C for 20 min. An autoradiogram is obtained after overnight exposure at –80°C.

3.7. Genetic Stability of Transplastomic Lines

1. As a control, germinate wild-type and T1 transplastomic seeds on ½-MS0 medium. The seeds should germinate normally.
2. Germinate the wild-type and transplastomic T1 seeds on ½-MS medium containing 300 mg/L spectinomycin in order to check the transmission of the spectinomycin resistance trait in seed progeny. The seedlings of transplastomic lines should be uniformly spectinomycin resistant.

3.8. Seed Germination under Salt Stress

1. Germinate the wild-type and T1 transplastomic seeds on ½-MS medium containing 300 mg/L spectinomycin and 200–300 mM NaCl. The transplastomic seeds, from say plants transformed with *otsB-A* operon, should grow normally, whereas wild-type seeds should not germinate.

4. Notes

1. It is strongly recommended that only purified plasmid DNA, free of proteins, is used to avoid formation of clumps with tungsten particles.
2. Spermidine solution, being very hygroscopic and oxidizable, should be aliquoted and stored properly only at –20°C.
3. Clump formation of plasmid DNA with tungsten particles needs to be avoided at any cost during bombardment for

successful plastid transformation. Clump formation greatly reduces transformation efficiency. Hence, the tungsten particle suspension stock should be stored for short periods only at −20°C and frequently vortexed before use.

4. Use plasmid DNA of concentration not less than 1 μg/μL. Mix in the following order: Plasmid DNA, 50 μL of 2.5 M $CaCl_2 \cdot 2H_2O$ and 20 μL of 0.1 M spermidine to 50 μL of 50 mg/mL tungsten particle suspension. This way, it helps efficient coating of plasmid DNA on tungsten particles.
5. To avoid contamination, it is recommended that macro-carriers, stopping screens, and rupture discs are sterilized thoroughly with ethanol and properly air-dried while working in the laminar flow.
6. The biolistic gene gun also needs to be thoroughly decontaminated by wiping it with 70% ethanol prior to every use, and is used inside a laminar air flow cabinet only.
7. It is recommended that the initial selection of the plastid transformants is performed with spectinomycin only because streptomycin selection delays shoots formation.
8. It is mandatory that shoots resistant to spectinomycin should be further subjected to another round of selection on MS medium containing 1 mg/L zeatin and 300 mg/L of both spectinomycin and streptomycin in order to achieve homoplasmy. Double selection is required to ensure elimination of antibiotic resistant shoots obtained as a result of spontaneous mutations in the 16S rDNA of the plastid genome.
9. To promote establishment of the transformants, plantlets should be maintained at high humidity for at least 1 week to allow proper hardening after transfer from in vitro to in vivo conditions.

References

1. Boyer JS (1982) Plant productivity and environment. Science 218:443–448
2. Varshney R, Bansal KC, Aggarwal PK, Datta SK, Craufurd P (2011) Agricultural biotechnology for crop improvement in a variable climate: hope or hype? Trends Plant Sci 16: 363–371
3. Clarke JL, Daniell H (2011) Plastid biotechnology for crop production: Present status and future perspectives. Plant Mol Biol 76: 211–220
4. Castiglioni P, Warner D, Bensen RJ, Anstrom DC, Harrison J, Stoecker M, Abad M, Kumar G, Salvador S, D'Ordine R, Navarro S, Back S, Fernandes M, Targolli J, Dasgupta S, Bonin C, Luethy MH, Heard JE (2008) Bacterial RNA chaperones confer abiotic stress tolerance in plants and improved grain yield in maize under water-limited conditions. Plant Physiol 147:446–455
5. Sharma RK, Bock R, Bansal KC (2005) Plastid transformation: safer alternative to transgenic plants. Physiol Mol Biol Plants 11:179–185
6. Quesada-Vargas T, Ruiz ON, Daniell H (2005) Characterization of heterologous multigene operons in transgenic chloroplasts: transcription, processing, and translation. Plant Physiol 138:1746–1762
7. Ruf S, Karcher D, Bock R (2007) Determining the transgene containment level provided by

chloroplast transformation. Proc Natl Acad Sci USA 104:6998–7002

8. Svab Z, Maliga P (2007) Exceptional transmission of plastids and mitochondria from the transplastomic pollen parent and its impact on transgene containment. Proc Natl Acad Sci USA 104:7003–7008
9. Sharma RK, Bansal KC (2003) Chloroplast transformation as a tool for prevention of gene flow from GM crops to weedy or wild relatives. Curr Sci India 84:1286–1287
10. Koop HU, Herz S, Golds TJ, Nickelson J (2007) The genetic transformation of plastids. In: Bock R (ed) Cell and molecular biology of plastids, vol 19, Topics in current genetics. Springer-Verlag, Berlin, pp 457–510
11. Singh AK, Verma SS, Bansal KC (2010) Plastid transformation in eggplant (*Solanum melongena* L.). Transgenic Res 19:113–119
12. Kumar S, Dhingra A, Daniell H (2004) Plastid expressed betaine aldehyde dehydrogenase gene in carrot cultured cells, roots and leaves confers enhanced salt tolerance. Plant Physiol 136:2843–2854
13. Lee SB, Kwon HB, Kwon SJ, Park SC, Jeong MJ, Han SE, Byun MO, Daniell H (2003) Accumulation of trehalose within transgenic chloroplasts confers drought tolerance. Mol Breeding 11:1–13
14. Martret BL, Poage M, Shiel K, Nugent GD, Dix PJ (2011) Tobacco chloroplast transformants expressing genes encoding dehydroascorbate reductase, glutathione reductase and glutathione-S-transferase, exhibit altered anti-oxidant metabolism and improved abiotic stress tolerance. Plant Biotechnol J 9: 661–673
15. Murashige T, Skoog F (1962) A revised medium for rapid growth and bioassay with tobacco tissue cultures. Physiol Plant 15: 473–497
16. Murray MG, Thompson WF (1980) Rapid isolation of high molecular weight plant DNA. Nucleic Acids Res 8:4321–4325

Chapter 24

Manipulating Expression of Tonoplast Transporters

Zhigang Li, Man Zhou, Qian Hu, Shane Reighard, Shuangrong Yuan, Ning Yuan, Bekir San, Dayong Li, Haiyan Jia, and Hong Luo

Abstract

Plant vacuoles have multifaceted roles including turgor maintenance, cytosolic pH and ionic homeostasis, plant protection against environmental stress, detoxification, pigmentation, and cellular signaling. These roles are achieved through the coordinated activities of many proteins in the tonoplast (vacuolar membrane), of which the proton pumps and ion transporters have been modified for improved abiotic stress tolerance in transgenic plants. Here we describe a method to manipulate vacuolar H^+-pyrophosphatase in turfgrass and evaluate the impact of the modified tonoplast on the phenotype, biochemistry, and physiology of the transgenics. Creeping bentgrass (*Agrostis stolonifera* L.) plants overexpressing an *Arabidopsis* vacuolar H^+-pyrophosphatase AVP1 exhibited improved growth and enhanced salt tolerance, likely associated with increased photosynthesis, relative water content, proline production, and Na^+ uptake. These transgenic plants also had decreased solute leakage in the leaf tissues and increased concentrations of Na^+, K^+, Cl^-, and total phosphorus in the root tissues. Similar strategies can be employed to manipulate other tonoplast transporters and in other plant species to produce transgenic plants with improved performance under various abiotic stresses.

Key words: Tonoplast, Vacuolar H^+-pyrophosphatase, Creeping bentgrass, Abiotic stress, Salt tolerance, Transformation, Electrolyte leakage

1. Introduction

Vacuoles, the membrane-bound compartments within the cytoplasm of a cell are multifunctional organelles. In plants, vacuoles transport and store many substances including minerals, nutrients, primary and secondary metabolites, proteins, and a variety of xenobiotic and toxic compounds. As such, vacuoles not only accumulate water by osmotic uptake, so maintaining the turgor pressure to provide structure support for cell expansion and plant growth, but

Sergey Shabala and Tracey Ann Cuin (eds.), *Plant Salt Tolerance: Methods and Protocols*, Methods in Molecular Biology, vol. 913, DOI 10.1007/978-1-61779-986-0_24, © Springer Science+Business Media, LLC 2012

also play several metabolic roles in intracellular digestion and in the recycling of waste cellular compounds, cytosolic pH and ionic homeostasis, protection against microbial pathogens and herbivores (through the release of relevant secondary metabolites and proteins), detoxification (through the sequestration of xenobiotics), pigmentation (through anthocyans, flavonoids accumulation), and cellular signaling (through storage and release of Ca^{2+}) (1, 2).

These multifaceted roles of vacuoles are achieved through the coordinated activities of a variety of proteins in the tonoplast (vacuolar membrane). This includes transporters and water channels (3–18). The tonoplast transporter family includes the following members: vacuolar proton ATPase (V-ATPase) and pyrophasphatase (V-PPase) (3–5), metal ion transporters such as Na^+/H^+ antiporters, Ca^{2+}-ATPases, Ca^{2+}/H^+ antiporters (6–10), sugar transporters as well as others (11–15). Genetic modification of various transport components of the tonoplast would lead to alterations in vacuolar content so changing plant development and response to environmental conditions. Manipulating the expression of tonoplast transporters to achieve enhanced plant tolerance to salinity stress is among the most important applications in agricultural biotechnology. Salinity is one of the major abiotic stresses, severely affecting agriculture production. Sodium sequestration into the vacuole is an effective mechanism that alleviates the sodium toxicity in the cytosol, while at the same time, enhancing cell water uptake and turgor maintenance by providing additional osmoticum (19–24).

Vacuolar compartmentation of Na^+ is operated by proton pumps via Na^+/H^+ antiporters (23). Genetic manipulation of Na^+/H^+ antiporters and proton pumps in transgenic plants could lead to modified subcellular compartmentalization of sodium, and consequently, the plant response to salt and drought stress (22–24). This has been demonstrated in many plant species. For example, the overexpression of either Na^+/H^+ antiporters (25–41) or vacuolar H^+-pyrophosphatase (26, 42–49) from different sources all resulted in enhanced salt and drought tolerance in transgenics of various plant species (25–41). Modified expression of vacuolar H^+-pyrophosphatase not only alters sodium sequestration, but also changes the transport and accumulation of auxin regulating organogenesis, thereby enhancing root and shoot growth via facilitation of auxin fluxes (50). Moreover, improved phosphorus uptake has also been achieved in transgenic plants of *Arabidopsis*, rice, and tomato overexpressing the *Arabidopsis* vacuolar proton pyrophosphatase AVP1 (51).

Here, we describe a method to manipulate expression of tonoplast transporters for enhancing plant resistance to salinity and drought stress, the two extremely important abiotic stresses. Specifically, we genetically engineered a perennial monocot turfgrass species, creeping bentgrass (*Agrostis stolonifera* L.), by the ectopic expression of an *Arabidopsis* vacuolar proton pyrophosphatase, AVP1. Transgenic plants overexpressing AVP1 exhibited improved plant growth and enhanced resistance to salt stress. The improved growth in transgenics was likely due to the enhanced plant photosynthesis; AVP1 overexpression led to increased leaf chlorophyll content in transgenic plants. The enhanced salt tolerance in transgenic plants was associated with increased relative water content (RWC), proline production, and Na^+ uptake as well as decreased solute leakage in leaf tissues. In addition, increased concentrations of Na^+, K^+, Cl^-, and total phosphorus were also observed in the root tissues of transgenic plants (49). Similar strategies of vacuolar proton pyrophosphatase modification can be employed to manipulate other tonoplast transporters in other plant species to produce transgenic plants with improved performance under adverse environmental conditions.

2. Materials

1. Gene coding for V-PPase: the cDNA coding for the *Arabidopsis* vacuolar H^+-pyrophosphatase gene *AVP1* (52) was used for chimeric gene construction (see Note 1).
2. Promoter: The promoter of the cauliflower mosaic virus 35S (CaMV 35S) gene was used to drive constitutive expression of *AVP1* in creeping bentgrass (see Note 2).
3. Plant materials: A commercial cultivar of creeping bentgrass (*A. stolonifera* L., cv. Penn A-4) supplied by HybriGene (Hubbard, OR, USA) was used for transformation (see Note 3).
4. The *Agrobacterium tumefaciens* strain LBA4404 (53) was used in all creeping bentgrass transformation experiments.
5. Plasmids: The binary vector pSB11 (54) was used to prepare the *AVP1*-expression chimeric gene construct, p35S-*AVP1*/Ubi-*bar*. The T-DNA region of this pSB11 derivative is shown in Fig. 1. The construct contains the CaMV 35S promoter driving the cDNA of the *Arabidopsis* vacuolar

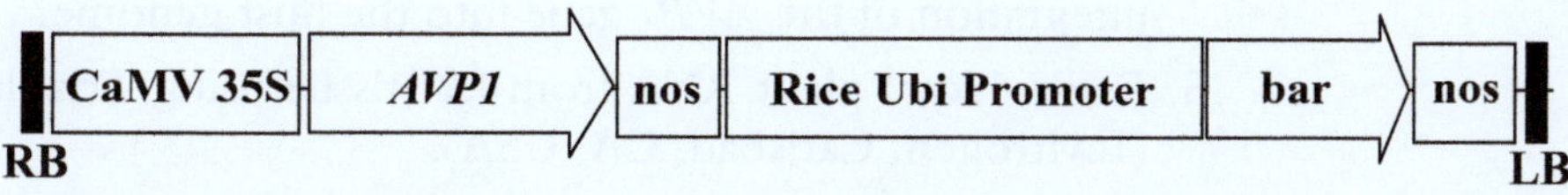

Fig. 1. Schematic diagram of the *AVP1*-expression chimeric gene construct, p35S-*AVP1*/Ubi-*bar*. The *AVP1* gene is under the control of the CaMV 35S promoter, and is linked to the herbicide resistance-conferring gene, *bar*, driven by a rice ubiquitin promoter.

H^+-pyrophosphatase gene *AVP1*. This is linked to the rice ubiquitin promoter (55) driving the *bar* gene for herbicide resistance as the selectable marker (see Note 4).

3. Methods

3.1. Plasmid Construction and Agrobacterium Transformation

1. Release the *AVP1* gene expression cassette from pRG389 (52) and cloned into pUbi-*bar* (Luo unpublished), generating the *AVP1*-expression chimeric gene construct, p35S-*AVP1*/Ubi-*bar*.
2. The construct was mobilized into the *A. tumefaciens* strain, LBA4404 by electroporation, which was then used for plant transformation.

3.2. Plant Transformation, Propagation, and Maintenance

1. Conduct *Agrobacterium*-mediated transformation of creeping bentgrass embryogenic callus initiated from mature seeds to produce transgenic lines stably expressing AVP1 as previously described (56, 57).
2. Transfer the regenerated transgenic plants from tissue culture to commercial potting soil (Fafard 3-B Mix, Fafard Inc., Anderson, SC, USA) or pure silica sand.
3. Maintain the plants in the greenhouse under 16 h photoperiods with supplemental lighting at 25°C in the light and 17°C in the dark.
4. Clonally propagate the AVP1-expressing transgenic creeping bentgrass plants (TG) and the control plants without AVP1 from stolons.
5. Grow the propagated plants in small cone-tainers (4.0 cm×20.3 cm, Dillen Products, Middlefield, OH, USA) (five individual stolons per cone-tainer) or big pots (15 cm×10.5 cm, Dillen Products, Middlefield, OH, USA) (50 individual stolons per pot) using pure silica sand (see Note 5).

3.3. Transgenic Analysis

3.3.1. Molecular Characterization of Transgenic Plants

1. Extract genomic DNA from leaf tissue of each independent transformation event as previously described (58).
2. Conduct polymerase chain reaction (PCR) and Southern blot hybridization as described by Sambrook et al. (59) to verify the integration of the *AVP1* gene into the host genome.
3. Extract total plant RNA from leaf tissue using Trizol reagent (Invitrogen, Carlsbad, CA, USA).
4. Conduct RT-PCR and Northern blot hybridization according to Sambrook et al. (59) to examine *AVP1* expression.

3.3.2. Plant Performance Evaluation Under Salt Stress

1. Maintain the AVP1-expressing transgenic plans and the control plants without AVP1 in growth rooms for 10 weeks.
2. Conduct plant salinity treatment experiments by watering plants daily with 10 mL (for cone-tainers) and 100 mL (for pots) of 200 ppm 20-10-20 Peat-Lite Special (Scotts) fertilizer (N:P:K = 20%:10%:20%) supplemented with 0, 100, 200, and 300 mM of NaCl, respectively for 2–3 weeks.
3. Harvest grass shoots (clippings) and roots every 4 days over 12 days and measure their biomass.
4. Analyze the collected samples to measure their mineral and proline contents as well as other biochemical properties.
5. Evaluate plant recovery from salt injury by watering plants daily with 200 ppm of fertilizer solution for 2–3 weeks (see Note 6).

3.4. Biochemical and Physiological Analyses of Salt-Tolerant Transgenic Plants

3.4.1. Measurement of Mineral Content

1. Harvest shoots (approximately 3 cm above silica sand) and roots from both the AVP1-expressing transgenic plants and the control plants of creeping bentgrass grown under normal or salt stress conditions.
2. Rinse the leaves in Millipore water for 30 s and rinse the roots thoroughly in Millipore water to eliminate the silica sand.
3. Incubate the samples at 80°C for 48 h and measure the samples for dry weights.
4. Determine the minerals and soluble chloride contents in leaves and roots using Spectro ARCOS ICP (Spectro, Mahwah, NJ, USA) following the protocols by Haynes (60) and Plank (61).

3.4.2. Measurement of Leaf Relative Water Content

1. Harvest leaves from both the AVP1-expressing transgenic plants and the control plants of creeping bentgrass grown under normal or salt stress conditions.
2. Weigh the harvested samples immediately for fresh weight (FW).
3. Cut the leaves into pieces and immerse them in Millipore water at 4°C for 16 h.
4. Measure the turgid weight (TW) at the end of the water immersion.
5. Incubate the leaves in an oven at 80°C for 24 h and weigh the samples for dry weight (DW).
6. Calculate leaf RWC using the following formula: $RWC = [(FW - DW)/(TW - DW)] \times 100\%$, where FW is fresh weight, DW is dry weight, and TW is turgid weight.

3.4.3. Measurement of Leaf Electrolyte Leakage for Cell Membrane Integrity

1. Harvest fresh leaves (0.2–0.5 g) from the AVP1-expressing transgenic plants and the control plants of creeping bentgrass grown under normal or salt stress conditions.
2. Cut the leaf samples into small segments and incubate in 20 mL Millipore water at 4°C for 16 h.
3. Measure the conductance of the incubation solution as the initial level of electrolyte leakage (EL) (Ci) using a conductance meter (e.g., AB30, Fisher Scientific, Suwanee, GA, USA). This measurement estimates the amount of the ions released from cells under normal or salt stressed conditions.
4. Kill the leaf tissue in the incubation solution by autoclaving it for 30 min.
5. Incubate the autoclaved leaf tissue and solution at room temperature for 24 h on a shaker.
6. Determine the conductance of the incubation solution containing the killed tissue (C_{max}). This measurement reflects the amount of the ions released from plant cells before and after heat killing (i.e., the total amount of ions contained in the leaf samples).
7. Calculate the relative EL using the following formula: $(Ci/C_{max}) \times 100$.

3.4.4. Measurement of Leaf Proline Content

1. Harvest 100 mg fresh leaves from both the AVP1-expressing transgenic plants and the control plants of creeping bentgrass grown under normal or salt stress conditions.
2. Grind the leaf samples in 2 mL of 3% sulfosalicylic acid.
3. Mix 200 μL of plant extraction with the same amount of acid ninhydrin (200 μL) and glacial acetic acid (200 μL) (see Note 7).
4. Incubate the mixture for 60 min at 100°C.
5. Stop the reaction in an ice bath.
6. Add 1,000 μL of toluene to the mixture and vortex to extract the proline.
7. Measure the absorbance of the toluene layer at 520 nm in a Thermo Spectronic BioMate 3 (Thermo Electron Corp., Waltham, MA, USA).
8. Determine the proline concentration from a standard curve and calculate on a fresh weight basis as follows: [μg proline/mL × (μL toluene/μL sample)/(g sample/10)]/115.5 μg/μmol = μmol proline/g of fresh weight material.

3.4.5. Measurement of Leaf Chlorophyll Content

1. Harvest 100 mg of fresh leaf from both the AVP1-expressing transgenic plants and the control plants of creeping bentgrass grown under normal or salt stress conditions.

2. Cut the leaf samples into small pieces with scissors.
3. Grind the samples for 5 min in 10 mL of 85% acetone in a mortar and pestle to extract the pigment.
4. Transfer the homogenate into a 15-mL Falcon tube and spin at 3,000 × *g* for 15 min.
5. Transfer the supernatant into a new 15-mL Falcon tube and make up to a total volume of 10 mL with 85% acetone.
6. Measure the optical density (absorbance) of the extract at both 663 and 644 nm with the Thermo Spectronic BioMate 3 (Thermo Electron Corp., Waltham, MA, USA).
7. Calculate the concentration of chlorophyll *a* and *b*, in milligram per gram of fresh weight (FW) tissue according to Arnon (62) and Koski (63) using the following formula:

$$\text{Milligram chlorophyll } a/\text{g FW} = 1.07(\text{OD}_{663}) - 0.094(\text{OD}_{644})$$

$$\text{Milligram chlorophyll } b/\text{g FW} = 1.77\ (\text{OD}_{644}) - 0.280(\text{OD}_{663})$$ (see Note 8).

3.4.6. Indole-3-Acid Extraction and Measurement by High-Performance Liquid Chromatography

1. Harvest 15 g of fresh tissues (leaves or roots) from both the AVP1-expressing transgenic plants and the control plants of creeping bentgrass grown under normal or salt stress conditions.
2. Grind the samples in fine powder in liquid nitrogen with a mortar and pestle.
3. Add 50 mL of methanol containing butylhydroxytoluene (1 mg/mL) and extract for 120 min under continuous shaking in the dark.
4. Collect the supernatant and filter through 0.22 μm nylon membrane filter (OSMONICS, Minnetonka, MN, USA).
5. Evaporate the filtrate in a vacuum rotary concentrator (room temperature) down to the aqueous phase.
6. Pass the filtrate again through a 0.22 μm nylon membrane filter.
7. Adjust the pH of the concentrated filtrate to 3.5 with glacial acetic acid (around 3 μL/mL filtrate).
8. Apply the filtrate to a Sep-Pak C-18 cartridge (500 mg, Waters, Milford, MA, USA), which is pre-equilibrated with 2 mL of methanol followed by 2 mL of 50 mM acetic acid.
9. Wash the cartridge with 2 mL of 50 mM acetic acid followed by 2 mL of water.
10. Elute the indole-3-acid (IAA) with 2 mL of methanol.
11. Concentrate the elution in a vacuum rotary concentrator (room temperature) to 200 μL.
12. Further purify the concentrated elution by passing through a 0.22 μm Cellulose Acetate Spin-X® Centrifuge Tube Filter (Corning Inc., Corning, NY, USA).

13. To quantify the IAA from plant tissue extraction by high-performance liquid chromatography (HPLC), connect a YMC-Pack-Pro C18 column (250 mm × 4.6 mm, S-5 μm, 12 nm, YMC Inc, Milford, MA, USA) to the LC-10AT HPLC system (Shimadzu, Kyoto, Japan) with a SPD-20A/AV detector (280 nm).
14. For each sample, inject 20–40 μL of the methanolic extract and elute with 1% (v/v) acetic acid/acetonitrile/(75/25, v/v) at a flow rate of 0.8 mL/min.
15. Quantify the levels of free IAA in samples using a calibration curve of the standards (0, 5, 25, 100, and 500 ppm of IAA) (see Note 9).

4. Notes

1. In the current protocol, *Arabidopsis* AVP1 was used for introduction into creeping bentgrass for ectopic expression. There are many other vacuolar H^+-pyrophosphatase genes cloned from various plant species, and these genes are also quite effective in rendering enhanced salt tolerance when overexpressed in transgenic plants. They all can be used to manipulate vacuolar proton pumps in the tonoplast.
2. Other strong, constitutive promoters can also be used for driving the vacuolar H^+-pyrophosphatase genes. For example, the maize (64) and rice (55) ubiquitin promoters have both been demonstrated to function very well in turfgrass species (49, 56, 57).
3. The seeds should be stored at 4°C until use. Seeds from any other genotypes of the creeping bentgrass can also be used for transformation.
4. The bacterial gene *bar*, which confers resistance to herbicide phosphinothricin (PPT), was used in the current protocol as a positive selectable marker for plant transformation. Other genes, such as the *hyg*, from the bacterium *Streptomyces hygroscopicus*, confers resistance to the antibiotic hygromycin, and can also be used as an effective selectable marker for creeping bentgrass transformation (49, 56, 57).
5. When repotting, all the stolons used for propagations were trimmed to the same size of 2 cm long for both shoots and roots. This ensures that the starting materials from both the AVP1-expressing transgenic and the control plants are in similar conditions.
6. The development of cool season creeping bentgrass at 25°C in the light and 17°C in the dark for 8–10 weeks is necessary

prior to conducting experiments for plant treatments. The illumination in plant growth room should be 350–450 μmol/m^2/s photosynthetically active radiation at canopy height.

7. The acid ninhydrin can be prepared by dissolving 0.3 g ninhydrin in 7.2 mL glacial acetic acid, 1.92 mL of 85% H_3PO_4, and bringing the volume up to 12 mL with water. The reagent remains stable for 24 h at 4°C.
8. Acetone will evaporate during sample grinding and centrifugation. It needs to be added to bring the final volume up to 10 mL.
9. The standards were treated by passing through the cartridge and spin column prior to HPLC. Samples were measured four times and the standard error was calculated.

Acknowledgements

This work was supported by the USDA grant CSREES SC-1700315, and in part by the USDA grants BRAG 2005-39454-16511, 2007-33522-18489, and 2010-33522-21656 as well as a grant from The Consortium for Plant Biotechnology Research, Inc. (CPBR). We are grateful to Dr. Roberto A. Gaxiola for providing plasmid pRG389, Dr. Xi Wang and Dr. Feng Chen for help in HPLC. Technical Contribution No. 5975 of the Clemson University Experiment Station.

References

1. Marty F (1999) Plant vacuoles. Plant Cell 11:587–600
2. Buchanan BB, Gruissem W, Jones RL (2000) Biochemistry and molecular biology of plants. American Society of Plant Physiologists, Rockville, MD
3. Maeshima M (2001) Tonoplast transporters: organization and function. Annu Rev Plant Mol Biol 52:469–497
4. Gaxiola RA, Palmgren MG, Schumacher K (2007) Plant proton pumps. FEBS Lett 581: 2204–2214
5. Fuglsang AT, Paez-Valencie J, Gaxiola RA (2011) Plant proton pumps:regulatory circuits involving H^+-ATPase and H^+-PPase. In: Geislet M, Venema K (eds) Transporters and pumps in plant signaling. Springer, Berlin
6. Apse MP, Blumwald E (2007) Na^+ transport in plants. FEBS Lett 581:2247–2254
7. Rodríguez-Rosales MP, Gálvez FJ et al (2009) Plant NHX cation/proton antiporters. Plant Signal Behav 4:265–276
8. Isayenkov S, Isner JC, Maathuis FJM (2010) Vacuolar ion channels: roles in plant nutrition and signaling. FEBS Lett 584:1982–1988
9. Isayenkov S, Isner JC, Maathuis FJM (2011) Rice two-pore K^+ channels are expressed in different types of vacuoles. Plant Cell 23:756–768
10. Pardo JM, Rubio F (2011) Na^+ and K^+ transporters in plant signaling. In: Eislet M, Venema K (eds) Transporters and pumps in plant signaling. Springer, Berlin
11. Neuhaus HE (2007) Transport of primary metabolites across the plant vacuolar membrane. FEBS Lett 581:2223–2226
12. Kühn C, Grof CPL (2010) Sucrose transporters of higher plants. Curr Opin Plant Biol 13:288–298
13. Ayre BG (2011) Membrane-transport systems for sucrose in relation to whole-plant carbon partitioning. Mol Plant 4:377–394
14. Klaumann S, Nickolaus SD, Fürst SH et al (2011) The tonoplast copper transporter COPT5 acts as an exporter and is required for

interorgan allocation of copper in *Arabidopsis thaliana*. New Phytol 192(2):393–404. doi:10.1111/j.1469-8137.2011.03798.x

15. Meyer S, Scholz-Starke J, De Angeli A et al (2011) Malate transport by the vacuolar AtALMT6 channel in guard cells is subject to multiple regulation. Plant J 67:247–257
16. Chrispeels MJ, Crawford NM, Schroeder JI (1999) Proteins for transport of water and mineral nutrients across the membranes of plant cells. Plant Cell 11:661–675
17. Tyerman SD, Hohnert HJ, Maurel C et al (1999) Plant aquaporins: their molecular biology, biophysics and significance for plant water relations. J Exp Bot 50:1055–1071
18. Bienert GP, Chaumont F (2011) Plant aquaporins: roles in water homeostasis, nutrition, and signaling processes. In: Geislet M, Venema K (eds) Transporters and pumps in plant signaling. Springer, Berlin
19. Blumwald E (2000) Sodium transport and salt tolerance in plants. Curr Opin Cell Biol 12:431–434
20. Blumwald E, Aharon GS, Apse MP (2000) Sodium transport in plant cells. Biochim Biophys Acta 1465:140–151
21. Jiang X, Leidi EO, Pardo JM (2010) How do vacuolar NHX exchangers function in plant salt tolerance? Plant Signal Behav 5:792–795
22. Flowers TJ (2004) Improving crop salt tolerance. J Exp Bot 55:307–319
23. Yamaguchi T, Blumwald E (2005) Developing salt-tolerant crop plants: challenges and opportunities. Trends Plant Sci 10:615–620
24. Møller IS, Tester M (2007) Salinity tolerance of *Arabidopsis*: a good model for cereals? Trends Plant Sci 12:534–540
25. Apse MP, Aharon GS, Snedden WA et al (1999) Salt tolerance conferred by overexpression of a vacuolar Na^+/H^+ antiport in *Arabidopsis*. Science 285:1256–1258
26. Brini F, Hanin M, Mezghani I et al (2007) Overexpression of wheat Na^+/H^+ antiporter TNHX1 and H^+-pyrophosphatase TVP1 improve salt-and drought-stress tolerance in *Arabidopsis thaliana* plants. J Exp Bot 58:301–308
27. Zhang HX, Blumwald E (2001) Transgenic salt-tolerant tomato plants accumulate salt in foliage but not in fruit. Nat Biotechnol 19:765–768
28. Zhang HX, Hodson JN, Williams JP et al (2001) Engineering salt-tolerant *Brassica* plants: characterization of yield and seed oil quality in transgenic plants with increased vacuolar sodium accumulation. Proc Natl Acad Sci USA 98:12832–12836
29. Ohta M, Hayashi Y, Nakashima A et al (2002) Introduction of a Na^+/H^+ antiporter gene from *Atriplex gmelini* confers salt tolerance to rice. FEBS Lett 532:279–282
30. Fukuda A, Nakamura A, Tagiri A et al (2004) Function, intracellular localization and the importance in salt tolerance of a vacuolar Na^+/H^+ antiporter from rice. Plant Cell Physiol 45:146–159
31. Wang J, Zuo K, Wu W et al (2004) Expression of a novel antiporter gene from *Brassica napus* resulted in enhanced salt tolerance in transgenic tobacco plants. Biol Plantarum 48: 509–515
32. Wu CA, Yang GD, Meng QW et al (2004) The cotton *GhNHX1* gene encoding a novel putative tonoplast Na^+/H^+ antiporter plays an important role in salt stress. Plant Cell Physiol 45:600–607
33. Xue ZY, Zhi D, Xue G et al (2004) Enhanced salt tolerance of transgenic wheat (*Tritivum aestivum* L.) expressing a vacuolar Na^+/H^+ antiporter gene with improved grain yields in saline soils in the field and a reduced level of leaf Na^+. Plant Sci 167:849–859
34. Yin XY, Yang AF, Zhang KW et al (2004) Production and analysis of transgenic maize with improved salt tolerance by the introduction of *AtNHX1* gene. Acta Bot Sin 46: 854–861
35. He C, Yan J, Shen G et al (2005) Expression of an *Arabidopsis* vacuolar sodium/proton antiporter gene in cotton improves photosynthetic performance under salt conditions and increases fiber yield in the field. Plant Cell Physiol 46:1848–1854
36. Lu SY, Jing YX, Shen SH et al (2005) Antiporter gene from *Hordum brevisubulatum* (Trin.) link and its overexpression in transgenic tobaccos. J Integr Plant Biol 47:343–349
37. Wu YY, Chen QJ, Chen M et al (2005) Salt tolerant transgenic perennial ryegrass (*Lolium perenne* L.) obtained by *Agrobacterium tumefaciens*-mediated transformation of the vacuolar $Na^+/^+$ antiporter gene. Plant Sci 169: 65–73
38. Zhao FY, Zhang XJ, Li PH et al (2006) Co-expression of the *Suaeda salsa* SsNHX1 and *Arabidopsis* AVP1 confer greater salt tolerance to transgenic rice than the single SsNHX1. Mol Breed 17:341–353
39. Chen H, An R, Tang J-H et al (2007) Overexpression of a vacuolar Na^+/H^+ antiporter gene improves salt tolerance in an upland rice. Mol Breed 19:215–225
40. Rodriguez-Rosales MP, Jiang X, Galvez FJ et al (2008) Overexpression of the tomato K^+/H^+ antiporter LeNHX2 confers salt tolerance by improving potassium compartmentalization. New Phytol 179:366–377

41. Duan XG, Song YJ, Yang AF et al (2009) The transgene pyramiding tobacco with betaine synthesis and heterologous expression of *AtNHX1* is more tolerant to salt stress than either of the tobacco lines with betaine synthesis or *AtNHX1*. Physiol Plantarum 135:281–295

42. Gaxiola RA, Li J, Undurraga S et al (2001) Drought- and salt-tolerant plants result form overexpression of the AVP1 H^+-pump. Proc Natl Acad Sci USA 98:11444–11449

43. Park S, Li J, Pittman JK et al (2005) Up-regulation of a H^+-pyrophosphatase (H^+-PPase) as a strategy to engineer drought-resistant crop plants. Proc Natl Acad Sci USA 102:18830–18835

44. Gao F, Gao Q, Duan XG et al (2006) Cloning of an H^+-PPase gene from *Thellungiella halophila* and its heterologous expression to improve tobacco salt tolerance. J Exp Bot 57: 3259–3270

45. Duan XG, Yang AF, Gao F et al (2007) Heterologous expression of vacuolar H^+-PPase enhances the electrochemical gradient across the vacuolar membrane and improves tobacco cell salt tolerance. Protoplasma 232:87–95

46. Asad S, Mukhtar Z, Nazir F et al (2008) Silicon Carbide Whisker-mediated embryogenic callus transformation of cotton (*Gossypium hirsutum* L.) and regeneration of salt tolerant plants. Mol Biotechnol 40:161–169

47. Li B, Wei A, Song C et al (2008) Heterologous expression of the *TsVP* gene improves the drought resistance of maize. Plant Biotechnol J 6:146–159

48. Lv SL, Zhang K-W, Gao Q et al (2008) Overexpression of an H^+-PPase gene from *Thellungiella halophila* (*TsVP*) in cotton enhances salt tolerance and improves growth and photosynthetic performance. Plant Cell Physiol 49:1150–1164

49. Li Z, Baldwin CM, Hu Q et al (2010) Heterologous expression of *Arabidopsis* H^+-PPase enhances salt tolerance in transgenic creeping bentgrass (*Agrostis stolonifera* L.). Plant Cell Environ 33:272–289

50. Li J, Yang H, Peer WA et al (2005) *Arabidopsis* H^+-PPase AVP1 regulates auxin-mediated organ development. Science 310:121–125

51. Yang H, Knapp J, Koirala P et al (2007) Enhanced phosphorus nutrition in monocots and dicots over-expressing a phosphorus-responsive type I H^+-pyrophosphatase. Plant Biotechnol J 5:735–745

52. Park S, Li J, Pittman JK et al (2005) Up-regulation of a H^+-pyrophosphatase (H^+-PPase) as a strategy to engineer drought-resistant crop plants. Proc Natl Acad Sci USA 102: 8830–18835

53. Hoekema A, Hirsch PR, Hooykaas PJJ et al (1983) A binary plant vector strategy based on separation of *vir*- and T-region of the *Agrobacterium tumefaciens* Ti-plasmid. Nature 303:179–180

54. Komari T, Hiei Y, Saito Y et al (1996) Vectors carrying two separate T-DNAs for co-transformation of higher plants mediated by *Agrobacterium tumefaciens* and segregation of transformants free from selection markers. Plant J 10:165–174

55. Huq E, Hirayama L, Hossain MA et al (1997) Characterization of a cDNA encoding a polyubiquitin gene in rice (accession no. U37687) (PRG97-004). Plant Physiol 113:305

56. Luo H, Hu Q, Nelson K et al (2004) *Agrobacterium tumefaciens*-mediated creeping bentgrass (*Agrostis stolonifera* L.) transformation using phosphinothricin selection results in a high frequency of single-copy transgene integration. Plant Cell Rep 22:645–652

57. Luo H, Kausch AP, Hu Q et al (2005) Controlling transgene escape in GM creeping bentgrass. Mol Breed 16:185–188

58. Luo H, van Coppenolle B, Seguin M et al (1995) Mitochondrial DNA polymorphism and phylogenetic relationships in *Hevea brasiliensis*. Mol Breed 1:51–63

59. Sambrook J, Fritsch EF, Maniatis T (1989) Molecular cloning: a laboratory manual. Cold Spring Harbor Laboratory Press, Cold Spring Harbor, NY

60. Haynes RJ (1980) A comparison of two modified Kjeldahl digestion techniques for multi-element plant analysis with conventional wet and dry ashing methods. Commun Soil Sci Plant Anal 11:459–467

61. Plank CO (1992) Plant analysis reference procedures for the Southern Region of the United States. South Cooper Bull 368, May 1992

62. Arnon DI (1949) Copper enzymes in isolated chloroplasts; polyphenol-oxidase in *Beta vulgaris*. Plant Physiol 24:1–15

63. Koski VM (1950) Chlorophyll formation in seedlings of *Zea mays* L. Arch Biochem Biophys 29:339–343

64. Christensen AH, Sharrock RA, Quail PH (1992) Maize polyubiquitin genes: structure, thermal perturbation of expression and transcript splicing, and promoter activity following transfer to protoplasts by electroporation. Plant Mol Biol 18:675–689

Chapter 25

Using Heterologous Expression Systems to Characterize Potassium and Sodium Transport Activities

Alonso Rodríguez, Begoña Benito, and Olivier Cagnac

Abstract

The expression of plant transporters in simple well-characterized cell systems is an irreplaceable technique for gaining insights into the kinetic and energetic features of plant transporters. Among all the available expression systems, yeast cells offer the highest simplicity and have the capacity to mimic the in vivo properties of plant transporters. Here, we describe the use of yeast mutants to express K^+ and Na^+ plant transporters and discuss some experimental problems that can produce misleading results.

Key words: Potassium transport, Sodium transport, Vacuole exchanges, Yeast mutants

1. Introduction

Salt toxicity is the detrimental result of a complex plant response to an aggressive environment, in which among other effects the excess of Na^+ in the soil perturbs plant K^+ homeostasis. Interestingly, the concentrations of Na^+ that produce toxic effects are not necessarily higher than the Na^+ concentrations that exist in the extracellular fluids of mammals, 150 mM, yet mammalian cells are able to maintain low, atoxic Na^+ contents. Therefore, identifying the transporters that mediate K^+ and Na^+ uptake in root cells, the distribution in cell organelles and its long distance transport will be key in understanding the effects of Na^+ in plants. However, this identification may not be sufficient if the kinetics and functional mechanisms of the transporters are not elucidated. The best approach to achieve this is to perform direct studies in the cells in which the transporters are expressed. To be successful, this approach requires the comparison of wild type and mutant plants and that the transporter

Sergey Shabala and Tracey Ann Cuin (eds.), *Plant Salt Tolerance: Methods and Protocols*, Methods in Molecular Biology, vol. 913,
DOI 10.1007/978-1-61779-986-0_25,

encoded by the disrupted gene does not have redundant systems within cells. However, these conditions do not always apply. Therefore, the use of an expression system that mimics the native function of the plant transporters and that is also devoid of endogenous systems of redundancy is ideal (1).

The energization of the plant plasma membrane and the mechanisms of K^+ and Na^+ uptake are similar in plant and fungal cells (2). Thus, fungi are convenient systems in which to express plant K^+ and Na^+ transporters. This conclusion applies to the many yeast species (3) that can be handled in the laboratory with simple microbiological techniques. Among these species, *Saccharomyces cerevisiae* is a model organism (4) that has been deeply investigated from many biological perspectives, including alkali cation transport (2). The first K^+ uptake (5) and Na^+ efflux (6) mutants in fungal species were identified in *S. cerevisiae*; and were used for the first identification of a plant K^+ uptake system (7, 8). In addition to all the knowledge accumulated on its physiology and molecular biology, *S. cerevisiae* was the first eukaryotic species to have its genome sequenced (9). This scientific information and technical facilitation has made *S. cerevisiae* a convenient system for the functional expression of plant genes encoding many different types of transporters including those mediating K^+ and Na^+ uptake, efflux and accumulation into the vacuole. Convenient expression mutants for each of these transporters are available.

In addition to yeast mutants, other heterologous expression systems have been used to express plant K^+ and Na^+ transporters. *Xenopus laevis* oocytes are an excellent system used to reveal the electrical properties of the transporters, which must be determined to understand the function of K^+ channels (10). *Escherichia coli* mutants are an excellent model system for examining transporter functions when there is no functional expression in yeast cells (11, 12). Recently, several mutants of the moss *Physcomitrella patens* have been constructed (13, 14). If a larger collection of *P. patens* mutants becomes available, the amenability of this species to molecular manipulation (15) gives it the potential to provide another excellent system for testing plant transporters.

For the functional tests of plant transporters in *S. cerevisiae* three types of mutants can be used: (1) *trk1 trk2* mutants in which K^+ and Na^+ uptake is defective and the function of plant K^+ and Na^+ uptake systems can be revealed with little interference of the endogenous systems (16–18); (2) *ena1-4 nha1* mutants, which are devoid of Na^+ efflux and reveal the function of plant Na^+ efflux systems (19); and (3) *nhx1*, *vnx1*, and *vcx1* mutants (20) for testing the function of vacuolar K^+ or Na^+/H^+ antiporters in tonoplast membrane vesicles. Although the functions of many plant transporters have been revealed using these strains, in some cases the protein is expressed but the transporter is not functional because *S. cerevisiae* cells do not activate it. In these cases, *the transformants* can be used for the selection of permanently activated mutant transporters (13, 21, 22). This would not only allow the study of transporter functions, but also provide information about the regulation of the transporters.

2. Materials

2.1. Biological Materials

1. The vector pYPGE15 (23) has been extensively used for cloning and functional expression of plant cDNAs in yeast cells, but any type of high-expression vectors could be used with similar results (see Note 1).
2. The *S. cerevisiae* mutants: W303Δ6 (*Mat*a *ade2-1 ura3-11 trp1-1 can1-100 trk1Δ::LEU2 trk2Δ::HIS3*) (24), defective in K^+ uptake; B31 (*Mat*a *ade2-1 ura3-11 trp1-1 can1-100 ena1ΔHIS3::ena4Δ nha1Δ::LEU2*) (25), defective in Na^+ efflux; and OC05 (*Mat*a, *ura3-1, leu2-3, trp1-1, his3-11,15, can1-100, nhx1Δ::HIS3, vnx1::KanMX6, vcx1Δ::HphNT1*) defective in tonoplast K^+ and Na^+/H^+ exchange (20, 26) are isogenic with strain W303.1A. Additionally, OC06 (*Mat*a; *his3Δ1; leu2Δ0; met15Δ0; lys2Δ0; ura3Δ0; vnx1Δ::KanMX4; vcx1Δ::KanMX4*) is defective in tonoplast K^+ and Na^+/H^+ exchange and isogenic with BY4741. All these strains can be obtained from the authors of this chapter.

2.2. Media and Buffers

1. Yeast extract, peptone, dextrose (YPD) medium: 1 % yeast extract, 2 % peptone, 2 % glucose.
2. Arginine phosphate (AP) medium: 8 mM H_3PO_4, 1 mM $MgSO_4$, 0.1 mM $CaCl_2$, a mix of oligoelements ($ZnSO_4$, 400 μg/mL; $CuSO_4$, 40 μg/mL; $MnSO_4$, 400 μg/mL; H_3BO_3, 500 μg/mL; KI, 100 μg/mL; $Na_2MoO_4{\cdot}2H_2O$, 200 μg/mL; $FeCl_3{\cdot}6H_2O$, 200 μg/mL), auxotrophic supplements as required (adenine, uracil, L-histidine, L-leucine, L-tryptophan, each at 20 μg/mL; or Amino Acid Drop-out mix, Sigma); the pH is adjusted to 6.5 with arginine. Sterile 2 % glucose and vitamins (nicotinic acid, pyridoxine, thiamine, calcium pantothenate, at a final concentration of 0.4 μg/mL; biotin 0.02 μg/mL) are added to the medium after autoclaving. The complete media are supplemented with K^+ as required (see Note 2).
3. Synthetic dextrose (SD) medium: 0.67 % Yeast Nitrogen Base (Difco or other trade marks) without amino acids and without ammonium sulfate, supplemented with 0.5 % $(NH_4)_2SO_4$, 2 % glucose and auxotrophic supplements as required (see AP medium).

4. LiAc/PEG solution: dissolve 40 g of polyethyleneglycol 4000 and 1.02 g Li^+ Acetate in 1 mL 1 M Tris–HCl pH 8, 0.2 mL 0.5 M Na^+-EDTA pH 8.0, and deionized water up to 100 mL; sterilized by autoclaving.
5. Cation uptake buffer: 10 mM 2-(*N*-morpholine)-ethanesulfonic acid (MES) adjusted to pH 6.0 with $Ca(OH)_2$ and supplemented with 2 % glucose; prepared with Milli-Q deionized water.
6. Sodium loading buffer: 10 mM trisodium citrate pH 8.5, 50 mM NaCl and 2 % glucose (27).
7. Sodium efflux buffer A: 10 mM MES adjusted to pH 5.5 with $Ca(OH)_2$ and supplemented with 2 % glucose, 0.1 mM $MgCl_2$ and 50 mM RbCl.
8. Sodium efflux buffer B: 10 mM *N*-Tris(hydroxymethyl)methyl-3-aminopropanesulfonic acid (TAPS) adjusted to pH 8.0 and supplemented with 2 % glucose, 0.1 mM $MgCl_2$, 50 mM RbCl, and 20 mM NH_4Cl.
9. Cell washing solution: 20 mM $MgCl_2$ solution prepared with Milli-Q deionized water.
10. Cation extraction solution: 0.1 M HCl, 10 mM $MgCl_2$ prepared with Milli-Q deionized water.
11. Spheroplast buffer: 50 mM Tris–HCl pH 7.5, 1 M sorbitol. Store at 4 °C.
12. Vacuole buffer A: 10 mM MES-Tris, pH 6.9, 12 % Ficoll-400, 0.1 mM $MgCl_2$ (see Note 3). Store at 4 °C (see Note 4).
13. Vacuole buffer B: 10 mM MES-Tris, pH 6.9, 8 % Ficoll-400, 0.5 mM $MgCl_2$. Store at 4 °C (see Note 4).
14. Vacuole buffer C: 5 mM MES-Tris, pH 6.9. Store at room temperature.
15. Zymolyase 20T: Zymolyase can be purchased from Seikagaku-America or AMS Biotechnology Europe Ltd.
16. 1 M DTT solution. Store at –20 °C.
17. Protease Inhibitor Cocktail (PIC) for plant cell and tissue extracts, DMSO solution (Sigma). It is used at 1/1,000 v/v unless otherwise stated.
18. Bafilomycin A1 solution: 1 μM Bafilomycin A1 in Ethanol 100 %. Make 20 μL aliquots and store at –80 °C.
19. Acridine Orange solution: 8 mM Acridine Orange hemi (zinc chloride). Store in the dark at room temperature.
20. Tris-ATP solution, 60 mM: Dissolve 1 g ATP in 25 mL Milli-Q water, add Tris powder with a spatula to adjust pH to 6.9, and adjust the volume to 30 mL. Make 1 mL aliquots and store them at –80 °C.

3. Methods

3.1. Yeast Transformation

1. Inoculate the yeast mutant strain in YPD medium and incubate with shaking overnight at 28 °C. For the growth of *trk1 trk2* mutants, YPD should contain 30 mM KCl (see Note 2).
2. Harvest 1–2 mL of the overnight culture in a sterile microcentrifuge tube at 1,000 × *g* for 5 min. Pour off the medium, resuspend the cells in 1 mL of sterile water and centrifuge again. Pour off the water, add 1 μg of the purified yeast expression plasmid used for transformation and resuspend the cells in 0.5 mL of the LiAc/PEG solution.
3. Incubate in the LiAc/PEG solution for a minimum of 4 h at room temperature. Sometimes longer incubation can improve the transformation efficiency. Pellet the cells at 1,000 × *g* for 5 min and remove the LiAc/PEG solution with a micropipette. Wash the cells with 0.5 mL of sterile water and centrifuge again. Pour off the supernatant.
4. Spread the yeast cells onto a plate of SD agar medium using 5–10 sterile glass beads per plate with gentle shaking. Pour off the glass beads and incubate the plates at 28 °C for 3–4 days until the appearance of the yeast transformant colonies is observed. The SD medium must be prepared including all of the yeast auxotrophic requirements except the one synthesized by the selectable marker of the plasmid. The plasmid pYPGE15 contains the *URA3* yeast selectable marker.

3.2. Growth Tests

Two different methods, drop tests and growth in liquid medium, can be used to test either the capacity of a plant K^+ transporter to suppress the defective growth of the *trk1 trk2* yeast strain in a medium with low K^+ or the capacity of a plant Na^+ efflux system to suppress the defective growth of an *ena1-4Δ nha1Δ* yeast strain at high Na^+.

1. For drop tests, culture the yeast cells in AP medium supplemented with KCl as required for the strain. Harvest and resuspend the cells in sterile water, adjusting the A_{600} to a value of approximately 1.0. Prepare tenfold serial dilutions of the cells in a microtiter plate containing 200–300 μL of the yeast suspension in each well. Spot 5 μL aliquots of each dilution on agar plates of AP medium containing different Na^+ and/or K^+ concentrations. For the efficient transfer of the cell suspensions to numerous agar plates, a 96 well replica-plating device may be used (see Note 5). Incubate the plates at 28 °C for 2–3 days and check growth (Fig. 1a).
2. For growth tests in liquid medium, prepare a suspension of yeast cells as described in step 1. Use this cell suspension to

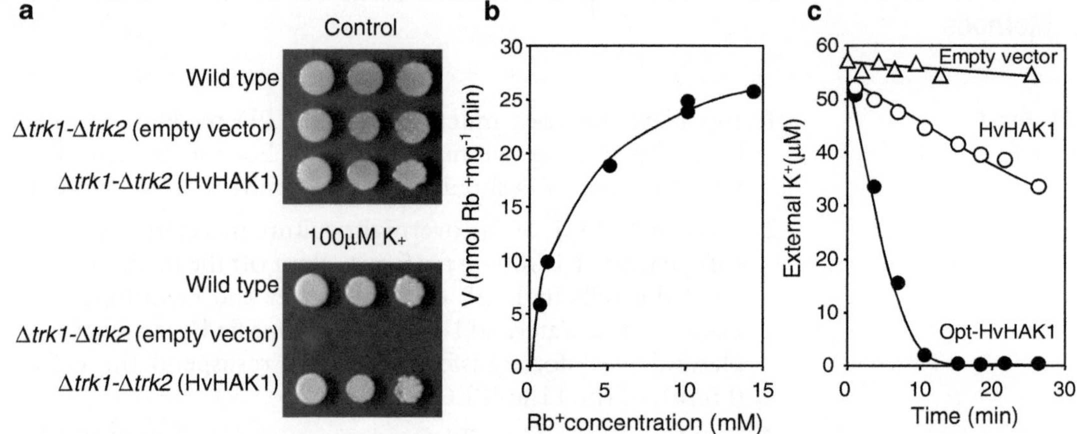

Fig. 1. Functional expression of a plant K^+ transporter in a *trk1Δ trk2Δ Saccharomyces cerevisiae* mutant deficient in K^+ uptake. The data shown correspond to the barley *HvHAK1* cDNA transformed into strain W303D3 (16). (**a**) Drop test assay: Suppression of the incapacity of the mutant to grow at low K^+ concentration. The yeast strains were inoculated in AP medium supplemented with either 30 mM KCl (control) or 100 μM KCl. (**b**) Kinetic analysis of Rb^+ uptake. The concentration dependence plot of Rb^+ influx was constructed by determining the initial rates of Rb^+ uptake from the time courses of the increase of cellular Rb^+ in cells exposed to different Rb^+ concentrations. (**c**) Enhancement of K^+ uptake by optimizing the ATG environment (see Note 1). Time course of the decrease of external K^+ in K^+-starved yeast cells exposed to 50–60 μM K^+. In the experiments shown the yeast mutant was transformed with the empty vector (*open triangles*), the native cDNA (*open circles*), or a construct in which the ATG environment was optimized (*closed circles*; see Note 1).

inoculate the AP medium with different K^+ or Na^+ concentrations at an initial cell density of $A_{600} = 0.05$. Fill the wells in the 96-well ELISA plate with 300 μL of the cell suspensions and load the microplate onto the microplate growth reader (see Note 6). Program the incubation at 28 °C for 3–4 days with periodic and intensive shaking of the microplate. The microplate reader will measure the absorbance values in each well at 600 nm at given intervals. The absorbance data are exported to Microsoft Excel for further processing.

3.3. Cation Uptake Tests

Cation uptake tests, including the kinetic analysis of the uptake, can be carried out in two different ways, either by measuring the increase of the cell cation content or by measuring the depletion of the cation in the external medium (Fig. 1c; see Note 7). In both cases, the first step is the preparation of K^+-starved yeast cells (see Note 8).

1. To prepare K^+-starved cells, inoculate the yeast strain in AP medium with 30 mM KCl and incubate with shaking overnight at 28 °C (see Note 2). At a cell density of A_{600} between 0.1 and 0.4, harvest the cells by centrifugation at $1,000 \times g$ for 5 min and pour off the medium; wash the cells twice with 100 mL of deionized water and resuspend the cells in 100 mL of AP medium. Incubate the yeast cells in this medium for 4 h at 28 °C with shaking. After K^+ starvation, centrifuge the cells at $1,000 \times g$ for 5 min.

2. To test the ion uptake by measuring the increase of the cell cation content, resuspend the K^+-starved cells in an appropriate volume of the cation transport buffer at a cell density of A_{600} of 0.1–0.2, determine and annotate the absorbance value, (see Note 9) add the cation and incubate at 28 °C with continuous shaking. At intervals, take alternating samples of 5 and 10 mL and collect the yeast cells on 0.8 μm pore Millipore membrane filters by filtration (see Note 10). Wash the cells rapidly with cell washing solution, transfer the filters with cells to a tube, and soak the filters with the cells in 5 mL of cation extraction solution. If the cation concentration is high (>1 mM) after the first filtration, transfer the cells by washing the Millipore membrane filter with cell washing solution to a second filter and wash again the cells on this filter. This second filter with the cells is then soaked in the cation extraction solution as described above. Keep the cells in the extraction solution for 12–24 h, centrifuge the extracted cells at 1,000 × g for 5 min, and measure the cation concentrations in the supernatant by atomic emission spectrophotometry. Divide the cation content of the supernatant by the dry weight of the sampled cells calculated from the absorbance of the culture and volume of samples. If the cells grow significantly during the experiment, measure the absorbance of the culture as many times as required to know the actual weight of cells at each sampling point. Kinetic analyses can be performed with the initial rates of uptake (Fig. 1b), which are determined from the time courses of cation accumulation.
3. To test the depletion of an external cation (Fig. 1c), suspend the K^+-starved cells in the testing buffer at a cell density of A_{600} of 1.0–2.0, at intervals, centrifuge 2 mL of the yeast suspension at 10,000 × g for 30 s, transfer 1 mL of the supernatant, which must be free of yeast cells, to a new tube, and measure the cation concentrations using atomic emission spectrophotometry.

3.4. Sodium Efflux Tests

The possibility that a certain plant cDNA encodes a sodium efflux system can be estimated by studying the suppression of the defective growth of the transformed *ena1-4Δ nha1Δ* yeast strain in a medium containing high Na^+ (Subheading 3.2, step 1; Fig. 1a). However, a Na^+ efflux system should be characterized either from the time course of Na^+ uptake (Subheading 3.3, step 2; Fig. 1b) or from the time course of the decrease in the cell Na^+ content when the cells are suspended in a buffer that does not contain Na^+ (Subheading 3.4, step 3; Fig. 1c). In the former test, the function of a Na^+ efflux transporter may be deduced from the time course of Na^+ accumulation with respect to the control yeast transformed with the empty plasmid (see Note 11). For the Na^+ efflux test, the first step is the preparation of Na^+-loaded yeast cells.

1. To prepare Na^+-loaded cells, inoculate the yeast strain in AP medium with 3 mM KCl and grow the cells at 28 °C up to a cell density of A_{600} between 0.1 and 0.4 (see Note 9). Harvest the cells using centrifugation at 1,000 × *g* for 5 min and wash them with 100 mL of deionized water. Resuspend the washed cells in 100 mL of sodium loading buffer and keep them in this medium for 1 h at 28 °C with shaking. After Na^+ loading, centrifuge the cells at 1,000 × *g* for 5 min.
2. To test the Na^+ efflux by measuring the decrease of the Na^+ cation content, resuspend the Na^+-loaded cells in sodium efflux buffer A or B at a cell density of A_{600} 0.1–0.2, annotate the cell density, and incubate at 28 °C with continuous shaking. At intervals, take and process samples described above (Subheading 3.3, step 2; see Note 12).

3.5. Vacuolar Cation/ H^+ Exchange Test

This protocol involves two steps: the preparation of the vacuoles and the cation/H^+ exchange test subsequently performed with the prepared vacuoles. For the latter test, the Acridine Orange fluorescence quenching method is used to monitor the establishment and dissipation of vacuolar inside acidic pH gradients that result from the activity of the vacuolar ATPase and cation/H^+, respectively.

1. Inoculate a single yeast colony containing the transgene into 10 mL of SD medium and grow it to saturation. The next day in the late afternoon, inoculate 2 × 1 L of SD medium in a 2 L flask with 5 mL of the saturated pre-culture. Incubate at 28 °C for 14–16 h to reach the post-diauxic-shift phase. Harvest the cells by centrifugation in 500 mL bottles for 5 min at 3,000 × *g*.
2. Concurrently with the centrifugation of the cells, add 50 mg of Zymolyase-20T and 100 μL of 1 M DTT solution (1 mM final concentration) to 100 mL of pre-warmed (28–30 °C) spheroplast buffer. Next, gently stir for 1–2 min until complete resuspension of the enzyme is achieved. After centrifugation, remove supernatant, gently resuspend the pellet with 100 mL of spheroplast buffer, and gently rotate the bottle at approximately 80 rpm at 30 °C for 1 h (see Note 13).
3. Check the spheroplast formation by mixing 10 μL of cells with 990 μL of water and spheroplast buffer. If spheroplast formation is sufficient, the water-cell mix should be clearer than the buffer-cell mix.
4. Collect the spheroplast by centrifugation, 10 min at 3,000 × *g* at 4 °C in a precooled rotor. With a wide-mouth transfer pipette, gently resuspend the pellet in 100 mL of cold spheroplast buffer at 4 °C, containing 1 % glucose (see Note 14). Centrifuge again as described above, remove supernatant with

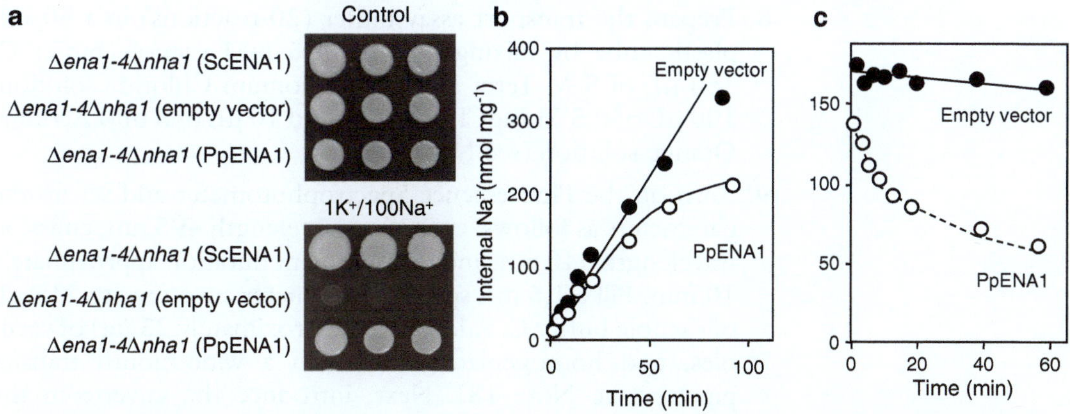

Fig. 2. Functional expression of a plant Na^+ efflux system in an *ena1-4Δ nha1Δ S. cerevisiae* mutant deficient in Na^+ and K^+ effluxes. The data shown correspond to the *Physcomitrella patens PpENA1* cDNA transformed into strain B31 (19). (**a**) Drop test assay: Suppression of the Na^+ sensitivity of the yeast mutant by *PpENA1*. The yeast strains were inoculated in AP medium supplemented with 1 mM KCl (control) or 1 mM KCl plus 100 mM NaCl. The growth of the mutant strain transformed with the endogenous *ScENA1* gene is shown for comparative purposes. (**b**) Time course of the increase of the Na^+ content in the mutant strain and in its *PpENA1* transformant. Yeast cells were transferred to AP medium with 1 mM K^+ and 100 mM Na^+. The data points in the *PpENA1* strain can be fitted to an equation that was formulated considering a constant influx and first order kinetic efflux (32) (fitted values: influx, 5.1 ± 0.5 nmol/mg/min; *k* of efflux, 0.025 ± 0.004/min, *n* = 3). (**c**) Time course of the Na^+ loss in the mutant strain and its *PpENA1* transformant. Na^+ loaded yeast cells were transferred to sodium efflux buffer B. The data points in the *PpENA1* strain can be fitted to a first order efflux kinetics (fitted value: *k* = 0.032 ± 0.003/min, *n* = 4).

a transfer pipette and gently resuspend the pellet in 25 mL of cold vacuole buffer A containing PIC (see Note 15). To increase cell lysis, transfer the mix to a pre-chilled Dounce Homogenizer and perform 5–6 strokes.

5. Transfer the cell lysate to pre-chilled centrifuge tubes and carefully add buffer A containing PIC up to 5 mm from the top of the tube without mixing the two layers. Centrifuge at 70,000 × *g* for 30 min at 4 °C (see Note 16). The vacuoles will float and form a white layer at the top of the tube.
6. Collect the top white layer of vacuoles with a spatula that has been previously moistened in vacuole buffer B. Dip the edge of the spatula along the side of the tube down to about 1 cm under the vacuole layer and lift up. Use a transfer pipette to resuspend the vacuoles in a Ultra-clear tube filled with up 20 mL of ice-cold vacuole buffer A with PIC. Then, carefully add buffer B with PIC up to 5 mm from the top of the tube without mixing the two layers. Centrifuge at 70,000 × *g* for 30 min at 4 °C.
7. Collect the floating vacuoles as previously described, and gently resuspend them in 500 μL of pre-chilled vacuole buffer C with PIC. Determine the protein concentration with the Bio-Rad protein assay kit. Adjust the concentration to 1 μg/μL by diluting the vacuoles with vacuole buffer C. Store on ice.

8. Prepare the transport assay buffer (20 reactions) in a 50 mL plastic tube by mixing 14 mL of 5 mM vacuole buffer C, 160 μL of 5 M Tetra Methyl Ammonium Chloride solution, 100 μL of 0.5 M $MgCl_2$ solution, and 10 μL of 8 mM Acridine Orange solution (see Note 17).
9. Turn on the Fluorescence Spectrophotometer and set up the parameters as follows: excitation wavelength 495 nm, emission wavelength 540 nm, and measurement duration approximately 10 min. Fill a 1.5 mL spectrophotometer cuvette with 715 μL of vacuole buffer C. Add 25 μL (approximately 25 μg) of vacuoles, and homogenize gently with a wide-mouth transfer pipette (see Note 18). Next, introduce the cuvette in the Fluorescence Spectrophotometer, and start the monitoring of the A_0 emission intensity at 540 nm (Fig. 3).

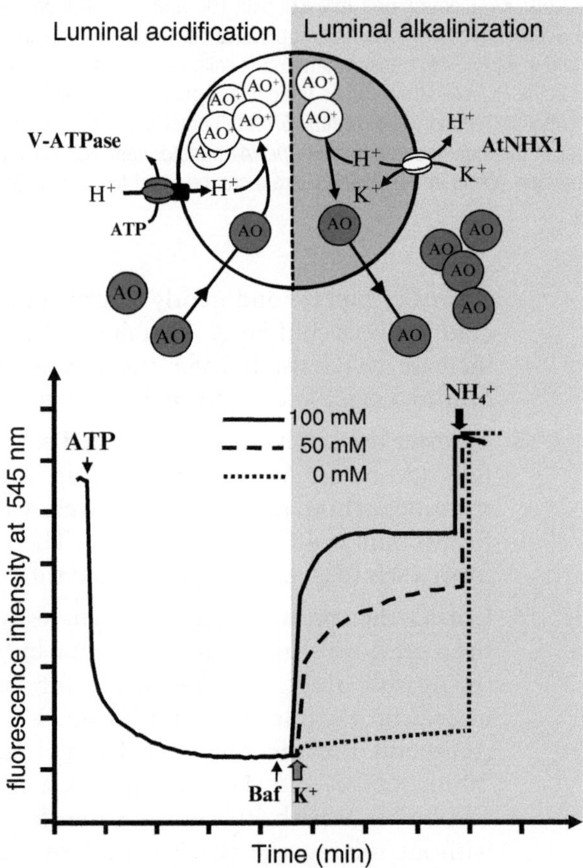

Fig. 3. Acidification of the lumen of purified intact vacuoles by the vacuolar Mg^{2+}-ATPase and its subsequent alkalinization by a cation/H^+ antiporter. Vacuole acidification was initiated by the addition of ATP (ATP *arrow*). After the steady-state acidic inside pH gradient was attained, the activity of the V-ATPase was partially inhibited by the addition of bafilomycin A1 (Baf *arrow*). AtNHX1 antiport was monitored as the recovery of fluorescence quench upon addition of various concentration of K^+ (K^+ *arrow*). At the indicated time (NH_4^+ *arrow*), 20 mM of $(NH_4)_2SO_4$ is added to collapse the pH across the vacuolar membrane to recover 100 % of the fluorescence.

10. After approximately 1 min, add 20 μL of 60 mM Tris-ATP solution to the cuvette and mix well (4–5 times) with a transfer pipette. A rapid quenching of the fluorescence within the first few seconds after the addition of the ATP should be observed; this is caused by the activation of the V-ATPase and the subsequent acidification of the vacuolar lumen (Fig. 3; see Note 19). After the minimum fluorescence point is reached (within 2 or 3 min), add 1 μL bafilomycin A1 solution (1 μM) and mix gently with a transfer pipette (see Note 20).
11. Wait approximately 1 min, then add 40 μL of the desired cation solution, and mix gently with a transfer pipette. Cation/H^+ exchange activity induces the alkalinization of the vacuolar lumen, the de-protonation of the Acridine Orange, and therefore a recovery of fluorescence at 540 nm (Fig. 3; see Note 21). Approximately 1 min after the addition of the cation, the intensity of fluorescence should level off and reach a plateau.
12. Wait approximately 2 min, then add 20 μL of 500 mM $(NH_4)_2SO_4$ solution to anneal the proton gradient and recover 100 % of the fluorescence. Alternatively, add 0.5 μL of a 5 mM monensin or nigericin solution. These are Na^+/H^+ and K^+/H^+ antiporter ionophores, respectively.

4. Notes

1. To optimize the expression in yeast cells, full-length cDNAs must be cloned, eliminating any ATG triplet in the cloning polylinker or in any sequence preceding the first ATG of the open reading frame. These ATGs originate from the ORFs in the untranslated region (upstream ORFs [uORFs]) of the expressed mRNA, and may substantially reduce the expression of the transporter (28). Moreover, it is also convenient if the sequence around the ATG is as similar as possible to the following sequence: (A/T)A(A/C)A(A/C)A**ATG**TC(T/C) (29).
2. The AP medium is free of K^+, Na^+, and ammonium, and is used supplemented with K^+ and Na^+ as required. If used with agar for solid media, special attention must be paid to the K^+ and Na^+ content of the agar. Commercially available purified agars may contain up to 15 μM K^+ and 750 μM Na^+. For growing purposes, the medium can be supplemented with 3 mM KCl for wild type and Na^+ efflux mutant strains and with 30 mM KCl for *trk1Δ trk2Δ* mutant strains. At lower K^+ concentrations, these mutants produce partial revertants with improved growth at low K^+. These partial revertants are mutants in genes that are independent of *TRK1* and *TRK2* (30).

3. The complete dissolution of the Ficoll-400 can take several hours with a magnetic stirrer or can be achieved manually in less than 30 min with a spatula.
4. For long-term storage, vacuole buffers A and B must be sterilized by filtration.
5. The replica-plating device can be sterilized with ethanol and flaming. In this case, the use of ethanol denatured with benzalkonium chloride should be avoided because yeast growth may be inhibited.
6. We use a microplate growth reader marketed by Growth Curves Oy, Finland, Microbiology Reader Bioscreen C Workstation.
7. The depletion of K^+ or Na^+ can be used exclusively in high-affinity transport systems because the cation taken up by the cells must decrease significantly the external concentration of the cation. This only occurs when the external concentration of the cation is low. Even when using high concentrations of yeast cells, 1–2 mg/mL, the maximum external cation concentration that can be used in these tests is approximately 0.5 mM. This excludes any uptake systems that show a low uptake rate at that concentration. Notably, in kinetic studies the influx at zero internal concentration must be used to calculate the kinetic constants rather than the rate of the uptake (17). However, this requirement is never fulfilled for K^+ and is difficult to fulfill for Na^+ (see Note 9). In contrast, it may be fulfilled for Rb^+ and Li^+, but the use of these cations as tracers of K^+ and Na^+ is inadvisable unless it has been previously demonstrated that the uptake system does not discriminate between the investigated cation and the putative tracer (2). The depletion of K^+ or Na^+ from the external medium can be followed by atomic emission spectrophotometry or with a cation selective electrode (17).
8. The use of K^+-starved cells is convenient for obtaining a rapid and long-lasting uptake of any cation. These cells have half the K^+ content, a low cytoplasmic pH (31) and can take up more than 150 nmol/mg of K^+ or Na^+ without decreasing the uptake rates. To test high-affinity Na^+ uptake systems, it must be taken into account that the Na^+ content of the starvation medium prepared with normal chemical components may be 30 μM or higher, and that in this medium the cells will be K^+ starved, but not Na^+ starved. The Na^+ content of these cells makes it impossible to determine the Na^+ influx without a Na^+ tracer.
9. For cation uptake experiments, yeast cells must be harvested early in the exponential growth phase to avoid excessive glucose consumption and ethanol production. This induces

changes in yeast metabolism and possibly in the internal pH. Cells growing at A_{600} of 0.1–0.4 in AP medium may be used with reproducible results.

10. The procedure to determine the increase of the cell cation content is very sensitive to a defective washing of the cation from the filter. Therefore, it is advisable to alternate two different volumes of samples, e.g., 5 and 10 mL, so that the different amounts of cells in each filter reveals any problems of contamination by producing significantly higher apparent cation contents in the smaller samples.
11. When yeast cells grow in a medium with Na^+ (i.e., 100 mM Na^+), they increase their Na^+ content at a rate that obviously depends on the Na^+ influx and efflux; Na^+ efflux is practically inexistent in the *ena1-4Δ nha1Δ* strain. Therefore, by following the time course of the increase of the cellular content of Na^+ (31) in a growing medium, the Na^+ efflux can be calculated (32, 33). However, these types of uptake tests at high Na^+ concentrations present technical and functional difficulties because washing the external Na^+ from the cells is difficult and because Na^+ influx may not be constant throughout the experiment. Therefore, it is normally easier to load the cells with Na^+ and to test Na^+ efflux in a medium without Na^+. For this test, loading the cells in AP medium with Na^+ is simple, but the Na^+ contents of the mutant strain (control) and the strains expressing the transporters will be completely different. This results in difficulties when making kinetic comparisons between different efflux systems. Alternatively, the cells can be loaded in Na^+ loading buffer as described in Subheading 3.4, step 2. This method produces cells with similar Na^+ contents, independently of the efflux systems that they are expressing.
12. Na^+ efflux can be normally tested in buffer A. However, some Na^+ efflux systems are inhibited by the acidification of the cytoplasm that is concomitant with the Na^+/H^+ exchange. In these cases, long testing periods must be carried out in sodium efflux buffer B that maintains a high cytoplasmic pH.
13. Do not exceed a 90 min digestion time, otherwise the quality of the vacuoles will be decreased. Galactose can be added in the spheroplast buffer to maintain the expression of heterologous genes driven by an inducible promoter without affecting the digestion efficiency.
14. Glucose is necessary to maintain the association between V_1 and V_0 subunits of the V-ATPase (34). The omission of glucose in the spheroplast buffer during the washing steps will result in an important decrease of the proton pumping activity necessary to energize cation/H^+ antiport activity.

15. The PIC is not stable in water solution and has to be added extemporaneously in vacuole buffers A and B.
16. The authors use a Beckman Ultracentrifuge, Beckman SW28 rotor, and Beckman Ultra-clear centrifuge tubes (25 × 89 mm).
17. Prepare the transport assay mix immediately before it is to be used and wrap the tube in aluminum foil to protect the Acridine Orange from light.
18. Before each transport assay, resuspend the vacuoles fraction with a wide-mouth pipette. It is preferable to pipette 25 μL of this fraction with a cut tip pipette in order to avoid breaking the vacuoles. For the same reason, do not use a magnetic stirrer and avoid air bubbles during mixing with the transfer pipette.
19. Acridine Orange (AO) is a permanent weak base in its neutral form, whereas its protonated form (AO^+) is trapped inside the vacuole (35). The protonation displaces the Acridine Orange emission wavelength towards a longer wavelength, consequently inducing a decrease in the fluorescence intensity at 514 nm. Therefore, the accumulation and protonation of Acridine Orange in the vacuoles produces a concomitant quenching of the 514 nm fluorescence (Fig. 3).
20. Bafilomycin A1 is used to slow down the vacuolar ATPase activity and therefore to establish a steady-state pH gradient (26, 36). This gradient can be visualized on the trace by a slight increase of the slope. Even with a standardized procedure, each vacuole preparation will vary in V-ATPase activity. The quantity of bafilomycin A1 given in this protocol is an average of the amount routinely used.
21. We successfully used this method to measure the transport of monovalent and bivalent cations, such as K^+, Na^+, Li^+, Rb^+, and Ca^{2+}. The transport of heavy metals (Cd^{2+}, Cu^{2+}, Zn^{2+}) cannot be monitored because of their inhibitory effect on the V-ATPase activity (20, 37).

Acknowledgements

Figure 2 reproduces two panels of Fig. 4 of ref (19). The authors would like to acknowledge the Plant Journal, the Society for Experimental Biology, and Wiley-Blackwell Publishing for permitting the reproduction.

References

1. Dreyer I, Horeau C, Lemaillet G et al (1999) Identification and characterization of plant transporters using heterologous expression systems. J Exp Bot 50:1073–1078
2. Rodríguez-Navarro A (2000) Potassium transport in fungi and plants. Biochim Biophys Acta 1469:1–30
3. Botha A (2011) The importance and ecology of yeast in soil. Soil Biol Biochem 43:1–8
4. Müller B, Grossniklaus U (2010) Model organisms—a historical perspective. J Proteomics 73: 2054–2063
5. Ramos J, Contreras P, Rodríguez-Navarro A (1985) A potassium transport mutant of *Saccharomyces cerevisiae*. Arch Microbiol 143:88–93
6. Haro R, Garciadeblas B, Rodríguez-Navarro A (1991) A novel P-type ATPase from yeast involved in sodium transport. FEBS Lett 291: 189–191
7. Sentenac H, Bonneaud N, Minet M et al (1992) Cloning and expression in yeast of a plant potassium ion transport system. Science 256:663–665
8. Anderson JA, Huprikar SS, Kochian LV et al (1992) Functional expression of a probable Arabidopsis thaliana potassium channel in *Saccharomyces cerevisiae*. Proc Natl Acad Sci U S A 89:3736–3740
9. Goffeau A, Barrell B, Bussey H et al (1996) Life with 6000 genes. Science 274:546–567
10. Miller A, Zhou J (2000) *Xenopus* oocytes as an expression system for plant transporters. Biochim Biophys Acta 1465:343–358
11. Uozumi N (2001) *Escherichia coli* as an expression system for K+ transport systems from plants. Am J Physiol Cell Physiol 281: 733–739
12. Senn ME, Rubio F, Bañuelos MA et al (2001) Comparative functional features of plant potassium HvHAK1 and HvHAK2 transporters. J Biol Chem 276:44563–44569
13. Garciadeblas B, Barrero-Gil J, Benito B et al (2007) Potassium transport systems in the moss *Physcomitrella* patens: pphak1 plants reveal the complexity of potassium uptake. Plant J 52:1080–1093
14. Fraile-Escanciano A, Kamisugi Y, Cuming AC et al (2010) The SOS1 transporter of *Physcomitrella* patens mediates sodium efflux in planta. New Phytol 188:750–761
15. Cove D, Bezanilla M, Harries P et al (2006) Mosses as model systems for the study of metabolism and development. Annu Rev Plant Biol 57:497–520
16. Santa-María GE, Rubio F, Dubcovsky J et al (1997) The HAK1 gene of barley is a member of a large gene family and encodes a high-affinity potassium transporter. Plant Cell 9: 2281–2289
17. Bañuelos MA, Garciadeblas B, Cubero B et al (2002) Inventory and functional characterization of the HAK potassium transporters of rice. Plant Physiol 130:784–795
18. Garciadeblás B, Senn ME, Bañuelos MA et al (2003) Sodium transport and HKT transporters: the rice model. Plant J 34:788–801
19. Benito B, Rodríguez-Navarro A (2003) Molecular cloning and characterization of a sodium-pump ATPase of the moss *Physcomitrella* patens. Plant J 36:382–389
20. Cagnac O, Aranda-Sicilia M, Leterrier M et al (2010) Vacuolar cation/H+ antiporters of *Saccharomyces cerevisiae*. J Biol Chem 285: 33914–33922
21. Rubio F, Santa-María GE, Rodríguez-Navarro A (2000) Cloning of Arabidopsis and barley cDNAs encoding HAK potassium transporters in root and shoot cells. Physiol Plant 109: 34–43
22. Quintero J, Martinez-Atienza J, Villalta I et al (2011) Activation of the plasma membrane Na/H antiporter salt-overly-sensitive 1 (SOS1) by phosphorylation of an auto-inhibitory C-terminal domain. Proc Natl Acad Sci USA 108:2611–2616
23. Brunelli JP, Pall ML (1993) A series of yeast *Escherichia coli* expression vectors designed for directional cloning of cDNAs and cre/lox-mediated plasmid excision. Yeast 9:1309–1318
24. Haro R, Rodríguez-Navarro A (2003) Functional analysis of the M2D helix of the TRK1 potassium transporter of *Saccharomyces cerevisiae*. Biochim Biophys Acta 1613:1–6
25. Bañuelos MA, Synchrová H, Bleykasten-Grosshans C et al (1998) The Nha1 antiporter of *Saccharomyces cerevisiae* mediates sodium and potassium efflux. Microbiology 144: 2749–2758
26. Cagnac O, Leterrier M, Yeager M et al (2007) Identification and characterization of Vnx1p, a novel type of vacuolar monovalent cation/H+ antiporter of *Saccharomyces cerevisiae*. J Biol Chem 282:24284–24293
27. Rodríguez-Navarro A, Benito B (2010) Sodium or potassium efflux ATPase. A fungal, bryophyte, and protozoal ATPase. Biochim Biophys Acta 1798:1841–1853
28. Bañuelos M, Haro R, Fraile-Escanciano A et al (2008) Effects of polylinker uATGs on the

function of grass HKT1 transporters expressed in yeast cells. Plant Cell Physiol 49:1128–1132

29. Hamilton R, Watanabe CK, de Boer HA (1987) Compilation and comparison of the sequence context around the AUG start codons in *Saccharomyces cerevisiae* mRNAs. Nucleic Acid Res 15:3581–3593
30. Gaber RF, Styles CA, Fink GR (1988) TRK1 encodes a plasma membrane protein required for high-affinity potassium transport in *Saccharomyces cerevisiae.* Mol Cell Biol 8:2848–2859
31. Ramos J, Haro R, Rodríguez-Navarro A (1990) Regulation of potassium fluxes in *Saccharomyces cerevisiae.* Biochim Biophys Acta 1029:211–217
32. Rodríguez-Navarro A, Asensio J (1977) An efflux mechanism determines the low net entry of lithium in yeast. FEBS Lett 75:169–172
33. Rodríguez-Navarro A, Ortega MD (1982) The mechanism of sodium efflux in yeast. FEBS Lett 138:205–208
34. Sherman F (1991) Getting started with yeast. Methods Enzymol 194:3–21
35. Clerc S, Barenholz Y (1998) A quantitative model for using acridine orange as a transmembrane pH gradient probe. Anal Biochem 259:104–111
36. Manolson M, Proteau D, Preston R et al (1992) The VPH1 gene encodes a 95-kDa integral membrane polypeptide required for in vivo assembly and activity of the yeast vacuolar H+-ATPase. J Biol Chem 267:14294–14303
37. Kakinuma Y, Ohsumi Y, Anraku Y (1981) Properties of H+-translocating adenosine-triphosphatase in vacuolar membranes of *Saccharomyces cerevisiae.* J Biol Chem 256: 10859–10863

Part V

Other Methods

Part V

Other Methods

Chapter 26

Isotope Techniques to Study Kinetics of Na^+ and K^+ Transport Under Salinity Conditions

Dev T. Britto and Herbert J. Kronzucker

Abstract

Radioisotopes (particularly ^{22}Na, ^{24}Na, ^{42}K, and ^{86}Rb) have been used for many decades to trace the fluxes and accumulation of sodium and potassium ions in plant tissues. In this article, standard procedures for the tracing of ion fluxes are described, with emphasis on special problems encountered when examining K^+ and Na^+ transport under salinity conditions. We focus in particular on unidirectional influx measurements, while also providing a brief introduction to compartmental analysis by tracer efflux.

Key words: Influx, Efflux, Salinity, Sodium, Potassium, Radiotracers

1. Introduction

Isotopic tracing of nutritional and toxicological processes in plant tissues has greatly enhanced our understanding of plant physiology, biophysics, and biochemistry in a wide range of contexts. Of these, the study of ion transport across the membrane barriers of the plant cell has particularly benefited from the tracer approach. Key advances in this area include the identification and characterization of the following:

- Nutritionally relevant high- and low-affinity ion transport systems (1–3).
- Uptake, release, and partitioning of toxicants in intact plants (4–6).
- Subcellular transport and accumulation of nutrients or toxic ions (6–9).
- Adaptations of plants to their edaphic environments (3, 9).

Sergey Shabala and Tracey Ann Cuin (eds.), *Plant Salt Tolerance: Methods and Protocols*, Methods in Molecular Biology, vol. 913, DOI 10.1007/978-1-61779-986-0_26, © Springer Science+Business Media, LLC 2012

The chief advantage of isotope methodology over other approaches is that it makes possible the measurement of unidirectional fluxes, which are more appropriate than net fluxes in determining kinetic flux parameters (e.g., K_M, V_{max}), and in enabling the study of specific transport capacities and their regulation. The need to isolate a flux in one direction is particularly important under conditions where the flux in the opposite direction is high, and the turnover of intracellular pools is rapid (10). In addition, because fluxes of one isotope are usually observed against a background of a different isotope of the same element, tracer measurement can be conducted under fairly high nutrient conditions, relative to other methods (but see caveat below). Interestingly, both of these conditions may apply to the study of Na^+ fluxes in plants. Isotopes can also be used in the measurement of net fluxes, but for this purpose a host of other well established methods can also be applied, including the following:

- Measurements of depletion in the external medium.
- Measurements of accumulation in the tissue.
- Use of ion-selective vibrating electrodes (e.g., MIFE, SIET).
- Use of ion-selective fluorescent dyes.

In the study of salt stress and tolerance in plants, the transport and accumulation of sodium and potassium are considered to be of crucial importance (11). Thus, measurements of both unidirectional and net fluxes of both ions are appropriate in this context. However, because of the unusual conditions imposed by salt stress (e.g., high sodium background, hyperosmolarity), special considerations need to be made when designing experimental protocols. In this article, we shall address general approaches to isotope-based measurements of influx, efflux, and net flux, and features that specifically affect the measurements of K^+ and Na^+ fluxes under salinity conditions. A note of caution is appropriate at the outset, regarding the interpretation of tracer fluxes in salt-stressed roots: one of the early impacts of salt stress is an osmotic one, which can result in cellular damage, particularly to membranes. This can have a significant effect on ion fluxes in plants suddenly exposed to high external $[Na^+]$, often suppressing influx and enhancing efflux (12).

2. Materials

1. Plant material (see Notes 1–3).
2. Glass or plastic vessels for pre-absorption, labeling, and desorption.
3. Equipment for aeration and/or stirring of solutions.

4. Dissecting tools.
5. Pre-absorption, labeling, and desorption solutions (see Notes 4 and 5).
6. Pipettes for specific activity samples.
7. Radiotracer (see Notes 6–8).
8. Safety equipment, including protective clothing, eyewear, and shielding (see Note 6).
9. Radiometric equipment (see Note 9).
10. Low-speed (clinical-type) centrifuge (see Note 10).
11. Drying oven (see Note 10).
12. Analytical balance.

3. Methods

1. Measure specific activity of uptake solution (see Note 11).
2. Pre-equilibrate plants, if appropriate (see Note 12).
3. Immerse roots in radioactive solution (see Note 13).
4. Remove plants from radioactive solution after appropriate labeling period (see Notes 14 and 15).
5. Desorb roots of extracellular tracer (see Note 16).
6. Detach roots from shoots, if appropriate (see Note 17).
7. Weigh plants, subsequent to centrifugation or drying (see Note 17).
8. Count radioactivity in plant samples (see Note 9).
9. Calculate the flux (see Notes 18–19).

4. Notes

1. Plants should be grown hydroponically, so that culture conditions can be precisely controlled, and roots are directly accessible for examination. Typically, seedlings are used once they have grown to a convenient size and age (e.g., 1 week for barley, 3 weeks for rice, 4–5 weeks for *Arabidopsis*), but it is important to consider the use of plants from a variety of developmental stages.
2. Excised roots are frequently used in tracer flux experiments, as they reduce complications arising from transpiration and translocation to the shoot. However, their use also entails tissue damage at the point of excision, which can be propagated to

the remainder of the root segment. Therefore, excised tissue should be aged for several hours prior to measurement, ideally in conjunction with an indicator of recovery. It may also be necessary to supplement this heterotrophic tissue with a source of energy (e.g., sucrose). Even when such precautions are taken, however, valuable information is inevitably lost when excised roots are used—e.g., information about the partitioning of a substance between organs of the plant, or the influence of transpiration as a potential driver of the flux. Thus, we recommend that, when possible, intact plants be used for tracer flux measurements.

3. Often several plants are bundled together at the shoot base and treated as a single replicate. This can improve statistics, and measuring accuracy when seedlings of low root mass are used, or when specific activity is low.
4. For steady-state investigations, labeling solutions should be identical to plant growth solutions, except for the addition of radiotracer. In addition, all other growth conditions, especially temperature, light, and humidity, should be maintained during experimentation. For non-steady-state conditions, the uptake solutions and/or ambient conditions are often modified, e.g., to include variations in substrate concentrations (as in the development of flux isotherms), or the provision of metabolic inhibitors.
5. Particularly in older studies, the uptake of a labeled substance has often been monitored against a background of only that substance, plus a small amount of Ca^{2+} (usually as $CaSO_4$), typically at about 100–200 μM, to maintain membrane integrity and basic membrane function. While this may reduce complications due to interactions between the traced substance and other materials in solution, it also may change the nature of the experimental system. For example, removing the K^+ provided during plant growth will likely result in electrical hyperpolarization of the plasma membrane, thus changing the driving force for ions across the membrane and therefore, quite possibly, their fluxes. In general, we recommend using complete nutrient solutions, unless the hypothesis guiding a particular study requires the removal or addition of solution components (e.g., see ref. (13)).
6. Often the choice between stable and radioactive isotope is a matter of convenience, and will depend upon the availability of appropriate isotopic material or instrumentation. In general, the processing time for radioactive counting is much shorter than that for measurement of stable isotopes by use of mass spectrometry. On the other hand, radioisotopes of some elements (e.g., N, O) are very short-lived and are only available to researchers working in close proximity to a production facility

such as a cyclotron. Tracers for K^+ and Na^+ tend to have slightly longer half-lives, on the order of several hours (see below). In all cases involving radiotracers, appropriate safety measures must be taken. This often involves a combination of shielding types, such as Plexiglas and/or lead.

7. While the short half-lives of some radiotracers may make them difficult to obtain, they are advantageous from the perspective of radioactive waste; for example, a sample of ^{13}N will have more or less completely decayed into stable ^{13}C within a few hours of its production. Nevertheless, longer-lived tracers are much more widely used due to their availability and longer handling period.

8. With respect to the tracing of plant K^+ and Na^+ fluxes under salinity conditions, radiotracers are much more frequently used than stable tracers. Perhaps surprisingly, the most widely used tracer for K^+ tracing is not potassium at all, but a radioisotope of its alkali-metal "analog," rubidium-86. However, there is substantial evidence that Rb is an imperfect substitute for K, for example with respect to its translocation to the shoot (14). ^{40}K, with its extremely long half-life (1.25 billion years), is rarely used as a tracer but rather as an environmental indicator (e.g., see ref. (15)). The other radioisotopes of K are all too short-lived to be of significant practical use, except for ^{42}K and ^{43}K (half-lives, respectively, of 12.36 and 22.3 h). Of these, ^{42}K is almost exclusively used as a tracer in plant systems, although ^{43}K has been successfully used, for example, in the microautoradiographic tracing of K^+ fluxes in stomatal cells (16). In the case of Na^+ tracing, only two isotopes of sodium are sufficiently long-lived to be useful: ^{22}Na and ^{24}Na (half-lives of 2.6 years and 14.96 h, respectively); both have been used extensively in plant systems. It is worth noting that the production of these isotopes typically involves bombardment of non-radioactive ^{39}K and ^{23}Na. Because these nuclear transformations are generally incomplete, researchers must consider the presence of (sometimes substantial) residual amounts of "cold" isotope in their preparation of experimental solutions.

9. Almost all radiotracing in biological systems involves the measurement of electromagnetic radiation associated with the decay of beta (plus or minus)-emitting nuclides. Liquid scintillation counters and gamma counters are used to measure most of these biologically important radioisotopes. Scintillation is the most widely used counting method, despite the need to use a scintillation "cocktail" in which the sample is dissolved or suspended; gamma counting has no such requirement. Regardless of counting device used, it is essential that it correct for radioactive decay, particularly when using short-lived tracers. As well, care must be taken to ensure that the geometry of the sample-counter system

is uniform from one sample to the next (or corrections are made to account for differences), since the position of the detector relative to the sample can influence the quantity of radiation measured. This phenomenon is not always fully acknowledged by providers of detection instrumentation and usually must be worked out by the individual researcher. Similarly, manufacturer claims about the effectiveness of detector shielding against ambient radiation (e.g., from nearby samples within the counter) are at times exaggerated, and such issues must also be worked out for individual measuring systems.

10. When fluxes are to be expressed on a dry weight basis, centrifugation is not required (see Note 17). Tissue is typically dried in an oven at around 50°C, or air-dried, for several days, or until weight becomes stable.

11. Prior to placing plants into an uptake solution, a small sample of solution (typically, 1 mL or less) must be removed to determine its specific radioactivity. Specific activity is usually expressed as cpm/µmol (cpm = counts per minute, which is related, by the counting efficiency of the detecting instrument, to the dpm, or disintegrations per minute), and is often denoted "SA," or S_o (the subscript "o" indicating the outside solution, bathing the roots).

12. When plants are to be measured under non-steady-state conditions, their roots are often first exposed to modified solution without radiotracer, for an equilibration period of 5–10 min, prior to their exposure to modified solution containing tracer.

13. When roots of intact plants are immersed in radioactive uptake solution, care must be taken to ensure that the aerial parts of the plant do not come into contact with the solution. This is particularly important when solutions are vigorously stirred or bubbled, or with plants having a short stem, such as *Arabidopsis*. A shielding collar is sometimes used to isolate stems and leaves from solution. If they are not isolated from radioactive solution, substantial translocation artifacts may be obtained (sometimes identifiable due to their high variability).

14. The duration of tracer absorption (and desorption; see below) by roots is an important issue when tracing influx at a high substrate concentration, because such "low-affinity" influxes can be extremely high, and are often associated with very high efflux rates and rapid turnover times (17). Under such conditions, a failure to consider the simultaneous efflux of a substrate over the course of influx measurement will result in an underestimate of the flux; such errors can be minimized by reducing the duration of labeling and desorption (10). Measurement of sodium influx under salinity conditions may be the most pronounced example of this situation, since external concentrations of Na^+ can be 100 mM or more, greatly

exceeding the naturally occurring, and experimentally provided, levels of other ions. Indeed, the rapid, futile cycling of Na^+ under salt stress has been reported many times (5, 18–20), and it has been recommended that the uptake period be as short as 2 min, to minimize the effects of simultaneous efflux (19). However, such reports and recommendations must be approached with caution, because the character and biological significance of Na^+ efflux from plant roots is still poorly understood. It is resistant, in many cases, to changes imposed by a wide range of experimental treatments (4), and appears not to have a commensurate respiratory burden associated with it (20, 21) despite its putatively active transport mechanism (11). This has led to speculation that there may be a large artifactual component to the excess tracer flux, which does not represent a cycling across cell membranes but may be apoplastic in nature (11, 18, 22). This flux, which has yet to be rigorously demonstrated, may be related to the well established (e.g., in rice) "apoplastic bypass" of sodium as it travels from root to shoot, independent of symplastic discrimination (22, 23). While apoplastic components of sodium transport in plants may nevertheless prove critical to the understanding of salinity stress, it is important not to conflate them with the measurement of membrane transport activity.

15. In practice, however, it can be very difficult to distinguish between extracellular events and those occurring across cell membranes. This presents a dilemma to the researcher: should one attempt to prevent underestimates of influx across the membrane caused by tracer efflux from the cell during measurement, by opting for very short labeling and desorption times? Alternatively, should one minimize artifacts associated with extracellular accumulation of tracer, by opting for the opposite? The answer to this depends at least partly on the intention of the study. Because the accumulation of sodium in plant tissues is central to the toxicology of salinity stress, the latter approach (basically a net-flux measurement) can be quite appropriate and satisfactory in many situations. On the other hand, if the study is to focus on a classical enzyme-kinetics evaluation of membrane transport systems (1), the requirement for unidirectional flux measurements is stringent, and thus the former approach would be recommended. However, veracity of the efflux component of the flux as a trans-membrane phenomenon must also be determined in the experimental system, e.g., by performing CATE analyses (see Note 19).

16. In all cases, care must be taken to clear (desorb) tracer from known apoplastic phases, once labeling is complete. These phases include the surface-water film of roots, and the electrostatically binding "Donnan" phase of cell walls, which consists

of fixed extracellular charges (mostly negative). Desorption is generally done by immersing roots in a solution identical to the uptake solution, except that it contains no radiotracer; this solution is sometimes chilled to 4°C to minimize loss of tracer from the symplast. The length of the desorption period is typically 5 min, but can often be more precisely determined for these phases by means of compartmental analysis by tracer efflux (CATE; see Note 19).

17. Once desorption is complete, roots are typically detached from shoots (if intact plants are used) for separate counting to estimate translocation rates. At this stage, roots can be weighed prior to counting if fluxes are to be normalized to root fresh weight. If so, a brief, low-speed centrifugation of root tissue (e.g., in a clinical centrifuge at 7,000 rpm) is required to remove surface and interstitial water; however, this water can sometimes contain tracer that originated within cells, especially when substrate concentrations are high, and this should be accounted for (24). If dry weight is the standard, radioactivity of samples may be counted before or after drying and weighing.

18. The influx or net flux into the plant can be calculated quite straightforwardly using the formula $\varphi = \frac{Q^*}{S_o w t_L}$, where φ is the flux (e.g., μmol/g/h), Q^* (cpm) is the quantity of tracer accumulated in tissue (usually root and shoot combined), S_o is the specific activity of the uptake solution (cpm/μmol), w is the root weight (g), and t_L is the labeling time (h). More sophisticated calculations can also be made, accounting for tracer efflux from root cells during labeling and desorption, and based on parameters obtained using CATE analysis (for details, see ref. (10); also see Note 19, and caveats above). The absolute quantification of transport to the shoot is more problematic, because (1) the specific activity of the translocating pool is difficult to estimate (25), and (2) a lag phase that retards the appearance in the shoot of some labeled ions, particularly K^+, is often observed (e.g., see ref. (26)). Thus, investigations of root-to-shoot transport may require longer labeling times and is sometimes expressed as % of total tracer absorbed that is found in the shoot. One additional issue is the possibility of apoplastic bypass flow of Na^+ to the shoot, well documented in species such as rice (23), which contributes to the non-cellular component of influx into the plant (see above).

19. CATE has been used extensively in biology and medicine (26) to quantify ion fluxes and metabolic pool sizes, including those of K^+ and Na^+ in plants examined under salinity conditions (5, 20, 27, 28). While its use in plant science has diminished somewhat in recent years, it remains an important methodology

in medical science, particularly pharmacokinetics (e.g., see ref. (29)). This method uses long labeling periods (typically one to several hours) followed by a partial washing out of tracer from roots by means of a timed series of non-radioactive eluates. When done correctly, CATE can provide a more comprehensive view of unidirectional fluxes than the procedure outlined above, as well as compartmentation data on both cellular and whole-plant scales. However, it is more labor-intensive, generally limited to steady-state conditions, and prone to its own set of heuristic problems. While a detailed exposition of CATE is beyond the scope of this chapter (but see refs. (7, 30–32) for rationale and procedures), a few points are worth mentioning. Exponential half-times of tracer release from intra- and extracellular phases of plant roots, as well as the ratio of unidirectional efflux to influx, can be estimated using CATE. This information is pertinent to the design of "direct-influx" protocols, because it facilitates the choice of (1) labeling time, during which tracer efflux from an absorbing and releasing root can be minimized (hence reducing underestimates of the flux); and (2) desorption time, to maximize the release of tracer from extracellular spaces. For details on this approach, see ref. (10). CATE can also be useful as an independent line of investigation against which direct-influx results may be compared. When discrepancies appear between the two systems of measurement, a comparison can yield useful insights, both biological and methodological. Requirements for the correct interpretation of CATE data, however, are stringent. Phases of tracer release (e.g., surface film, cell wall, cytosol, vacuole) must be correctly identified (31, 32), and should also be sufficiently distinct kinetically from one another to be resolved (33). In addition, steady state must be ensured, limiting the use of CATE for some investigations, such as those involving flux isotherms.

References

1. Epstein E (1966) Dual pattern of ion absorption by plant cells and by plants. Nature 212:1324–1327
2. Siddiqi MY, Glass ADM, Ruth TJ et al (1990) Studies of the uptake of nitrate in barley. I. Kinetics of NO_3^- influx. Plant Physiol 93: 1426–1432
3. Kronzucker HJ, Siddiqi MY, Glass ADM (1997) Conifer root discrimination against soil nitrate and the ecology of forest succession. Nature 385:59–61
4. Cheeseman JM (1982) Pump-leak sodium fluxes in low-salt corn roots. J Membrane Biol 70:157–164
5. Wang B, Davenport RJ, Volkov V et al (2006) Low unidirectional sodium influx into root cells restricts net sodium accumulation in Thellungiella halophila, a salt-tolerant relative of Arabidopsis thaliana. J Exp Bot 57:1161–1170
6. Britto DT, Siddiqi MY, Glass ADM et al (2001) Futile transmembrane NH_4^+ cycling: A cellular hypothesis to explain ammonium toxicity in plants. Proc Natl Acad Sci USA 98: 4255–4258
7. Walker NA, Pitman MG (1976) Measurement of fluxes across membranes. Part A. In: Lüttge U, Pitman MG (eds) Encyclopedia of plant physiology, vol 2. Springer-Verlag, Berlin

8. Lee RB, Clarkson DT (1986) ^{13}N studies of nitrate fluxes in barley roots. I. Compartmental analysis from measurements of ^{13}N efflux. J Exp Bot 37:1753–1767
9. Memon AR, Saccomani M, Glass ADM (1985) Efficiency of potassium utilization by barley varieties—The role of subcellular compartmentation. J Exp Bot 173:1860–1876
10. Britto DT, Kronzucker HJ (2001) Can unidirectional influx be measured in higher plants? A mathematical approach using parameters from efflux analysis. New Phytol 150:37–47
11. Kronzucker HJ, Britto DT (2011) Sodium transport in plants: A critical review. New Phytol 189:54–81
12. Britto DT, Ebrahimi-Ardebili S, Hamam AM et al (2010) ^{42}K analysis of sodium-induced potassium efflux in barley: Mechanism and relevance to salt tolerance. New Phytol 186: 373–384
13. Horie T, Costa A, Kim TH et al (2007) Rice OsHKT2;1 transporter mediates large Na+ influx component into K+ -starved rice roots for growth. EMBO J 26:3003–3014
14. Britto DT, Kronzucker HJ (2008) Cellular mechanisms of potassium transport in plants. Physiol Plant 133:637–650
15. Brownridge JD (1984) The radial distribution of ^{137}Cs and ^{40}K in tree stems. J Plant Nutr 7:887–896
16. Maier-Maercker U, Jahnke A (1980) Microautoradiography with ^{43}K: a method for the reliable tracing of ion transport in stomata. Zeitschr Pflanzenphysiol 100:35–42
17. Britto DT, Kronzucker HJ (2006) Futile cycling at the plasma membrane: A hallmark of low-affinity nutrient transport. Trends Plant Sci 11:529–534
18. Roslyakova TV, Molchan OV, Vasekina AV et al (2011) Salt tolerance of barley: Relations between expression of isoforms of vacuolar Na^+/H^+-antiporter and $^{22}Na^+$ accumulation. Russ J Plant Physiol 58:24–35
19. Essah PA, Davenport R, Tester M (2003) Sodium influx and accumulation in Arabidopsis. Plant Physiol 133:307–318
20. Malagoli P, Britto DT, Schulze LM et al (2008) Futile Na^+ cycling at the root plasma membrane in rice (*Oryza sativa* L.)—kinetics, energetics, and relation to salinity tolerance. J Exp Bot 59:4109–117
21. Britto DT, Kronzucker HJ (2009) Ussing's conundrum and the search for transport mechanisms in plants. New Phytol 183:243–246
22. Oh D-H, Leidi E, Zhang Q et al (2009) Loss of halophytism by interference with SOS1 expression. Plant Physiol 151:210–222
23. Yeo AR, Yeo ME, Flowers TJ (1987) The contribution of an apoplastic pathway to sodium uptake by rice roots in saline conditions. J Exp Bot 38:1141–1153
24. Szczerba MW, Britto DT, Kronzucker HJ (2006) The face value of ion fluxes: the challenge of determining influx in the low-affinity transport range. J Exp Bot 57:3293–3300
25. Lazof D, Cheeseman JM (1988) Sodium and potassium compartmentation and transport in the roots of intact lettuce plants. Plant Physiol 88:1279–1284
26. Jacquez JA (1996) Compartmental Analysis in Biology and Medicine, 3rd edn. U Michigan Press, Ann Arbor
27. Kronzucker HJ, Szczerba MW, Moazami-Goudarzi M et al (2006) The cytosolic Na^+:K^+ ratio does not explain salinity-induced growth impairment in barley: A dual-tracer study using $^{42}K^+$ and $^{24}Na^+$. Plant Cell Env 29:2228–2237
28. Hajibagheri MA, Flowers TJ, Collins JC et al (1988) A comparison of the methods of X-ray microanalysis, compartmental analysis and longitudinal ion profiles to estimate cytoplasmic ion concentrations in two maize varieties. J Exp Bot 39:279–290
29. Ashworth S, Rabiner EA, Gunn RN et al (2010) Evaluation of ^{11}C-GSK189254 as a novel radioligand for the H_3 receptor in humans using PET. J Nucl Med 51:1021–1029
30. Siddiqi MY, Glass ADM, Ruth TJ (1991) Studies of the uptake of nitrate in barley. III. Compartmentation of NO_3^-. J Exp Bot 42:1455–1463
31. Kronzucker HJ, Siddiqi MY, Glass ADM (1995) Analysis of $^{13}NH_4^+$ efflux in spruce roots: A test case for phase identification in compartmental analysis. Plant Physiol 109:481–490
32. Britto DT, Kronzucker HJ (2003) Trans-stimulation of $^{13}NH_4^+$ efflux provides evidence for the cytosolic origin of tracer in the compartmental analysis of barley roots. Func Plant Biol 30:1233–1238
33. Cheeseman JM (1986) Compartmental efflux analysis—An evaluation of the technique and its limitations. Plant Physiol 80:51–64

Chapter 27

Trait Dissection of Salinity Tolerance with Plant Phenomics

Bettina Berger, Bas de Regt, and Mark Tester

Abstract

The phenotypic response of plants to high external levels of NaCl is very dynamic so the methods used for phenotyping should be able to capture those dynamics. Non-destructive high-throughput plant imaging enables researchers to measure salt stress-induced changes in plant growth over time. In addition, the color information of the digital images allows the quantification of leaf senescence induced through long-term exposure to NaCl. In combination with destructive leaf sampling to measure leaf [Na^+], this approach allows dissection of salinity tolerance into its individual components.

Key words: Plant imaging, Osmotic tolerance, Tissue tolerance, Leaf [Na^+], Growth analysis

1. Introduction

The response of plants to high external concentrations of NaCl occurs in two distinct phases (1). The osmotic response occurs immediately, once the salt concentration in the growth medium reaches a critical level (1) and affects the growth rate of plants by slowing down the expansion of young leaves and inhibiting the initiation of new leaves. The ion-specific response occurs at a later stage when Na^+ and/or Cl^- has accumulated to toxic levels in the plant tissue. This leads to premature senescence of older leaves. Plants have evolved different ways to cope with high levels of NaCl and overall tolerance can be split into three distinct mechanisms, (i) osmotic tolerance, (ii) Na^+ exclusion from the shoot and (iii) tissue tolerance. High-throughput measurements of Na^+ exclusion can be determined fairly easily by measuring the tissue Na^+ concentration in a defined leaf at a given time point. This has resulted in the discovery and characterization of many genes involved in this process, genes that are being incorporated into breeding programs.

Sergey Shabala and Tracey Ann Cuin (eds.), *Plant Salt Tolerance: Methods and Protocols*, Methods in Molecular Biology, vol. 913, DOI 10.1007/978-1-61779-986-0_27, © Springer Science+Business Media, LLC 2012

Osmotic and tissue tolerance on the other hand are very dynamic processes that affect plant growth and development and require continuous, non-destructive measurements over time. Automated imaging setups that are now becoming available allow such measurements, making it possible to dissect the salt stress response of plants into its individual components (2).

Plants that are osmotic tolerant display only minor reductions in relative growth rate upon exposure to NaCl when compared with osmotic sensitive plants that show a severe growth reduction. Changes in growth can be measured by daily imaging of plants, using the size of the plant in the images as a surrogate for plant biomass (3). Using this approach, a reduction in relative growth rate due to salt stress can be observed within days of salt application, before Na^+ has accumulated to toxic levels in the leaves. This makes it possible to study the osmotic component of salt stress independent of the ion-specific stress (2). An alternative approach of measuring osmotic tolerance is by infrared thermography and is presented in this volume in the chapter by James and Sirault.

Plants that are tissue tolerant accumulate high concentrations of shoot Na^+ while still maintaining plant health. They are therefore green. In contrast, sensitive plants will not be able to effectively compartmentalize Na^+ away from the cytosol, resulting in increased cell death and an early onset of senescence in older leaves. Since senescent leaves change color, the onset and degree of senescence can be quantified through the color information of digital color images. Together with measurements of Na^+ accumulation in the same plant, this allows quantification of the tissue tolerance of that plant.

The aim of this chapter is to describe how digital color imaging and growth analysis in combination with tissue Na^+ measurements can be used to quantify osmotic tolerance, Na^+ exclusion, and tissue tolerance for hydroponically and soil grown plants. We have found that the methods described to measure osmotic tolerance and Na^+ exclusion are applicable to a range of plants. However, the measurement of tissue tolerance through quantifying leaf senescence and leaf [Na^+] can currently only be used for highly diverse germplasm, such as Triticum monococcum, where there is large variation in the degree of salt-induced senescence. However, we hope that advances in 3D plant modeling will soon allow the quantification of tissue tolerance in plants such as bread, wheat, and barley that have a lower degree of salt-induced senescence.

2. Materials

2.1. Seed Germination

1. Uniformly sized seeds (see Note 1).
2. Petri dishes.

3. Filter paper or paper towel.
4. 70% (v/v) ethanol.
5. 3% (v/v) sodium hypochlorite (see Note 2).

2.2. Growth Solutions and Hydroponics Setup

1. Stock solution A: 0.04 M NH_4NO_3, 1 M KNO_3. Add about 2.5 L deionized water to a 5 L measuring beaker. Weigh 12.8 g NH_4NO_3 and 404.4 g KNO_3 and add to the beaker. Mix and adjust to a final volume of 4 L. Store at 4°C.
2. Stock solution B: 0.4 M $Ca(NO_3)_2$. Add about 2.5 L deionized water to a 5 L measuring beaker. Weigh 377.8 g $Ca(NO_3)_2{\cdot}4H_2O$ and add to water. Mix and adjust to a final volume of 4 L. Store at 4°C.
3. Stock solution C: 0.4 M $MgSO_4$, 0.02 M KH_2PO_4. Add about 2.5 L deionized water to a 5 L measuring beaker. Weigh 394.4 g $MgSO_4{\cdot}7H_2O$ and 10.8 g KH_2PO_4 and add to water. Mix and adjust to a final volume of 4 L. Store at 4°C.
4. Liquid $NaSiO_3$stocks can be purchased and diluted accordingly to reach a final concentration of 0.5 M. (see Note 3). Store at 4°C.
5. Stock solution E: 0.05 M NaFe(III)EDTA. Dissolve 14.7 g NaFe(III)EDTA in about 0.5 L of RO water, mix and adjust to 0.8 L. Store at 4°C.
6. Micronutrient stock F: 50 mM H_3BO_3, 5 mM $MnCl_2$, 10 mM $ZnSO_4$, 0.5 mM $CuSO_4$, 0.1 mM Na_2MoO_4. Dissolve 2.47 g H_3BO_3, 0.79 g $MnCl_2{\cdot}4H_2O$, 2.3 g $ZnSO_4{\cdot}7H_2O$, 0.10 g $CuSO_4{\cdot}5H_2O$, 0.02 g $Na_2MoO_4{\cdot}2H_2O$ in about 0.5 L of RO water and adjust to 0.8 L. Store at 4°C.
7. Polycarbonate pellets, approximately 3 mm in diameter (see Note 4).
8. PVC tubes (4.5 cm diameter, 28 cm height) fitted with mesh at the bottom to prevent pellets from falling through.
9. 80 L storage tank.
10. Two 50 L growth tubs with grid to hold PVC tubes in upright place.
11. Aquarium pumps that are able to lift water at least 1.2 m at a minimum of 300 L per hour.

2.3. Growth in Potting Mix

2.3.1. Measurement of Field Capacity of Potting Mix

1. Sintered glass funnel.
2. 1.3 m silicon or clear plastic tubing with diameter to fit the funnel outlet.
3. Retort stand and clamp.
4. Large beaker or bucket as water reservoir.

2.3.2. Pot Preparation and Plant Growth in Potting Mix

1. Draining plastic pots with a capacity of about 3 L (see Note 5 & subheading 3.5.6) and saucers big enough to enclose the bottom third of the pot.
2. Potting mix.

2.4. Measurement of Third Leaf Na^+ and K^+ Concentration by Flame Photometry

1. Hot block for acid digest (e.g. Environmental Express, Mount Pleasant, USA).
2. 1% HNO_3 stored in a dark glass bottle with dispenser.
3. Flame photometer (e.g. model 420, Sherwood, Cambridge, UK).

2.5. Image Acquisition

1. Industry grade digital color camera with automated software control (e.g. LemnaTec 3D Scanalyzer system, LemnaTec GmbH Germany).
2. Automated setup to move plants to the camera or vice versa. If manual systems are used, experiments are usually limited to about 150–200 plants per experiment.
3. Adequate computer hardware for image storage (see Note 6).
4. Adequate illumination equipment.
5. Optional a color reference card and/or ruler for calibration purposes (e.g. RHS Colour Chart; ColorChecker, X-Rite,USA).

2.6. Image Analysis

1. Adequate computer hardware for high-throughput image processing.
2. Image analysis software package, included with imaging system like LemnaGrid (LemnaTec GmbH, Germany) and/or stand-alone software such as MATLAB (Mathworks, USA), Halcon (MVTec Software GmbH, Germany) or Labview (National Instruments, USA). An open source alternative is ImageJ (http://rsbweb.nih.gov/ij).

3. Methods

3.1. Seed Germination

1. Surface sterilize uniformly sized seeds (see Note 7) for 1 min in 70% (v/v) ethanol followed by 5 min in 3% (v/v) sodium hypochlorite.
2. Rinse the seeds several times in deionized water.
3. Place the seeds evenly on several layers of moist filter paper or paper towel in petri dishes.
4. Germinate the seeds for 2–4 days at room temperature in the dark until the coleoptiles are about 1.5–2 cm long.

3.2. Growth in Supported Hydroponics

There are numerous ways how to design a supported hydroponics system. One example is presented in this volume in the chapter by James and Sirault. The method presented here follows the protocol described by Genc et al. (4), which also contains an image of the hydroponic setup).

1. Two 50 L opaque plastic tubs are mounted on a trolley and connected to an 80 L storage tank with nutrient solution. Each tub holds 42 PVC tubes filled with polycarbonate pellets. The pellets should be filled to 1 cm above the maximum height of the nutrient solution during the filling cycle. Each tub is filled and drained with 25 L of nutrient solution every 20 min (see Note 8).
2. Fill the storage tank with about 50 L of RO water then add each 400 mL of stock solutions A, B, and C and each 80 mL of stock solutions D, E, and F. Adjust the pH to 6.0 with 10% (v/v) HCl and make up to a final volume of 80 L.
3. To each PVC tube, transplant one, uniformly germinated seedling at a depth of 1 cm, ensuring that the roots reach the nutrient solution during the filling cycle.
4. After 1–2 days, once the seedlings have established, gently fill up the PVC tubes another 1–2 cm with polycarbonate pellets to support the seedlings and reduce the growth of algae at the surface.
5. Monitor the pH of the nutrient solution daily and adjust if necessary.
6. Change the nutrient solution weekly.
7. Start the salt application at the emergence of leaf three by adding 25 mM NaCl and 1.67 mM $CaCl_2$ (see Note 9) in increments once or twice per day to avoid osmotic shock. The final concentrations typically range from 100 mM to 300 mM NaCl, depending on the plant species and duration of the treatment.

3.3. Growth in Potting Mix

3.3.1. Measurement of Field Capacity of Potting Mix

When working in pots, it is important to carefully consider the watering to avoid water-logging and hypoxia (5). Many experiments will adjust watering to "water holding capacity" or "pot capacity", which is the volumetric water content of a free draining pot. However, this value greatly depends on the height of the pot and might often result in hypoxia, especially with fine potting mixes or field soil. In our experiments, we measure "field capacity", defined as the volumetric water content of the potting mix or soil at 1 m suction.

The setup described here to measure this parameter is comparable to the one shown in Figure 2 of Passioura (5).

1. Attach the silicon tubing to the funnel outlet.
2. Mount the funnel with tubing on a retort stand about 1 m above the water reservoir (see Note 10).

3. Add about 2 L of water to the water reservoir below the funnel.
4. Fill the funnel and silicon tube with water ensuring that all air bubbles are removed.
5. Add the soil/potting mix to be tested into the funnel and let settle it settle. About half to two-thirds of the funnel should be filled with soil.
6. Once the water has drained to just above the soil level, cover the funnel with clingfilm to avoid evaporation from the surface.
7. To ensure hydraulic conductivity, there should be no air bubbles present between the filter plate, tubing and water reservoir.
8. Adjust the position of the filter to obtain a height of 1 m from the sintered filter plate down to the water level in the reservoir.
9. Let the soil/potting mix equilibrate for several days up to 1 week, ensuring that no air bubbles form.
10. Take out the wet soil from the funnel and record the wet weight (WW).
11. Dry the soil in an oven at 105°C until constant weight is reached.
12. Record the dry weight (DW).
13. The volumetric field capacity is given by the equation (WW – DW)/DW.

3.3.2. Plant Growth in Potting Mix and Salt Application

There are numerous possibilities of how to design salt stress experiments in soil or potting mix. In several studies, the salt is pre-mixed into the bottom part of the soil (6) to simulate a saline subsoil environment. Here, we present salt application through watering with saline solution. This has the advantage that the time point of salt application can be chosen precisely and the changes in growth immediately after salt application can be measured.

1. Fill a pot to about 4 cm below the rim after gentle tapping and then weigh it.
2. Use the same weight to fill up all remaining pots.
3. Include several spare pots to monitor water evaporation from the soil during the experiment and at least two pots to determine the oven dry weight of the soil.
4. Once all pots are filled, add enough water for germination.
5. Plant three to four seeds per pot, about 1 cm deep and cover them with soil.
6. Use the soil dry weight (see subheading 3.4.3) to calculate the target weight of a pot at field capacity as determined by Subheading 3.3.1.

7. Adjust the watering level of each pot to field capacity about 2–3 times per week and record the water use.
8. Once the seedlings are about 8–10 cm long, thin out to one seedling per pot.
9. Daily image acquisition should start once the second leaf is expanding and should continue until about 1 week after salt application. Then the imaging frequency can be reduced to about three times per week.
10. At the time of third leaf emergence, start salt application by adding saline solution into the saucer.
11. The amount and concentration of the saline solution added will depend on the amount of dry soil in the pot, the field capacity of the potting mix and the desired final NaCl concentration. For wheat and barley, we observe noticeable reductions in growth from around 75 mM NaCl.
12. When choosing the volume and concentration of saline solution to add, one has to make a compromise between avoiding osmotic shock due to high NaCl concentrations and overwatering and hypoxia due to a too larger volume added.

As an example, if using pots with 2 kg of dry potting mix, a field capacity of 20% (w/w) and a desired final NaCl concentration of 100 mM, one could add 150 mL of a 267 mM NaCl solution. This would temporarily bring the soil moisture above field capacity (27.5% at 2.55 kg wet weight) and pots would have to be dried down to field capacity (20% at 2.4 kg) over the following days.

3.4. Measurement of Third Leaf Na^+ and K^+ Concentration by Flame Photometry

Measuring the tissue Na^+ and K^+ concentration is an easy way to determine the ability of a plant to exclude Na^+ and to maintain a high K^+/Na^+ ratio. Both are important traits contributing to overall salt tolerance.

1. Sample the third leaf 10 days after start of salt application (see Note 11) from the soil or hydroponically (see Note 12) grown plants.
2. Record the leaf fresh weight and place leaves in 50 mL Falcon tubes.
3. Dry the leaves at 65°C in a drying oven until constant weight is reached.
4. Record the dry weight of each leaf (see Note 13).
5. To each sample, add 10 mL of 1% HNO_3.
6. Digest the leaves for 4 h at 85°C in a hot-block.
7. Ensure that the whole leaf is submerged during the digestion and gently shake the samples every 30 min (see Note 14).
8. Let the samples cool down to room temperature after the digest. The samples can be stored for several weeks at room temperature.

9. Prepare a 0 μM NaCl and 0 μM KCl lower standard and a top standard of 500 μM NaCl and 500 μM KCl (see Note 15).
10. Turn on Flame Photometer, model 420 (Sherwood, UK) and set top and bottom standards for both Na and K.
11. Dilute samples with appropriate dilution factor so the readings are within the two standards (see Note 16).
12. For each sample record the reading, then use the dilution factor of the sample, the standard curve, volume of liquid the sample was digested in and the weight of the plant material to convert the reading into either [μmole·g^{-1}] based on dry weight or [mM] based on the tissue water content.

3.5. Image Acquisition to Monitor Plant Growth and Leaf Color

How images are acquired will greatly depend on the hard- and software available to the researcher and the trait to be measured. There are complete systems available from LemnaTec (LemnaTec GmbH, Germany) that combine plant handling, imaging hardware, and the control software. Other institutes might have the capability to build their own automated in-house solutions (7, 8) or use a fairly simple camera setup and manual handling of plants. We will therefore only present aspects of image acquisition that are generally applicable and important for any type of setup.

The aim of any imaging setup should always be to obtain the best possible image of the plants for measuring the trait of interest. Image acquisition should be done as consistent as possible. This will greatly facilitate the image analysis and ideally allow the generation of automated image analysis algorithms that require minimum user input.

In general, there are two methods for image acquisition.

1. The plants are stationary and the camera is moved to the plant. This is most commonly used for plants with a simple architecture, such as, for example, Arabidopsis, where a single image from the top often provides sufficient data.
2. The plants are moved to a stationary camera setup. This is of advantage for plants with a complex morphology, such as wheat and barley, where images from several angles will greatly increase the quality of data obtained through imaging. In addition, the imaging environment, such as background and illumination, is easier to control.
3. Illumination conditions should be as uniform as possible, both over time and throughout the field of view. It is important to pre-heat the lamps until constant illumination is reached before the first images are taken. Hunter et al. (9) give detailed information on how to achieve optimal lighting and avoid shadows and reflections.

4. Use of a color card and ruler allow calibration of the imaging setup. If both are present in an image, it is possible to normalize the recorded colors and calibrate for the zoom factor used. This allows comparisons between different imaging setups that differ in lighting conditions and the cameras used.
5. The imaging background should be chosen carefully to facilitate the identification of the plant in subsequent analysis. Backgrounds, such as white or blue are preferable, since the green of the plant will be easy to differentiate.
6. Green and gray should be avoided as pot colors. White, blue, and black are suitable for most plant types and white has the advantage of keeping the soil cooler than darker colors. Materials with a flat finish reduce undesired reflections.
7. The soil surface can become challenging in the image analysis, since sandy or drying soils can have very similar colors to senescent leaves. Colored plastic mulch or white gravel on the surface can reduce this problem and have the further advantage of reducing water loss from the soil surface.
8. Many plants, especially wheat and barley, will need some sort of support when grown in pots, such as carnation frames. Again, they should not be green and if they are out of metal it needs to be tested if they can be easily eliminated in the image analysis. In some cases, it might be easier to get color-coated frames to avoid problems in the automated image analysis.
9. When choosing the exposure for the images, it is generally better to have a lower exposure. Overexposure will lead to white spots and thus a loss of color information that cannot be compensated for by image analysis.
10. The file format for storing the images should not lead to loss of image information, such as done by JPG or BMP. PNG or TIFF are commonly used formats and do not lead to loss of information through compression.

3.6. Image Analysis for Plant Size and Senescence Measurements

Since plant imaging allows daily recordings, already simple image analyses, such as plant size measurements, yield valuable information about plant growth and performance. Nevertheless, basic image analysis also requires the use of specialized software, computing infrastructure, and database management if it is to be performed at high-throughput.

Depending on the software solution used, different levels of prior knowledge in image analysis and programming are necessary to develop image analysis algorithms and collaboration with scientists experienced in that area is advisable.

MATLAB (MathWorks, Massachusetts, USA) is possibly the most commonly used and powerful software to develop image analysis algorithms and offers solutions for automated image acquisition. Halcon (MVTec Software GmbH, Germany) is a fairly comprehensive application for image analysis and it is compatible with common programming languages such as C, C#, and .NET. ImageJ (http://rsbweb.nih.gov/ij) presents a Java based solution for image analysis that is open source, so it is easily accessible. However, all three softwares require a certain amount of programming skills to use to write and implement analysis algorithms. The in-built image analysis solution of LemnaTec setups, LemnaGrid (LemnaTec GmbH, Germany), was designed to allow researchers without prior programming knowledge to create algorithms for image analysis through drag-and-drop software where individual operators can be connected to create a processing pipeline. Unfortunately, algorithms can only be shared among LemnaTec users and the functionalities are not as comprehensive at those of specialized image analysis software.

Since the specific algorithms will depend on the software used and the imaging setup, we will only discuss general steps common to digital image processing (10) that are necessary to measure the traits of interest for osmotic and tissue tolerance studies, the size of the plant and its degree of senescence.

1. Image retrieval. Recorded images need to be loaded into the software from a database or storage folder. Images may need to be cropped or a Region Of Interest (ROI) may need to be set to shorten the computing time and/or to remove unnecessary parts of the image that can become a source of noise.
2. Image pre-processing. The application of filters to minimize noise or increase sharpness can improve the outcome of the subsequent analysis steps. However, there is a possibility of losing information that cannot be retrieved in later steps. If thresholding is used to make a binary image in the next step, the color image needs to be converted into a gray scale image by transforming the 3D RGB color information into a single channel.
3. Image segmentation. The next step is the segmentation of the image into objects of interest, parts of the plant, and objects that will later be discarded, such as the background, pot, carnation frame, or soil. Depending on the composition of the image, there are several options to produce a binary image. Classification by color with a supervised nearest neighbor algorithm or thresholding of a gray scale image are commonly used. In both instances, the result is a binary image, where pixels that belong to the object of interest are set to a value of 1, all others to 0.
4. Noise reduction. Morphological operations such as erosion-dilation steps or filling holes can be used to correct for unavoidable imperfections in the binary image, that result from noise

from image acquisition or difficulties in distinguishing between parts of the object and background that have similar colors.

5. Image composition. Leaves can often become fragmented in earlier steps due to curling of the leaves and the individual fragments need to be merged to create one single object, the plant.
6. Image description. Features of the identified object, such as area, height, width, convex hull, or compactness are quantified. The features mostly consist of mathematical characteristics calculated from the object.
7. Color classification. The identified object, the plant, can now be extracted from the original RGB image. Based on the color information of the original image, damaged leaf area can be identified and quantified using supervised nearest neighbor color classification.

3.7. Growth Analysis for Measurement of Osmotic Tolerance

It is beyond the scope of this chapter to discuss plant growth analysis in detail. For in depth information on this subject, please consult the publications by Hunt (11, 12).

The purpose of the two methods presented here is to measure the reduction in plant growth in response to the osmotic stress component of NaCl immediately after salt application, before symptoms of Na^+ toxicity become visible (see Note 17). Both methods use the projected shoot area of the plant (A) extracted from the images as a measure of plant size and the time of salt application is used as zero time point.

1. Exponential growth model. An exponential growth model of the form $A(t) = A.e^{Rt}$ is frequently used to describe the growth of biological systems. Here, t is the time, $A(t)$ is the size or biomass at time t, A the size at time 0, and R the relative growth rate (RGR). This model assumes that the RGR is constant and does not change over time. This assumption is valid for many systems, such as bacterial growth, and can be used as an approximation in early stages of plant growth, before the plants change to reproductive development. An advantage of using an exponential growth model is its ease of use. Most standard spread sheet programs, such as Microsoft Excel (Microsoft Cooperation, USA), will allow fitting an exponential growth curve through a given set of data points.
   1.1. Transform the time values of your data to "days after salt application".
   1.2. Perform a log transformation (ln) of your projected shoot area measurements (A) obtained from the image analysis.
   1.3. The RGR can now be calculated as the slope of a linear regression curve through the transformed data (time after salt application t as x-values, ln of projected shoot area

ln(A) as *y*-values). The RGR indicates the growth of the plant per day, independent of its initial size and has the unit [d^{-1}].

1.4. To obtain an index for osmotic tolerance, use the equation: *osmotic tolerance = RGR in salt/RGR in control* (see Note 18).

2. Spline curve or polynomial growth model. Spline curves and higher order polynomials have the advantage that they make no prior assumptions about the growth of the plant, so give a more realistic representation of plant growth, especially when growth is monitored over longer periods of time and when the plants shift from vegetative to reproductive growth. In both cases, RGR is not assumed to be constant, but changes over time. Consequently, close monitoring of growth dynamics becomes possible. Since higher order polynomials can also be calculated with many basic spreadsheet programs, we will use a polynomial growth curve as an example.

   2.1. Transform the time values of your data to "days after salt application".

   2.2. Use statistical software to fit a higher order polynomial through your raw data (time *t* as *x*-values, projected shoot area *A* as *y*-values).

   2.3. The differential of the polynomial function describes the absolute growth rate (*AGR*) over time with *AGR = dA/dt.*

   2.4. The function describing the RGR can be calculated as *RGR = dA/dt1/A.* Since in this case RGR is a function of time, a simple index for osmotic tolerance cannot be calculated as compared to an exponential growth model. However, the dynamic changes in RGR upon salt application can be monitored using this approach. This might reveal details obscured when using exponential curve fitting.

3.8. Measurement of Tissue Tolerance

Tissue tolerance refers to the ability of a plant to tolerate high levels of Na^+ within its tissues or organs. To measure tissue tolerance we need to relate the levels 3rd leaf [Na^+] as determined by flame photometry (see subheading 3.4) with the level of salt-induced senescence measured through color classification of the images (see subheading 3.6.7). We found that this approach works well for germplasm that shows large variation in the level of salt-induced senescence, such as *T. monococcum* (2). However, modern wheat and barley cultivars show relatively low levels of salt-induced leaf senescence, with a relative area of senescence generally below 10%. It may therefore be necessary to determine the life span of individual leaves as an alternative measure of tissue tolerance for plants such as wheat and barley. This will hopefully become possible in high-throughput experiments with the more sophisticated plant image analysis that is currently being developed.

For now, the method described here is only suitable for plants with large variation in tissue tolerance and follows the previously described protocol by Rajendran et al. (2). The values used to determine the tissue tolerance index should be extracted from images of plants that were exposed to high salinity for several weeks, when clear symptoms of Na^+ toxicity are visible.

1. Calculate the relative natural senescence (*N*) of plants grown in control conditions of the same age as *N* = *senescent area/total projected shoot area.*
2. The relative salt-induced senescence (*S*) can be determined as *S* = *(senescent area in salt/total projected shoot area in salt)—N.*
3. Plants with a low level of salt-induced senescence in the presence of high [Na^+] are considered to be tissue tolerant. The index for tissue tolerance can therefore be calculated as *Tissue tolerance* = *3rd leaf* $[Na^+] \cdot (1 - S)$.

4. Notes

1. We have used the described methods for wheat (bread wheat, du\rum wheat, and Einkorn wheat), barley, and rice.
2. The sodium hypochlorite solution can be prepared using a household product such as Domestos®, when taking into account the lower active concentration of Cl^- compared to a lab grade solution.
3. It is easiest to get a homogenous solution by first adding the water and then the sodium silicate stock due to its high viscosity.
4. Polycarbonate pellets can be obtained from plastic manufacturers that use them as a starting product for molding. If using other types of plastics, ensure their density is higher than water.
5. The color of the pot should allow an easy distinction from the plants in the image-processing step, preferably white or blue. Black is possible, but it leads to an increased soil temperature. Standard green nursery pots should not be used.
6. Since the assay presented for osmotic tolerance is based on growth rates, it is extremely important that the seeds and seedlings used are as uniform as possible. If sufficient seed is available, one should always germinate excess amounts to be able to select for evenly sized seedlings. If it is known that the used

lines germinate at different rates, the germination should be staggered to have evenly sized seedlings at the time of transplanting. In the case of limited seed stocks and uneven germination, the faster growing seedlings can be placed at 4°C to slow down growth until the other seedlings have reached the same size. However, this should be a last resort.

7. We generally take three images per plant (two from the side at 90° rotation and one from the top) at about 15–20 time points throughout an experiment. With a file size of about 4 MB, this amounts to 4 MB × 3 images × 20 time points = 240 MB per plant. Even a smaller scale experiment with 200 plants will therefore need 47 GB of storage.
8. To avoid excessive growth of algae, light exposure to the solution should be kept at a minimum and the space between the PVC tubes should be covered.
9. An increased concentration of Na^+ in the solution will decrease the Ca^{2+} activity and can lead to Ca^{2+} deficiency if no additional Ca^{2+} is supplied. There are software programs such as Visual MINTEQ (www2.lwr.kth.se/English/OurSoftware/vminteq) that can be used to calculate the Ca^{2+} activity. As an approximation a ratio of 15:1 for Na^+: Ca^{2+}can be used.
10. If no large retort stand is available, a smaller one can be placed on a table with the water reservoir on the ground.
11. The third leaf should be harvested once it is fully elongated and mature, but before senescence starts to occur. In general, 10 days after salt application is a good time interval under most growth conditions.
12. Plants grown in hydroponics might need to be rinsed with RO water and wiped dry to avoid salt contamination from the growth solution.
13. By subtracting the dry weight from the fresh weight, it is possible to determine the tissue water weight. This is useful if wanting to express the Na^+ or K^+ concentrations on a tissue water basis.
14. After the acid digestion, the leaves should become translucent.
15. Depending on salt treatment, plant sample, and dilution it may be necessary to adjust the concentrations of the top standard. For wheat and barley, we typically use 500 μM NaCl and 500 μM KCl. For extremely salt stressed plants it may be necessary to use a top standard of 500 μM NaCl and 100 μM KCl.
16. We try to adjust the dilution factor so that all readings are between 10 and 100 units. Readings below 10 should be avoided.

17. We generally treat the first 5–7 days after salt application as the osmotic phase.

18. If the experiment does not allow growing control plants in parallel, a ratio of *RGR after salt application/RGR before salt application* can be used instead. However, this index will be sensitive early vigor of the seedlings and extra care should be taken when selecting even sized seedlings.

References

1. Munns R, Tester M (2008) Mechanisms of salinity tolerance. Annu Rev Plant Biol 59:651–681
2. Rajendran K, Tester M, Roy SJ (2009) Quantifying the three main components of salinity tolerance in cereals. Plant Cell Environ 32(3):237–249
3. Golzarian MR, Frick RA, Rajendran K et al (2011) Accurate inference of shoot biomass from high-throughput images of cereal plants. Plant Methods 7. doi:2.10.1186/1746-4811-7-2
4. Genc Y, McDonald GK, Tester M (2007) Reassessment of tissue Na^+ concentration as a criterion for salinity tolerance in bread wheat. Plant Cell Environ 30:1486–1498
5. Passioura JB (2006) The perils of pot experiments. Funct Plant Biol 33(12):1075–1079
6. Harris BN, Sadras VO, Tester M (2010) A water-centred framework to assess the effects of salinity on the growth and yield of wheat and barley. Plant Soil 336(1–2):377–389
7. Granier C, Aguirrezabal L, Chenu K et al (2006) PHENOPSIS, an automated platform for reproducible phenotyping of plant responses to soil water deficit in *Arabidopsis thaliana* permitted the identification of an accession with low sensitivity to soil water deficit. New Phytol 169(3):623–635
8. Jansen M, Gilmer F, Biskup B et al (2009) Simultaneous phenotyping of leaf growth and chlorophyll fluorescence via GROWSCREEN FLUORO allows detection of stress tolerance in *Arabidopsis thaliana* and other rosette plants. Funct Plant Biol 36(10–11):902–914
9. Hunter F, Biver S, Fuqua P (2007) Light-science & magic: an introduction to photographic lighting. Focal Press
10. Gonzalez RC, Woods RE (2006) Digital image processing, 3rd edn. Prentice-Hall, NJ
11. Hunt R (1978) Plant growth analysis. Edward Arnold Ltd
12. Hunt R, Causton DR, Shipley B, Askew AP (2002) A modern tool for classical plant growth analysis. Ann Bot 90(4):485–488

Chapter 28

Measuring Soil Salinity

Marcus Hardie and Richard Doyle

Abstract

Soil salinity is a form of land degradation in which salts accumulate in the soil profile to an extent that plant growth or infrastructure are negatively affected. A range of both field and laboratory procedures exist for measuring soil salinity. In the field, soil salinity is usually inferred from apparent electrical conductivity (EC_a) using a range of devices, depending on the required depth of analysis, or size of the survey area. Field measurements of EC_a require calibration to the actual salt content by laboratory analysis. In the laboratory, soil salinity is usually assessed by determining either the total soluble salts by evaporation of a soil water extract (TSS), or by determining the electrical conductivity (EC) of either a 1:5 distilled water:soil dilution, or a saturated paste extract. Although procedures for measuring soil salinity appear relatively straightforward, differences in methodology have considerable influence on measured values and interpretation of results.

Key words: Electrical conductivity, Electromagnetic induction, Soil texture, Saturated paste, Salt

1. Introduction

Soil salinity refers to the presence of soluble salts in soil or soil water at levels that adversely affect plant growth. Globally salinity is predicted to affect 3,230,000 km^2 (1). Salinity can be natural or induced by human activities including clearing of deep rooted vegetation and use of irrigation. Soil salinity reduces plant growth by a combination of (1) reducing the osmotic potential of the soil solution which limits water uptake by the plant, (2) specific ion toxicity from ions such as sodium, chloride, and boron, and (3) reduced uptake of nutrients. In order to monitor changes in the extent of salt affected land over time or predict the effect of soil salinity on plant growth or crop yield, it is necessary to measure soil salinity in a reliable, repeatable, cost effective manner.

A variety of techniques exist to measure soil salinity. In the field, soil salinity is commonly measured by determination of the

Sergey Shabala and Tracey Ann Cuin (eds.), *Plant Salt Tolerance: Methods and Protocols*, Methods in Molecular Biology, vol. 913, DOI 10.1007/978-1-61779-986-0_28,

apparent electrical conductivity (EC_a) at multiple locations in order to generate a map of apparent electrical conductivity. Laboratory methods for determining soil salinity involve either the determination of total soluble salts (TSS), electrical conductivity (EC) of a soil suspension, or determination of the type and proportion of salt species using spectrophotometers. Advantages and disadvantages of each approach are discussed, as well as procedures for reporting EC values, and procedures for converting $EC_{1:5}$ values to an equivalent saturated paste value (EC_{eq}).

1.1. Field Techniques

Soil salinity may be indirectly measured using a range of devices of which electromagnetic induction (EM) is the most popular (2). Electromagnetic induction (EM) uses a transmitting coil to create a magnetic field that penetrates into the soil. The magnetic field creates an electrical current that in turn creates a secondary magnetic field which is received by the meters receiver coil (3). The ratio of the secondary to primary magnetic field is in linear proportion to the soil's apparent electric conductivity (EC_a). Consequently, EC_a represents the electrical conductivity of both the soil and water, in which EC_a is influenced by salt content, soil porosity, permeability, moisture content, soil temperature, and clay composition (2, 4). Thus, interpretation of EC_a values in terms of salt content is non-unique and depends on other assumptions about soil properties. Studies in Australia typically report that between 75% and 90% of EC_a values can be explained by total soluble salt content (2). Provided that total soil moisture content is greater than 10%, electromagnetic approaches are usually a reliable means for rapid determination of soil salinity (5, 6). In topsoils or within the crop root zone EC_a is usually determined by EM 38 which operates to a maximum depth of approximately 1.5 m, whilst subsoil EC_a is usually measured by EM 31, or down boreholes by EM 39. EM 38 and EM 31 devices may be operated manually or mounted to a vehicle with precision GPS to enable rapid paddock or farm scale mapping. Interpretation of soil salinity from EC_a maps requires calibration of EC_a values to laboratory determined measurements of electric conductivity ($EC_{1:5}$, EC_{sp}) from multiple locations and soil depths. Electromagnetic devices may also be mounted behind fixed wing aircraft or helicopters (AEM). Combined with other devices such as LIDAR, radiometrics, or digital elevation models, AEM enables rapid catchment scale mapping of multiple soil attributes and their position in the landscape. Calibration of AEM surveys requires considerable calibration data usually from down borehole induction (EM 39) or electrical depth sounding. Interpretation of AEM surveys requires sophisticated processing, statistical and image processing software. AEM has advantages over standard EM surveys due to its lower cost per hectare, rapid survey of large areas, and the ability to resolve spatial variations in EC_a at discrete depth intervals.

1.2. Laboratory Techniques

Soil salinity may be determined in the laboratory by measuring the (1) electrical conductivity (EC) of soil water extracts, (2) soluble ion concentrations of soil water, (3) mass of total dissolved solids (TDS), or total dissolved ions (TDI).

Soil salinity is most commonly measured as electrical conductivity (EC) of a 1:5 soil:distilled water suspension following 1 h of end-over-end mixing. Electrical conductivity is a surrogate measure of salinity in which the salt content of a soil water suspension or a soil water extract is determined by the ability of the extract/suspension to conduct electricity between two metal electrodes accessed as based on the ability of the extract/suspension to conduct electricity. The more salt in the soil extract or suspension, the more current is conducted between the two electrodes, resulting in higher EC values. Electrical conductivity values increase with increasing temperature and this must be corrected for if not measured at 25°C. An approximate correction can be made by increasing the values by 2% for each degree that the ambient temperature is below 25°C, and decreasing when the temperature is above 25°C (7).

Soil salinity may also be determined as the total dissolved salts or TDS (mg/L). This is measured by evaporating a known filtered volume of water to dryness, then weighing the remaining solid residue. This approach is prone to error resulting from incomplete filtration of clay platelets and inclusion of non-saline dissolvable organic compounds in the evaporite (8). There is no exact relationship between EC (μS/cm) and TDS (ppm) as the form of the salt species present influences conversion values. However, in water with a high proportion of sodium chloride, one can estimate TDS(ppm) by multiplying EC (μS/cm) by 0.5 – 0.67 (7).

Approaches such as EC and TDS are not able to determine the species of ions that contribute to soil salinity. Soil salinity may result from a range of soluble salts, typically the cations Na^+, Mg^{2+}, and Ca^{2+} and the anions Cl^-, SO_4^{2-}, CO_3^{2-}, and HCO_3^-. Soluble fertilizer also contributes other ions such as K^+, NH_4^+, and NO_3^-. In order to determine the types and relative importance of the different ions that contribute to soil salinity, the soluble anion and cation concentrations of soil water extracts need to be determined by approaches such as flame-atomic absorption spectroscopy, colorimetric methods, ion chromatography, and titrimetric methods (9).

2. Materials

1. Field techniques require access to appropriate electromagnetic devices depending of the scale and depth of the required investigation. Typical EM devices include airborne electromagnetic (AEM), hand held or vehicle mounted EM 38, EM 31, and

down borehole apparatus (EM 39). These are usually provided by specialist consultants or may be purchased directly from suppliers such as Geonics Ltd., Fugro Airborne Surveys, Geophex Ltd., etc.

2. A calibrated EC meter, with temperature sensor or temperature compensation. A number of manufacturers supply suitable EC meters including Hanna instruments, HM Digital, Extech Instruments, Milwauke meters, etc.
3. Calibration solution. Commercially available solutions are typically available at 1.413 dS m^{-1} and 12.880 dS m^{-1}, or can be made directly from potassium chloride.
4. Deionized or Distilled Water. The water must have an electrical conductivity of $<10^{-4}$ dS m^{-1}, and have a CO_2 concentration not more than that in equilibrium with the atmosphere.
5. 100 mL vials with lid (depending on the quantity of soil required).
6. Shaker, preferably end-over-end shaker.
7. The saturated paste approach will also require a small beaker or cup for mixing saturated paste and either Buchner funnel with highly retentive filter paper and vacuum apparatus or centrifuge.

3. Methods

3.1. $EC_{1:5}$ Soil/Water Extract

In the laboratory, soil salinity is commonly measured as the electrical conductivity at 25°C from an unfiltered 1:5 soil:distilled water suspension ($EC_{1:5}$) (see Note 2).

1. Prepare a 1:5 soil/water suspension. For example, weigh 20.0 g air-dry soil into a vial and add 100 mL deionized water.
2. Mechanically shake (end-over-end preferred) at 25°C in a closed system for 30 min to dissolve soluble salts.
3. Allow about 15 min for the soil to settle.
4. Calibrate the conductivity cell and meter in accordance with manufacturers' instructions.
5. If standard solutions are not available, make a 0.010 M potassium chloride reference solution by dissolving 0.7455 g potassium chloride (KCl; dried at 110°C for 2 h) and make volume to 1.0 L with CO_2-free, distilled or deionized water. This solution has an electrical conductivity of 1.413 dS m^{-1} at 25°C.
6. Dip the conductivity cell into the supernatant, moving it up and down slightly without disturbing the settled soil. Take the reading with the cell stationary when the system has stabilized. Rinse the cell with deionized water between samples and

remove excess water. Complete EC measurements within 3–4 h of obtaining the aqueous supernatant.

7. If readings become erratic, clean electrodes by soaking them in an acid-dichromate cleaning solution overnight (7).
8. Report $EC_{1:5}$ (dS m^{-1}) at 25°C on an air-dry basis (see Note 1).

3.2. EC_{se} Saturated Paste Extract

This extract is obtained by making a saturated soil paste with distilled water and then extracting the liquid from the soil with a centrifuge or a suction device after allowing time for the soil to equilibrate (see Note 2).

1. Determine moisture content of air dried soil by drying at 105°C for 24 hours.
2. Weigh 200–400 g of air dried soil ground to <2 mm of known soil moisture content into a container with lid.
3. Add deionized water while mixing the ground soil sample to saturation. Saturation is said to exist when, the soil paste glistens, flows slightly when the container is tipped, slides cleanly from the spatula and readily consolidates after a trench is formed upon jarring the container.
4. Allow the soil to equilibrate for at least 4 h. Check to ensure saturation criteria is still met. If free water has accumulated on the surface, add more soil and remix. If soil has stiffened or does not glisten, add distilled water and mix thoroughly.
5. Allow the soil paste to stand for a further 4 h or preferably overnight.
6. Transfer soil paste to Buchner funnel fitted with highly retentive filter paper and apply vacuum to collect extract until air passes through the filter.
7. Alternatively extract soil solution by centrifuge (10).
8. Store extracts at 4°C until analyzed for EC.
9. Calibrate the conductivity cell and meter in accordance with manufacturers' instructions.
10. If standard solutions are not available, make a 0.010 M potassium chloride reference solution by dissolving 0.7455 g potassium chloride (KCl; dried at 110°C for 2 h) and make volume to 1.0 L with CO_2-free, distilled or deionized water. This solution has an electrical conductivity of 1.413 dS m^{-1} at 25°C.
11. Dip the conductivity cell into the supernatant, moving it up and down slightly without disturbing the settled soil. Take the reading with the cell stationary when the system has stabilized. Rinse the cell with deionized water between samples and remove excess water. Complete EC measurements within 3–4 h of obtaining the aqueous supernatant.
12. If readings become erratic, clean electrodes by soaking them in an acid-dichromate cleaning solution overnight (7).

13. Report EC_{eq} (dS m^{-1}) at 25°C on an air-dry basis (see Note 1).
14. If required, saturation percentage can be calculated from the weight of added water, soil moisture content, and mass of soil (9).

4. Notes

1. Reporting results
   Considerable confusion surrounds the reporting of soil salinity values. It is proposed that the following nomenclature be adopted for reporting soil salinity. Determination of electrical conductivity by unfiltered soil:water 1:5 dilution as $EC_{1:5}$, determination by saturated paste filtered extracts as EC_{sp}, and determination by saturated paste equivalent in which $EC_{1:5}$ values are multiplied by a texture coefficient as EC_{eq} (see Note 3). The nomenclature EC_{eq} is preferred to EC_e in order to remove confusion as to whether determination was conducted by texture coefficient or saturated paste method.

   Historically, electrical conductivity has been measured in micro-mhos per centimeter (μmho/cm), or in milli-mhos per centimeter (mmho/cm), today electrical conductivity is reported in deci-siemens per meter (dS/m), in which one dS/m is equivalent to 1 mmho/cm or 1,000 μS/cm. Salinity may also be measured by TDI in units of mg/L (milligrams/liter) or ppm (parts per million) (8), in which 1 dS/mis approximatly equal to 670 ppm or mg/L of salt depending on the species of ion present in the soil solution. Conversions between units are presented in Table 1.

Table 1
Conversion (multiply by) of electrical conductivity units and approximate salt concentrations

From/to	dS/m	mS/m	μS/m	mS/cm	μS/cm	TDI mg/L	Meq/L
dS/m	1	100	100,000	1	1,000	667	10
mS/m	0.01	1	1,000	0.01	10	6.7	0.1
μS/m	0.00001	0.001	1	0.00001	0.01	0.0067	0.0001
mS/cm	1	100	100,000	1	1,000	667	10
μS/cm	0.001	0.1	100	0.001	1	0.67	0.01
TDI mg/L	0.0015	0.15	150	0.0015	1.5	1	0.015
Meq/L	0.1	10	10,000	0.1	100	66.7	1

TDI total dissolved ions (note conversion is approx and based on the composition of seawater), *Meq/L* milliequivalent per Liter

Table 2
Cost comparison of selected methodology based on Spies and Woodgate (2)

Approach	Symbol	Approximate cost	Comment
Electric conductivity of a 1:5 dilution	$EC_{1:5}$ and EC_{eq}	\$10–20	Costs vary between laboratories
Electrical conductivity of a saturated paste extract	EC_{sp}	\$20–100	May be difficult finding a laboratory willing to do saturated paste extracts
Total soluble salts of saturated paste extract	TSS	\$100–200 per sample	Basic cations and anions. Cost differ depending on laboratory and prewash conditions
Hand held EM38 or EM 31 survey	EC_a	Starts from approximately \$1,500 a day	Coverage ranges from 5 to 50 ha/day depending on spacing and measurement interval
Vehicle mounted EM38 or EM 31 survey	EC_a	Starts from approximately \$1,500 a day	Coverage ranges from 50 to 500 ha/day depending on spacing. Consider additional calibration and mapping costs
Borehole EM 39	EC_a	Approximately \$2,000 per day for up to 10 bores	Number per day depends on access and depth of bores.
AEM	EC_a	Less than \$1/ha	Large area required to justify cost of aircraft. Considerable additional calibration and data management costs

Note: Cost differs considerably between laboratories. Costs as per AU\$ in 2005

2. Advantages and disadvantages of soil water ratios, $EC_{1:5}$, EC_{sp}, EC_{eq}

Determination of soil salinity from the 1:5 soil:distilled water suspension has been widely used in Australia. Consequently a considerable data base has accumulated for comparison between soils. The 1:5 suspensions are relatively straightforward, quick, inexpensive, and can be done in the field or laboratory (Table 2). Furthermore it produces a large quantity of solution for easy measurement. However, due to the high water to soil dilution of the 1:5 suspension the approach is a poor representation of the actual soil water salinity experienced by plants. Consequently it can be difficult to predict plant response to $EC_{1:5}$ values (10). The 1:5 dilution is also prone to poor setting of solids, specifically dispersed clays in which suspended clay particles contribute to the electrical conductivity. In soils with sparingly soluble salts, the 1:5 ratio results in higher EC values than would otherwise be expected at lower dilution ratios more commonly experienced in the actual soil water. Furthermore, in soils that contain more than 1% gypsum, EC is likely to be underestimated in 1:5 dilutions as the soil solution will approach saturation at approximately 2.0 dS m^{-1} as gypsum will not dissolve completely in the 1:5 dilution (7).

Determination of soil salinity by saturated paste (EC_{sp}) is more meaningful for predicting plant response to soil salinity because the dilution factor is closer to field water content (10). Consequently EC_{sp} rather than $EC_{1:5}$ is considered a more realistic measure of soil water environment that plants experience. The saturated paste approach is however more laborious, requiring extraction and analysis of small amounts of soil water from soil samples near their liquid limit. Furthermore, difficulty exists with replication of samples between researchers and technicians as the end point for dilution is visually determined. The extent to which results represent the soil water environment experienced by plants is also questioned because the procedure results in a structureless "gel" that is influenced by the degree of soil grinding as opposed to 'true' soil water extracted from the pores of an intact soil. The approach is also prone to errors associated with absorption of atmospheric CO_2 during vacuum filtration or centrifuge (9, 10).

3. Conversion of $EC_{1:5}$ to EC_{sp} and EC_{eq}

Given the relative simplicity of the $EC_{1:5}$ procedure, the ability to convert $EC_{1:5}$ values to EC_{sp} is desirable. However, conversion of $EC_{1:5}$ to saturated paste equivalent EC_{eq} is not straightforward due to the effect of texture, clay content, and soil porosity on electrical conductivity. Conversion factors have been developed for different texture classes to convert $EC_{1:5}$ values to saturated paste equivalent EC_{eq}. Concern however exists with the universal application of such coefficients as little literature exists to support regional differences in conversion factors, presented in Table 3.

Shaw (10) details a procedure by which $EC_{1:5}$ values can be converted to saturation paste equivalent EC_{eq} from the air dried moisture content (ADMC) and saturation percentage,

$$EC_{eq} = EC_{1:5}\left(\frac{500 + 6\text{ADMC}}{\text{SP}}\right)^{b}$$

ADMC is the moisture held in the soil following 24 h of air-drying at no more than 40°C expressed on an oven-dry basis (mass of soil following drying at 105°C for 24 h), SP is

Table 3
Variations in texture coefficients used to convert $EC_{1:5}$ to a saturated paste equivalent EC_{eq}

	Tasmania. surface soil (11)	Tasmania subsoil (11)	WA (2006): web site[a]	Qld: factsheet (8)	NSW: salinity notes (12)	NSW dryland Salinity (13)	Saltland pastures (14)	Saltland solutions (15)	Australian Journal of Soil Research (16)
Sand	14	17	15	23	17	17	14	15	22.7
Loam	12	15	10	10	9.5	10	11	9.5	9.5
Sandy loam			12	14	13.8				13.8
Clay loam			8	9	8.6	9			8.6
Clays							8	6.5	
Light clay	11	13	6	7.5	8.6	8			8.6
Medium clay			6		7	7			7.5
Heavy clay	6	8		6	7	7			5.8
Very heavy clay	4	5							

[a]http:\\www.agric.wa.gov.au/content/lwe/salin/smeas/salinity_units.htm

Table 4
Australian salinity class based on saturated paste equivalent with corresponding 1:5 dilution values (15, 17)

	Saturated paste EC_{eq} or EC_{sp}	1:5 Dilution $EC_{1:5}$		
Salinity class	**All soils (dS/m)**	**Sand (dS/m)**	**Loam (dS/m)**	**Clay (dS/m)**
Non-saline	0–2	0–0.14	0–0.18	0–0.25
Low	2.0–4.0	0.15–0.28	0.19–0.36	0.26–0.50
Moderate	4.0–8.0	0.29–0.57	0.37–0.72	0.51–1.00
High	8.0–16.0	0.58–1.14	0.73–1.45	1.01–2.00
Severe	16.0–32.0	1.15–2.28	1.46–2.90	2.01–4.00
Extreme	>32.0	>2.28	>2.90	>4.00

the saturation percentage, and the power term b has values $1 \geq b \geq 0$. The latter represents the value for the mix of the totally soluble and sparingly soluble salts in the solution and needs to be determined experimentally.

4. Interpretation of salinity values

   Salinity class based on saturated paste equivalent EC_{eq} and $EC_{1:5}$ are presented in Table 4. Note salinity classes for $EC_{1:5}$ vary according to which soil field texture coefficient is selected from Table 3.

References

1. Brinkman R (1980) Saline and sodic soils. In: Land reclamation and water management. International Institute for Land Reclamation and Improvement (ILRI), Wageningen, The Netherlands, pp 62–68
2. Spies B, Woodgate P (2005) Salinity mapping methods in the Australian context. Department of the Environment and Heritage; and Agriculture, Fisheries and Forestry, Australia
3. McNeill JD (1980) Electromagnetic terrain conductivity measurements at low induction numbers. Geonics Limited, Canada
4. Cook PG, Walker GR, Buselli G, Potts I, Dodds AR (1992) The application of electromagnetic techniques to groundwater recharge investigations. J Hydrol 130:201–229
5. Williams BG, Baker GC (1982) An electromagnetic induction technique for reconnaissance surveys of soil salinity hazards. Aus J Soil Res 20:107–118
6. McFarlane DJ, George RJ (1992) Factors affecting dryland salinity in two wheatbelt catchments in Western Australia. Aus J Soil Res 30:85–100
7. Rayment GE, Higginson FR (1992) Australian laboratory handbook of soil and water chemical methods. Inkata Press, Melbourne
8. Watling K (2007) Measuring salinity, in facts, land series, L137. Natural Resources and Water, Queensland
9. Miller JJ, Curtin D (2008) Chapter 15: electrical conductivity and soluble ions. In: Carter MR, Gregorich EG (eds) Soil sampling and methods of analysis, 2nd edn. Canadian Society of Soil Science, CRC Press, Boca Raton, pp 161–171
10. Shaw RJ (2001) Soil salinity—electrical conductivity and chloride. In: Peverill KI, Sparrow LA, Reuter RJ (eds) Soil analysis, an interpretation manual. CSIRO Publishing, Melbourne, pp 129–145

11. McMahon S, Bell I (1992) Saltpak Tasmania: an information package, Department of Primary Industry and Fisheries Tasmania, National Soil Conservation Program, Landcare Tasmania
12. Gibbs S (2000) How to texture soils and test for salinity, in salinity notes, p 4, NSW Agriculture—Salt Action, No 8, October
13. Taylor S. (1993) Dryland salinity—introductory extension notes. Department of Conservation and Land Management, Bathurst, New South Wales
14. Barrett-Lennard EG (2003) Saltland pastures in Australia: a practical guide, 2nd edn. Land, water and wool sustainable grazing on saline lands sub-program
15. Land and Water Australia (2009) Saltland solutions, options for saltland restoration, Future Farm Industries CRC, Land Water and Wool SGSL, Canberra, ACT
16. Slavich PG, Petterson GH (1993) Estimating the electrical conductivity of saturated paste extracts from 1:5 soil, water suspensions and texture. Aus J Soil Res 31:73–81
17. Nichols P, Barrett-Lennard EG, Bennett S (2010) Pasture legumes and grasses for saltland, in Farmnote. Department of Agriculture and Food, Western Australia

INDEX

Sergey Shabala and Tracey Ann Cuin (eds.), *Plant Salt Tolerance: Methods and Protocols*, Methods in Molecular Biology, vol. 913, DOI 10.1007/978-1-61779-986-0, © Springer Science+Business Media, LLC 2012

D

E

F

G

H

I

K

L

M

MIX
Papier aus verantwortungsvollen Quellen
Paper from responsible sources
FSC® C105338

If you have any concerns about our products,
you can contact us on
ProductSafety@springernature.com

In case Publisher is established outside the EU,
the EU authorized representative is:
Springer Nature Customer Service Center GmbH
Europaplatz 3, 69115 Heidelberg, Germany

Printed by Libri Plureos GmbH
in Hamburg, Germany